Gerdina de Jong (Ed.)

Population Genetics and Evolution

With 70 Figures

Springer-Verlag
Berlin Heidelberg New York
London Paris Tokyo

Dr. GERDINA DE JONG
Department of Population and Evolutionary Biology
University of Utrecht
Padualaan 8
NL-3584 CH Utrecht
The Netherlands

The drawing of Drosophila melanogaster on the front cover was made by Mr. Bas Teunissen, of the Afdeling Beeldprocessing & Vormgeving, Faculteit Biologie, Universiteit Utrecht.

ISBN-13: 978-3-642-73071-9   e-ISBN-13: 978-3-642-73069-6
DOI: 10.1007/978-3-642-73069-6

Library of Congress Cataloging-in-Publication Data. Population genetics and evolution/ Gerdina de Jong (ed.). p. cm. Papers presented at a symposium at Woudschoten, Utrecht, The Netherlands, Sept. 7-13, 1986. Includes index. 1. Population genetics – Congresses. 2. Evolution – Congresses. I. Jong, Gerdina de. 1945– . QH455.P67 1988 575.1′5 – dc19 87-32118

© Springer-Verlag Berlin Heidelberg 1988
Softcover reprint of the hardcover 1st edition 1988

2131/3130-543210

Preface

At least since the 1940s neo-Darwinism has prevailed as the consensus view in the study of evolution. The mechanism of evolution in this view is natural selection leading to adaptation, working on a substrate of adaptationally random mutations. As both the study of genetic variation in natural populations, and the study of the mathematical equations of selection are reckoned to a field called population genetics, population genetics came to form the core in the theory of evolution. So much so, that the fact that there is more to the theory of evolution than population genetics became somewhat obscured. The genetics of the evolutionary process, or the genetics of evolutionary change, came close to being all of evolutionary biology.

In the last 10 years, this dominating position of population genetics within evolutionary biology has been challenged. In evolutionary ecology, optimization theory proved more useful than population genetics for interesting predictions, especially of life history strategies. From developmental biology, constraints in development and the role of internal regulation were emphasized. From paleobiology, a proposal was put forward to describe the fossil record and the evolutionary process as a series of punctuated equilibria; thus exhorting population geneticists to give a plausible account of how such might come about. All these developments tend to obscure the central role of population genetics in evolutionary biology.

Some re-examination of the role of population genetics seemed called for. At a symposium at Woudschoten, Utrecht, the Netherlands, from September 7 to 13, 1986, a start was made in that direction. The symposium was made possible by the financial support of the Royal Dutch Academy of Sciences (Koninklijke Nederlandse Academie van Wetenschappen), the University of Utrecht and the Biology Faculty of the University of Utrecht. This symposium was one in a series that started in Berlin (1981), and was continued by meetings organized by the Evolutionary Biology Departments of the Universities of Pavia (1982), Tübingen (1983) and Aarhus (1985). These meetings have as one of their goals the strengthening of evolutionary biology in Europe. Proceedings were published of the 1982 meeting: "Evolution and the Genetics of Populations", eds. S.D. Jayakar and L. Zonta, Suppl. Atti Ass. Genet. Ital. Vol. XXIX, and of the 1983 meeting: "Population Biology and Evolution", eds. K. Wöhrmann and V. Loeschcke, Springer-Verlag.

Two major questions can be posed: Is population genetics necessary and sufficient for a theory of evolution? With regards to necessity, the obvious, historical answer is no. Darwin clearly did without. Evolutionary ecologists who maintain that optimization provides satisfying answers to evolutionary problems are Darwinists rather than neo-Darwinists. The outstanding theory of phenotypic evolution is quantitative genetics, a statistical theory with a genetic background. To maintain that the answer to the question whether population genetics is necessary for the theory of evolution is yes, is to maintain that adding population genetics provides a level of understanding that is necessary for an evolutionary theory of up to date sophistication. In the contributions in the first section of this book several points of view on this question are defended or illustrated.

In the second section of the book some approaches are made with regard to the second question: Is population genetics sufficient for a theory of evolution? The answer might be: yes, if you are content with an inadequate theory of evolution, and this might be the answer that the proponents of a decoupling of macro-evolution from micro-evolution, or the proponents of a decisive role of development in evolution give. The answer might be yes, too, for anyone who takes a broader view of population genetics, including the study of the origin of developmental constraints and species differences. The authors attempt to show how far one can go in the direction of a satisfying theory of evolution, starting from population genetics. The Chapters are varied in content, from one that shows the strength of a genetic approach in unravelling development to a Chapter dealing with the possible rates of evolution.

This volume is addressed both to evolutionary biologists who like population genetics and to those who dislike it. Thus, they might find the strengths, or the weaknesses, of a now classical approach in evolutionary biology.

I would like to thank Dr. J.M. van Damme for his help by providing closing comments on the meeting, and Dr. F.R. van Dijken for organizational assistance. The help of the department B & V of the Biology Faculty in Utrecht is gratefully acknowledged for their polishing of figures, photographic work and cover design. Many thanks are due to the providers of financial support, the Royal Dutch Academy of Sciences (KNAW), the University of Utrecht and the Biology Faculty. Most thanks are due to the participants, in their contributions to making this an excellent meeting.

Utrecht, January 1988   Gerdina de Jong

Contents

Development and Selection

Is Population Genetics in Its Present Scope Sufficient for a Theory of Evolution?

Adaptation

Contents

Population Structure

Developmental Constraints

Extrapolations

Contributors

You will find the addresses at the beginning of the respective contribution

Introduction: The Place of Population Genetics in Evolutionary Biology

G. DE JONG and W. SCHARLOO

Since evolution is a change in the genetic composition of populations, the mechanisms of evolution constitute problems of population genetics.

Th. Dobzhansky, 1937

1 Population Genetics and Evolution

Without the theory of evolution there would be no science called "biology". Biology would not be one discipline, but a catchall name for a very varied set of disciplines, that all, in some way or other, studied the properties of living beings. Many of those disciplines contributed to the synthetic theory of evolution. This theory derives its strength from its roots in so many diverse fields; as it was possible to build one theory of evolution from blocks from several fields, evolutionary biology came to occupy a central and unifying place among the (sub)disciplines. The unification of all of biology into one science, by the providing of a major explanatory framework, forms the unassailable strength of the synthetic theory.

The synthetic, or neo-Darwinian, theory of evolution is very much a genetic theory. The principles governing adaptive evolution can of course be explained by simply assuming that some form of heridity exists: Darwin did just that. The investigation of the extent of genetic variation in nature, and the presence of a recent and fully fledged mathematical theory of selection at the time of the emergence of the synthetic theory gave evolutionary biology a strong slant towards the details of the genetic explanation of evolutionary change. It is questionable whether a definition of evolution as a gene frequency change would have had any attraction to Darwin, or to anyone else before, about, 1940. In the enthusiasm over the fit of genetics into evolutionary theory, neo-Darwinism amounted to a genetic takeover.

The definition of evolution as gene frequency change (Dobzhansky 1937) prevailed widely, but was never totally accepted. The reduction of the process of selection to "the aridities of neo-Darwinian algebra" (Waddington 1972) was occasionally countered by an emphasis on organic evolution. However unclearly defined and hazy Waddington's epigenetic landscape might be, it provided a metaphor to start thinking about development and evolutionary genetic change.

Yet it was the view of evolutionary theory as "gene frequency change is all there is to evolution" that provoked a reaction, as being too much an outline or an axiom, and not enough of an explanation. The dissatisfaction occasionally took the shape of an attack on neo-Darwinism and on the predominating position of population genetics in evolutionary theory (Gould 1980, Ho and Saunders 1984). One of the controversial

Population Genetics and Evolution
G. de Jong (ed.)
© Springer-Verlag Berlin Heidelberg 1988

subjects in evolutionary biology today is which factors determine the direction of evolutionary processes. Is it really natural selection using genetic variation that is present in every character and expressed randomly with respect to the direction of evolution? Or is the expression of genetic variability biased by processes internal to organisms, to the organization of developmental and physiological systems?

It seemed therefore to be the right time for population geneticists and some evolutionary minded ecologists to examine what they saw, from their own work, as the connection between population genetics and evolution.

2 Necessity

The contributions of this symposium are arranged around two main questions: Is population genetics necessary and sufficient for the theory of evolution? Twenty or 30 years ago, the answer might have been yes and yes; some would now say: no and no. To give a reasoned answer, several viewpoints have to be taken. With regards to the necessity, the question pertains to how far evolutionary theory can do without explicit genetics. With regards to sufficiency, the question as to which contributions population genetics can make, and which phenomena are definitely outside its scope, cannot be dealt with at all. Less technically, the questions are whether we can do without population genetics, and how far we can go with it.

2.1 Population Genetics as a Core Theory

In the first section of this book, five chapters are concerned with the place of population genetics. Dhondt, in a review of field data of birds, probes the several strategies employed to argue for micro-evolution in field populations. The field ecologist can provide circumstantial evidence from comparisons with other species that observed phenotypic changes reflect an underlying genetic change. This might not be very satisfactory, but occasionally it is as far as one can go. A stronger argument for micro-evolution in natural populations can be made when heritability from experimental populations is known for a character that shows changes in a field population, or differences between field populations, of the same species. For the field biologist on the lookout for micro-evolution, the best situation is, however, when knowledge of heritabilities or Mendelian segregation is supplemented by knowledge on the process of natural selection of that character. The genetic data are considered crucial for an interpretation of any change as micro-evolutionary. An example of this approach is given in another section by Brakefield. Pásztor argues for the primacy of the functional explanation, in her case a functional explanation of life history strategies within the context of different ecological situations. She outlines a model that is able to provide the mechanisms of both ecological change and evolutionary optimization. Many published life history models can be assessed as subsets of this general model. It is interesting to see what the optimal life history strategies should be, given a specified ecological background. It is interesting, too, that this model is concordant with some simple genetic models of the life histories it deals with. Both Hoekstra and

Boomsma provide examples of the dimension that is added when a problem is approached from a population genetics angle. Hoekstra distinguishes between the functional explanation and the ultimate evolutionary scenario. He adduces that the functional explanation for the evolution of anisogamy has never changed. What has changed over time is the attention given in models to the mechanisms of the evolution of anisogamy, shifting to more complete and more genetic explanations. Hoekstra argues that introducing population genetics into any evolutionary explanation aims at completeness of the explanation. Boomsma uses population genetics to choose between two competing scenarios for the evolution of the sex ratio within colonies of ants. The advantages of a population genetic approach over explanations based upon optimization (local mate competition) or upon kin selection (asymmetrical genetic relatedness) are that population genetics is able to provide detailed observations and predictions.

That theoretical population genetics does not absolutely mean the reduction of selection processes to changes in gene frequencies, is shown in the contribution of Gregorius. Selection on genotypes need not be reduced to averaged selection on alleles. Gregorius shows an algebra where selection is actually on genotypes; such an algebra can account for the vagaries of the mating system and of Mendelian segregation. It gives more generally valid equations, and comes nearer to the notion that selection is on organisms. A general condition for the maintenance of polymorphism by heterozygote advantage exemplifies the strengths of his method.

2.2 Quantitative Genetics and Evolution

Quantitative genetics provides the theory of phenotypic evolution. The well-known formula for the change R of the phenotypic mean of a quantitative character under selection, $R = h^2 S$, can be written as $R = h^2 \cdot \text{cov}(w,P)/\overline{w}$, given the definition of the selection differential S. This states that for any change in the mean of a quantitative character to occur, there should be additive genetic variation for the character ($h^2 = V_A/V_P > 0$) and there should be a covariance other than zero between the character P and fitness w within the population. The expression $R = h^2 \cdot \text{cov}(w,P)/\overline{w}$ gives a summary in mathematical notation of Darwin's argument, the nearest we have got to a mathematical expression for the theory of natural selection. Other than the classical population genetics model with three genotypes at one locus, each genotype having its fitness value, this expression does not emphasize fitness, but rather the importance of the covariance between phenotype and fitness. Perhaps we should have had less of a debate on natural selection being a tautology if this central prediction of the response to selection had been more generally regarded as the basic model of the theory of natural selection.

When quantitative genetics became a branch of population genetics (Fisher 1918), the question as to the necessity of population genetics to evolutionary theory was almost immediately answered. Key results in quantitative genetics not only hold for very many loci, or for normally distributed characters, but are already valid in a one-locus population genetics of a quantitative character. An example of such a result is the single-parent/offspring covariance; according to both the definition of the breeding value and a one-locus treatment, this is $1/2 V_A$. Falconer (1981) fairly routinely presents both approaches, the high level approach based on definitions and probability theory and the low level approach based on single-locus population genetics. Bürger and Gabriel in this volume choose the high level approach. Cheverud, and Zonta and Jayakar, choose the alternative approach, population genetics of a quantitative character, based upon a few loci. These four contributions are all concerned with the development of quantitative genetics theory. Dingle shows how effective quantitative genetics techniques can be in analyzing genetic variation in ecologically important characters in natural populations.

Bürger outlines a quantitative genetics model at a higher level, freed from the restrictions of multi-locus theory. He aims at an exact and general solution for the equilibrium variance in a (diploid, additive allelic effects) mutation selection balance model. Since estimates for an equilibrium variance only make sense if the corresponding equilibrium is stable, this point is first dealt with. An upper boundary for the equilibrium variance under stabilizing selection is found for this model by analytical methods; the upper boundary agrees with some other, less general, models and previous simulations. Gabriel starts with the expected genetic variance under a balance of mutation and selection, too, but in a parthenogenetic species. He then shows the force of quantitative genetic models in evolutionary ecology by applying them to evolutionary problems of cyclically parthenogenetic species. Cyclical parthenogenesis, by accumulation of mutations within each parthenogenetic line, stores a large amount of genetic variation between lines; the hoarded variation can be brought to light on sexual reproduction. Somewhat contrary to expectation, rates of evolution might be as high in a cyclically parthenogenetic species as in a sexual species.

Dingle stresses the importance of co-adapted complexes of life history traits. He asks whether the pattern of genetic variance and covariance is the same in two populations of the milkweed bug *Oncopeltus fasciatus*, a migratory and a non-migratory population. There are important differences that can be sensibly interpreted as a selected response to a colonizing and to a non-migratory life history. On the one hand, co-adaptation of traits in nature clearly exists; on the other hand, these observations must warn theorists not to assume too easily a constant genetic variance/covariance pattern in their models. Cheverud concerns himself specifically with the dynamics of the additive genetic variances of and covariances between quantitative traits as a consequence of selection. Equal effects at a number of loci, and equal selection on two traits, lead in a stabilizing selection model not to the theoretically predicted genetic correlation of −1; this must mean that in his model linkage disequilibrium proves to play an appreciable role, the net effect of gametes determining the system. Zonta and Jayakar show in a two-locus model where both loci influence mean and variance of a character, that cyclical selection can maintain genetic variation. Stable polymorphisms with three or all four of the possible gametes present can be found. This means that it is at least possible that a fluctuating environment leads to a wider niche.

2.3 Development and Selection

The formulation of the theory of natural selection as $R = h^2 \cdot cov(w,P)/\bar{w}$ focusses attention on the covariance of phenotype and fitness, as the factor expressing selection. This immediately implies that for the study of natural selection it is of as much importance how fitness differences come about as what the consequences of fitness differences are. The causes of fitness differences are to be found in the development and ecology of the individual, and very much in the interaction of the development of the individual and the ecology it perceives. Fitness is not a fixed, invariable character of the genotype: fitnesses, and fitness differences are specific to the environment, shaped by the course of individual development. This process of the causation of fitness differences by genotype-specific developmental pathways related to the environment, has as much claim to the name selection as the consequences of the fitness differences, i.e. the dynamics of allele frequency change. In the discussion on the units of selection this two-fold meaning of selection is often not recognized. It might need discussion, not perhaps what the units of selection are, but what we mean by selection. For anyone who thinks of selection as the process causing fitness differences, the unit of selection seems to be the individual, as a matter of course; the allele frequencies represent a bookkeeping of past events, but it is the individual by whose development in a certain environment the process of selection takes shape. But for anyone who thinks of selection as the dynamics of allele frequencies as caused by given fitness differences, the gene or allele itself might seem the unit of selection. This distinction, of selection as the cause or the consequence of fitness differences, might seem reminiscent of Sober's (1984) conceptual distinction between selection on (individuals) and selection for (a character). Reminiscent, but not identical; though both distinctions are useful in any discussion of the units of selection, Sober too seems to use fitness as an invariant property of a genotype, and only speaks of selection as the consequence of fitness. A more fertile way of referring to selection might be, however, to speak of it as the cause of fitness differences between genotypes. Development and ecology would thus become fully integrated into population genetics.

Two conceptual and two experimental studies are presented, the two conceptual papers contrasting in point of view and the two experimental papers providing examples of the practice of analyzing the causes of fitness differences. Tuomi, Vuorisalo and Laihonen thoroughly analyze a causal model of the selection process as a unifying framework for development, ecology and population genetics. They point out that when in the emergence of theoretical population genetics and its incorporation into the synthetic theory, fitness became a technical tool for quantifying natural selection in terms of its consequences, selection was redefined as a consequence of genetic variation in fitness. The causal process dropped out of sight; yet, it is this causal process, called selection, that integrates with development and ecology. Gouyon and Gliddon analyze selection in terms of information and vehicles for information, called (by them) avatars. It is information that in their view is selected for. Even the gene is an avatar and, therefore, the selection of information does not mean that the gene is the unit of selection. They see information at many hierarchical levels and, correspondingly, selection on many hierarchical levels. Van Noordwijk presents a simulation of the growth of nestling Great Tits, keeping close to field data on weights of nestlings. He is interested in the causation of any weight differences at the age of fledging, and the influence of sib competition on weights. In his model, a higher environmental variance and a lower heritability comes about under restricted food circumstances, and when sib competition exists. Under such circumstances environmentally caused differences in weight at fledging age might override genetic information for weight to be achieved at 15 days. Templeton and Johnston explore the character "abnormal abdomen" *(aa)* in *Drosophila mercatorum* in a natural population. The overall climate first sets an upper limit of about a week to survival of adult flies under desiccation. This fixed threshold dictates an advantage for any flies with high early fecundity. Abnormal abdomen adults never live much longer, even under humid circumstances; under humid circumstances *aa* is selected against, by its later maturation and lower total fecundity. But under the imposed threshold of survival under very dry conditions, their higher early fecundity leads to a higher frequency of *aa* flies. Moreover, the primary effect of the *aa* gene can be analyzed. This set of field observations and molecular studies gives a fairly full account of how development and ecology interact to cause fitness differences.

3 Sufficiency

The contributions in the second part of the book deal in some way with the question whether population genetics is sufficient for evolutionary theory. The chapters in this part are more often experimental than the works in the first part of the book. The four sections deal with the study of adaptation, the effects of subdivided populations, the tool provided for the study of development and developmental constraints by population genetics and with two chapters approaching a wider field: speciation and the trajectory of the phenotypic mean of a character over space or time.

3.1 Adaptation

The four chapters in this section analyze the adaptive significance of known genetic differences in natural populations. The approach is bottom-up: a known genetic difference is chosen as a starting point for study, and likely ecological factors are screened for their influence. In the case of allozyme differences, the search starts by looking for a direct connection between the genotypes at one locus and fitness differences in particular environments. The way in which such a search is conducted is intended to show that the several genetic variants found in natural populations are adaptations to factors likely to be encountered in the environment. This involves a functional analysis of the genetic variants; the point is to show that the genetic variants are indeed the building stones of adaptive evolution.

Yet, in all four chapters, further study reveals complexities which are essential to our understanding of evolution. The complexities are a consequence of the fact that genes cannot adapt in isolation. They act in the physiological context of the whole organism, and it emerges readily that we do not have enough knowledge of the interplay of genetic variation in reaching functioning phenotypes. Here, population genetics becomes a tool for studying the relation between the physiology and the ecology of the species.

Brakefield discusses the well-known case of industrial melanism in *Biston betularia*. Next to visual selection by predators, some non-visual selection seems necessary to explain observed geographic patterns. Moreover, predation is not as straightforward as thought, as the life moths do not rest on tree trunks, but on narrow branches in the canopy, behaviourally choosing to resemble lichen. The actual degree of crypsis has to be determined, especially in areas where the degree of pollution is decreasing. David observes that even if environmental alcohol is a strong selective force in nature, and alcohol dehydrogenase a key enzyme for adaptation to an alcohol-rich environment, the role of the polymorphism of the *Adh* locus in *Drosophila melanogaster* in this adaptation remains vague. The tolerance level is different for different geographic populations, being highly correlated with latitude. The frequency of the *Adh* alleles too is correlated with latitude. However, changes in allele frequencies are at most only a partial explanation. Tolerance differs between similar *Adh* genotypes from different regions, and can be selected for within one *Adh* genotype. Van Delden places single characters in the context of multigene studies. In the plantain *Plantago major* ecotypic differentiation in quantitative, life history characters is found in the two subspecies. Many of these ecotype-specific adaptations can be localized in one linkage group, together with particular enzyme loci such as *Pgm-1*. Such a clustering is specific to the species *Plantago major*, and is not found in the related species *Plantago lanceolata*. In *Drosophila melanogaster*, the alleles of the locus *Adh* are found to be embedded in multigene complexes, too. Therefore, it is the phenotype as a whole that is selected and adaptive, and not the single locus genetic variant. Klarenberg's contribution reminds us that beyond the differences at structural loci, genetic variation in the regulatory loci are important for understanding selection on allozymes, in this case amylase in *Drosophila melanogaster*.

3.2 Population Structure

The next four chapters have to do with phenomena in subdivided populations; otherwise, the topics are varied, from local adaptation and geographic differentiation to a possible ESS strategy for allocation of patch time in a parasitoid searching for patchily distributed hosts. Noer discusses the micro-evolutionary capabilities of parthenogenetic species, as exemplified by the triploid parthenogenetic form of the woodlouse *Trichoniscus pusillus*. He shows that evolution within clones seems to be possible, as some of his clones might be monophyletic, *i.e.* pairs of clones differ in electrophoretic mobility of one enzyme. Monophyletic clones tend to share a similar ecology, while clones of different descent tend to differ in ecology. The interesting point is that clones from different monophyletic groups are found to co-exist, but monophyletic clones exclude each other. Seitz studied the microlepidopteran species *Argyresthia mendica*, a rather common species living in hedges on *Prunus spinosa*. Over the extensive study area, there is a weak but significant correlation between the topographic distance and the genetic distance between populations, but more importantly, a better correlation between altitudinal distance and genetic distance. This contrast between altitudinal distance and topographic distance seems to be the result of an interplay of local adaptation to altitude and gene flow in the direction of the prevailing wind. Tomiuk and Wöhrmann estimated the degree of genetic heterogeneity in natural populations of the rose aphid *Macrosiphum rosae* and looked for correlations between genetic heterogeneity and climate. Van Alphen reviews the ideas on allocation of patch time by parasitoids searching for patchily distributed hosts. He points out that two distinct and contrasting views emerge. On the one hand, models developed from a population dynamics point of view stress the importance of parasitoid aggregation; on the other hand, optimal foraging models for individual parasitoids lead us to expect that parasitoids should be distributed according to an equal free distribution. Van Alphen, however, reminds us that parasitoids do not search as isolated individuals. They play an evolutionary game against other parasitoids. An ESS analysis might well predict an advantage for the individual parasitoid in aggregation, and be more in line with the dynamic models.

3.3 Developmental Constraints

Today, there is some controversy over the relative importance of natural selection and developmental constraints in evolution. Some argue that the expression of genetic variability is biased by processes internal to the organisms, that genetic variability and adaptation are secondary to the organization of developmental systems. Adaptation might not be possible in all directions, or to all environments, but only insofar as it is compatible with the determination by the developmental system. Or, genetic variation can only be expressed in characters that are of minor importance to development, and natural selection can never do more than fine-tuning to a system determined by developmental rules.

While there is a great deal of truth in this view — it is rather difficult to select for a higher number of abdominal legs in insects — there are still worthwile approaches, both environmental and theoretical, on the selection of developmental constraints and on selection shaping or altering the developmental system. The next two chapters show a population genetics approach analyzing the interplay of selection and development.

Wagner relates the concept of developmental constraint to quantitative genetic theory. In a model where the quantitative characters are not under any constraints, fitness functions that seem adequate to describe the evolution of functionally integrated organisms easily lead to an upper limit on the possible rate of evolution. He concludes that constraints do not always have a negative effect on the selection response. He even suggests that developmental constraints might facilitate the evolution of functionally integrated phenotypes if they lead to an appropriate allocation of phenotypic variance. Scharloo concludes that artificial selection on quantitative characters has shown the existence of biased expression of genetic variation, but that such constraints can be established or removed by artifical stabilizing or disruptive selection. Canalization can be created by stabilizing selection. The central concept in the study of the developmental system by way of quantitative genetics is the genotype-environmental factor/phenotype-mapping function. This function, relating the phenotype to the internal environment as the reaction norm relates the phenotype to the external environment, can be changed by natural selection; this alters the developmental properties of the system. Scharloo suggests too that more fundamental constraints can be revealed by artifical selection on morphological patterns.

3.4 Extrapolations

In evolutionary biology, the proposal that evolution proceeds by a series of punctuated equilibria has generated some debate; not so much by its paleontological aspects, as by its associated emphasis on the role of developmental constraints during stasis and on morphological change associated with random speciation followed by species selection. In the previous section something was said about developmental constraints and evolution. Here, two works are presented that attempt to show, by studying present-day phenomena, how one might be able to form a judgement on the mechanisms proposed for the punctuations. One study is on speciation, one presents a model of change in the phenotypic mean.

Fontdevila, in a genetic analysis of hybrid sterility, emphasizes the importance of middle repetitive DNA for speciation. The biological similarities between intraspecific hybrid dysgenesis and interspecific hybrid sterility are too clear to escape notice. In crosses of *Drosophila serido* females and *D. buzzatii* males, hybrid offspring are produced; the male offspring are sterile and the females fertile. These hybrid females are used in backcrosses to obtain strains that are introgressed with different chromosomal segments. Hybrid sterility of the offspring in these backcrosses proves to depend upon the total length of introgressed chromosomal segments, not on their location within the genome. De Jong is concerned with the patterns traced by the mean phenotypic value in time or over an environmental cline. Starting with present-day ecological patterns, we should look for ecological situations manifesting gradual changes or punctuation in the phenotypic mean. The ingredients of a population genetics model are the reaction norms of a character coded for by a number of loci, and the change of the optimum phenotype with the environment. Stabilizing selection on a character that evinces the developmental constraints of reaction norms causes the optimum phenotype to be tracked in a population with a supply of genetic variation. But any pattern of changes in the phenotypic mean can be generated by a combination of the tracking of the optimum phenotype and the constraint of the reaction norm. No contradiction in mechanism between gradual change and punctuations exists.

4 Concluding Remarks

Evolution has been defined both as descent with modification and as change in allele frequencies. At the time the latter definition was introduced (Dobzhansky 1937) it might have been supposed that it included all of the former. Though a strong statement, it does fairly well; one has to exclude reaction norms in a changing environment, but otherwise descent with modification implies gene frequency change. This is fairly well agreed upon; the reverse implication has evoked more unease. Not that it is often contested that gene frequency change implies descent with modification (though again there is an exceptional case with reaction norms in a changing environment, where gene frequency change is necessary to keep the phenotypic mean more or less constant), but what causes discontent is the feeling that just everything interesting in evolution is left out of this definition. There is too little biology there; it is too abstract, too abstruse, too theoretical. It might be true, but if it is, it is not informative.

Nor is the opposition recent. It has long been felt that not only a theory of gene frequency change was needed, but moreover a theory of adaptive genome organization. "It is doubtful, however, whether even the most statistically minded geneticists are entirely satisfied that nothing more is involved than the sorting out of random mutation by the natural selective filter" (Waddington 1942).

If one opens any book with "population genetics" in the title, the chance is that one finds population genetics theory; most often, theory only. Moreover, it is a particular type of theory; not about the evolution of certain characteristics but about selection on and evolution of any characteristic. The emphasis is not on the character that is selected, but on the general properties of selection models. The adaptive landscape can be regarded as the pinnacle of insight in the processes of evolution; but it is very much a metatheory. To invoke it directly in any practical selective situation risks the charge of "dogmatism": as the explanation is on the wrong level for the actual situation.

Experimental population genetics often has played a subsidiary role to theoretical population genetics. This is clearly evident in all work on fitness differences of electrophoretic variants. Allozymes at one locus seemed at first to correspond clearly to the situation supposed in the simple models, and in the 1960s expectations of quickly finding the elementary building blocks of evolution were high. But for those loci that have been subjected to intense scrutiny, the allelic differences proved enmeshed in a genomic network (van Delden, David, Klarenberg, this Vol.). Single characters resisted direct analysis. The exception in this volume is the character *"abnormal abdomen"* in *Drosophila mercatorum* (Templeton and Johnston). Although here too some modifiers are present, and it is coded for by two very closely linked loci on the X-chromosome, *aa* seems to behave as a single mutation with a large effect; a far larger effect than the allozyme differences of (wild-type!) flies, and seemingly less encapsulated or buffered by the background genome.

The contrast between the visible mutant *aa* and allozyme differences remains if one looks at their role in the ecology and development of the species. In the search for the adaptive significance of allozyme differences, one looks for "true" adaptations shown as such by a functional analysis, not for a serendipitous advantage. It would be stretching the meaning of the word adaptation to call the character *aa* in *D. mercatorum* an adaptation. Templeton and Johnston carefully point out that it is not an adaptation to desiccation, the direct selective factor involved. The higher fitness of *aa* flies under desiccation seems a coincident effect. The abnormal abdomen flies have a good claim to be "hopeful monsters". While adaptation has to arise by natural selection, natural selection can be an autonomous process, not leading to any adaptation. The covariance in $R = h^2 \text{ cov}(w,P)/\overline{w}$ is reduced to the covariance of fitness with itself.

These two points: the fact of the integrated genome and the ecological analysis of adaptation separate experimental population genetics from population genetics theory. Both are forced upon us by any further study of genetic variants under natural conditions. What the balance of studies on electrophoretic variants has shown is that it is futile to look at a single allozyme difference in isolation or as if it had any innate fixed fitness value. This very genetic approach is akin to population genetics theory; allozyme studies are often "pure" population genetics, unalloyed with ecology or development. But "pure" population genetics worked out to be not the most profitable or informative approach. Not that this makes population genetics theory redundant: as Hoekstra, Boomsma and Van Alphen point out, application of (a lower level of) population genetics theory is necessary or useful, to understand mechanisms, or to chose between rival explanations in evolutionary ecology, or to integrate a population dynamics problem with an evolutionary problem. But in the majority of the works presented here, the attitude towards population genetics is as towards a tool to analyze and integrate ecological and developmental aspects of the characters of individuals within a population. Genetic differences within a population provide the material to link molecular biology, development and ecology in a single study. The aim is to connect the differences at the levels of organization: the molecular level, the organismal level and the fitness level, tracing the causes of fitness differences. It is in this guise of a tool that population genetics is most productive. As in the work of Scharloo, experimental model systems can be generated to study the selective origin of developmental constraints. Would Noer's parthenogenetic woodlice suffice as a model system for species selection? It is by the use of population genetics as a tool that the neo-Darwinian theory of evolution can produce a synthesis between the biological disciplines.

References

Dobzhansky Th (1937) Genetics and the origin of species. Columbia Univ Press

Falconer DS (1981) Introduction to quantitative genetics. Longman, London

Fisher RA (1918) The correlation between relatives on the supposition of Mendelian inheritance. Trans R Soc Edinbourgh 52:399–433

Gould SJ (1980) Is a new and general theory of evolution emerging? Paleobiology 6:119–130

Ho M-W, Saunders PT (1984) Beyond new-Darwinism. Academic Press, London New York

Sober E (1984) The nature of selection. MIT, New York

Waddington CH (1942) Canalization of development and the inheritance of acquired characters. Nature (London) 150:563–565

Waddington CH (ed) (1972) Epilogue. In: Towards a theoretical biology. 4. Essays. IUBS Symp, pp 283–287

Is Population Genetics Necessary for the Theory of Evolution?

Population Genetics as a Core Theory

The Necessity of Population Genetics for Understanding Evolution: An Ecologist's View

A. A. DHONDT[1]

Evolutionary changes are changes in gene frequencies in a population over time. For these changes to be adaptive they must be caused by natural selection acting through differential survival or reproduction of individuals which differ in genotype. Changes in gene frequencies that cannot be related to changes in phenotype may illustrate that natural selection has operated and that (micro)evolutionary changes have taken place. They do not, however, help us to understand how evolution proceeds (changes in iso-enzyme frequencies would fall into this category). The knowledge of population genetics unrelated to phenotype is thus not really sufficient to understand how evolution proceeds.

Alternately, a population ecologist may observe changes in some phenotypic character. This change may or may not be a micro-evolutionary one, depending on whether the underlying genotypes underwent parallel changes. The question then is how far can and must the ecologist delve downwards to satisfy himself that micro-evolution has taken place. An extreme example is provided by Templeton in these proceedings. Usually, however, one cannot unravel the complete interrelation between genotype, morphology, life-history parameters and environment. What can one do?

A preliminary, though not very satisfactory approach is to provide circumstantial evidence that the observed phenotypic change reflects an underlying genotypic change. Thus, I argued (Dhondt 1986) that an increase in the per capita growth rate after a population crash caused by an extreme winter in populations of small resident birds in England was in fact a micro-evolutionary change. I assumed, thereby, that some life-history parameter such as reproduction or adult survival had a genetic basis. I could not provide evidence that this was actually so in the species observed. I could, however, refer to studies in other bird species where non-zero heritabilities in reproductive rates had been found (Van Noordwijk 1980). I could also think of examples from laboratory work on invertebrates in which a similar change had been observed (Bergmans 1983). Such an example is suggestive but will only convince the convinced.

A better example is one whereby one can reason through strong inference. Berthold and co-workers did some extremely nice work with the blackcap *Sylvia atricapilla*, a

1 Department of Biology, University of Antwerp, V.I.A., Universiteitsplein 1, B-2610 Antwerpen (Wilrijk), Belgium

Population Genetics and Evolution
G. de Jong (ed.)

small Old World warbler (Berthold, in press). The migratory habits of this species, which breeds over most of Europe, vary with latitude. Northern populations are migratory, southern populations mainly resident, central ones are partially migratory. Since the blackcap migrates at night, nocturnal restlessness gives a precise idea of its migratory urge. Populations differ as to the percentage of birds that show migratory restlessness, and as to the number of nights and hours per night this is observed. To give repeatable results experiments, however, must be carried out with birds born in captivity. Berthold and Querner (1981) showed (1) that birds from different populations differ in the amount of nocturnal restlessness, (2) that hybrids between populations show quantitatively intermediate amounts of migratory activity and (3) that strains can be bred from any population, through artificial selection in captivity, showing high/low amounts of migratory urge. The migratory behaviour of the blackcaps thus has a strong heritable component. This result is found without further unravelling of the genetics of this animal. Knowing that differences in migratory behaviour reflect difference in genome, resulting from different selective forces operating in different populations is interesting, but is not really unexpected. Many similar examples have shown that in captivity disruptive artificial selection yields strains differing in the character selected. One question of direct relevance is how rapidly will populations respond in natural situations to changes in the environment? The stonechat *Saxicola torquata* has a more limited distribution in Europe, but is also a partial migrant. As can be seen in Fig. 1 the number of birds observed in Belgium in winter depends on the temperature of the *previous* winter. In winters following cold winters fewer stonechats are observed than in winters following warm ones. Assuming that the migratory behaviour of the stonechat would reflect genetic differences, these data strongly suggest that natural selection causes rapid changes in the gene frequency of such populations. These data, by themselves, are inconclusive. The combined data of the blackcap and the stonechat (combined with similar experiments performed on the European robin *Erithacus rubecula*, Biebach 1983) clearly suggest that in partial migrants differences in migratory behaviour between individuals reflect differences in genes and that rapid changes in gene frequencies can occur as a result of changes in the environment which caused differences in the selection pressures.

Another way in which can ecologist can try to relate phenotypic changes to evolutionary changes is by using quantitative genetics. My involvement with this kind of genetics, which some consider to be part of population genetics, came from the need to evaluate field data. Dhondt et al. (1979) found that great tits had become about 7% smaller in wing length over the period 1963-1971. Using heritability estimates from the Oxford great tit population (Garnett 1981), which showed body size measurements to be highly heritable, it was assumed that a micro-evolutionary change had taken place in their population at Ghent. They explained the change as having been caused by a change in the strength of intraspecific competition after the beginning of their study, caused by an increase in the number of nest sites through the provision of nest boxes. I wanted to test the idea that competition would be a potent evolutionary force causing changes in body measurements. I therefore set up populations of blue tits *Parus caeruleus* near Antwerp at high and low population densities. The differences in densities were caused by manipulating the level of interspecific competition of great tits *Parus major* (Dhondt and Eyckerman 1980; Dhondt et al. 1982).

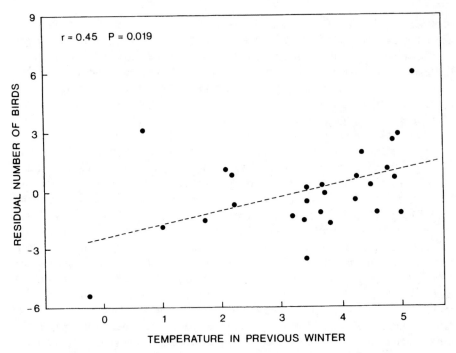

Fig. 1. Stonechats are occasionally observed in Belgium during winter. The number reported increased significantly over the period 1937–1963, but showed considerable variations around the trend. The residual number of birds not explained by the long-term trend is shown against the mean winter temperature of the previous winter. As winter temperature increases the number of overwintering stonechats observed in the following winter increases, suggesting that natural selection operates rapidly (cf. Dhondt 1982 for a fuller discussion)

Results of measurements of 15-day-old nestlings are shown in Fig. 2. At day 15 the tarsus of young tits had reached adult size and body weight had reached fledging weight. The three populations, although fluctuating at very different densities, tend to show parallel variations in body measurement. Rather than assume that body size had a non-zero heritability from the Oxford data on the related great tit, and aware of the fact that heritabilities are a ratio in which environmental variation is also included, I made repeated estimates of the heritability of tarsus length in the same population (cf. Table 1). Furthermore, I tried to correct for a possible inflation of this value, caused by bird/environment interaction through cross-fostering experiments (Dhondt 1982). Although tarsus length and body weight are correlated, the average tarsus length of the nestlings born in 1980 was smaller than that measured in 1979, although a substantial increase in mean body weight was found. This was observed in the three populations. To me, the obvious interpretation is that the smaller size of the birds in 1980 compared to 1979 reflects a micro-evolutionary change. The birds in 1979 were unusually large, rather than the birds in 1980 unusually small, as a result of the extreme winter 1978–79. This is confirmed by the observation that nestlings were large again in 1985 compared to 1984, after the cold winter 1984–85, although

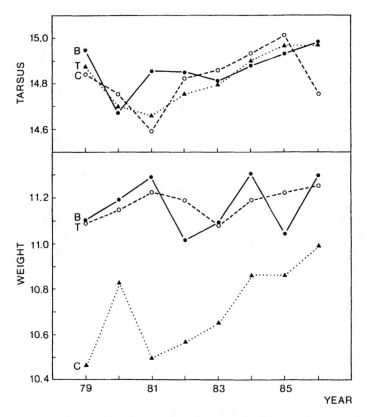

Fig. 2. Variations in 15-day nestling measurements between years in three blue tit populations subjected to different levels of intra- and interspecific competition. Mean measurements of first brood young only. Note that (1) weight is always lower in *plot C*; (2) tarsus length does not differ between plots; (3) variations in tarsus length tend to occur in parallel between the plots; (4) between 1979 and 1980 parallel changes occurred in the three plots, whereby tarsus length decreased but weight increased. This suggests that a micro-evolutionary change towards smaller birds was being observed (for more details, see text)

Table 1. Repeated heritability estimates of tarsus length in the blue tit population in plot C (Fig. 2) through mid-parent/mid-offspring regression reveal that heritabilities are significantly non-zero in all years, but differ significantly between years (Analysis of covariance comparing heritabilities: F = 3.054, df = 5,209, P < 0.01)

Year	h²	± SE
1981	0.51	0.159
1982	0.64	0.169
1983	0.65	0.209
1984	0.38	0.112
1985	0.48	0.099
1986	1.07	0.131

body weight was relatively light. This also fits well with the data presented by Dhondt et al. (1979) for the great tits at Ghent: their data series started in 1963, after the coldest winter on record this century. In order, however, to understand precisely how natural selection operates, I still need to analyze in great detail and year by year what birds survived and which birds died.

This example illustrates the idea that for an evolutionary ecologist, who is interested in studying evolution during his lifetime in the field, it may, in the beginning anyway, be useful to simply assume that some of the observed phenotypic variation has a heritable basis. Once changes are observed that could be interpreted as micro-evolutionary changes, however, he will have to try to substantiate this. The knowledge of the underlying genetic mechanisms combined with the observation of the operation of natural selection through differential survival of the adults or differential recruitment into the breeding population of the offspring will provide fairly good insight into the evolutionary processes.

Given the actual status of population genetics it will, in many cases, be impossible to provide a foundation for the assumptions. In that case the ecologist will have to reason from first principles.

References

Bergmans M (1983) Population biology of the harpacticoid copepod *Tisbe furcata* (Baird 1837). Thesis Vrije Univ Brussel

Berthold P (in press) The control of migration in European warblers. Proc XIX Congr Int Ornithol

Berthold P, Querner U (1981) Genetic basis of migratory behavior in European warblers. Science 212:82–85

Biebach H (1983) Genetic determination of partial migration in the European robin *(Erithacus rubecula)*. Auk 100:601–606

Dhondt AA (1982) Heritability of blue tit tarsus length from normal and cross-fostered broods. Evolution 36:418–419

Dhondt AA (1983) Variations in the number of overwintering stonechats possibly caused by natural selection. Ring Migrat 4:155–158

Dhondt AA (1986) The effect of an extreme winter on per-capita growth rates in some resident bird populations: an example of r-selection? Biol J Linn Soc 28:301–314

Dhondt AA, Eyckerman R (1980) Competition between the great tit and the blue tit outside the breeding season in field experiments. Ecology 61:1291–1296

Dhondt AA, Eyckerman R, Hublé J (1979) Will great tits become little tits? Biol J Linn Soc 11: 289–294

Dhondt AA, Schillemans J, De Laet J (1982) Blue tit territories in populations at different density levels. Ardea 70:185–188

Garnett MC (1981) Body size, its heritability and influence on juvenile survival among great tits *Parus major*. Ibis 123:31–41

Noordwijk AJ van (1980) On the genetical ecology of the great tit *(Parus major)*. D Thesis Rijksuniv Utrecht

Unexploited Dimensions of Optimization Life History Theory

E. Pásztor[1]

1 Introduction

Darwin's theory of evolution was not a genetic theory; it could not be that. The idea that a comprehensive theory of evolution must be genetic was a new and strong paradigm of the synthetic theory only. As the synthetic theory dominated so strongly, those evolutionary biologists who do not follow a genetic approach (e.g. Oster and Wilson 1978) should give better arguments for their point of view. I am a heretic of this kind, too, and I would like to show that it is possible to explore some unexploited dimensions of life history theory by the optimization method and that a non-genetic but systematic approach can have some consequences for the general theory of evolution.

Although the theoretical and empirical studies of ecology and evolution of life histories constitute a quickly developing branch of evolutionary ecology, its development seems to have slowed down in the 1980s. On the one hand, there have been considerable technical achievements, e.g. optimization methods of economy were applied to life history (Taylor et al. 1974; Goodman 1982; Schaffer 1983), statistical and empirical techniques were refined to measure the cost of reproduction (for a review, see Reznick 1985) and other allocation patterns (e.g. Pritts and Hanckok 1983) and quantitative genetic methods were developed further to analyze life history and other age-specific traits in field and experimental populations (e.g. Cheverud et al. 1983; van Noordwijk 1984; Baldwin and Dingle 1986).

On the other hand, the same ideas are discussed and the same theories are tested nowadays as 10 years ago. Twenty years have past since the publication of such "bestsellers" of evolutionary ecology as William's book on the nature of adaptation (1966) and MacArthur and Wilson's book on island biogeography (1967). Some particular arguments of these books might be refined today, but their guiding role in evolutionary ecology is the same. It has turned out that nothing is as simple as it was thought in the 1960s, however, we do not know yet how to construct a new theory from recent developments which would be as attractive and coherent as the previous one.

---

1 Department of Genetics, Eötvös Loránd University, Muzeum körut 4/a, H-1088 Budapest, Hungary

Population Genetics and Evolution
G. de Jong (ed.)
© Springer-Verlag Berlin Heidelberg 1988

Strong guiding ideas such as the allocation principle or r, K-selection have not been developed recently.

It is true even today that the study of life history evolution needs a "more comprehensive theory that makes more readily falsifiable predictions" (Stearns 1976). In the present chapter a consciously reductionist and mechanistic but systematic approach is presented towards such a more comprehensive theory. Following a brief summary of the structural characteristics of life history theory and the main problems connected to it, some of the classic questions will be discussed on a modified conceptual base. It will be shown that changes in life history strategy can be discussed systematically as the outcome of optimization processes in specified ecological situations.

2 Guiding Ideas and Problems Left Open

"Which life histories are appropriate to which ecological circumstances?" (Sibly and Calow 1983) is the often mentioned basic problem of the theory of life history evolution. Expressed in a more technical form, the general life history problem is "to compute a schedule of age-specific reproduction and mortality which, given environmental constraints and the biology of the species in question, is likely to be favored by natural selection" (Schaffer 1983). Neither the question nor the generalization is self-evident. In Table 1 a brief and arbitrary summary is given of those ideas which made it possible

Table 1. Main guiding ideas of the theory of life history evolution

Paradigm: Life History is an Adaptation		
The ideas	Some notes	Originating from
Cluch size is optimized	n-s Trade-off Field studies	Lack 1947
Fitness set is bounded Principle of allocation exists	Graphic analysis Fitness set Adaptive function Heterogeneous environment	Levins 1962, 1968
Reproduction is costly	n-p Trade-off Reproductive effort Cost of reproduction	Williams 1966
Population density has a key role in determining the optimal strategy	Density-dependent selection r,K-Dichotomy	MacArthur 1962
Age specificity of the environmental fluctuations has a key role in determining the optimal strategy	n-p Trade-off Demographic studies	Murphy 1966 Schaffer 1974
Allocation patterns can be studied in the field	Field studies of biomass, energy and time allocation	Harper 1967 Tinkle 1969
Tactics can be characterized by covariation of life history traits	Multivariate statistical studies on correlation structures	Pianka 1970 Stearns 1976

to treat specific life history characteristics as adaptations, to ask theoretical and practical questions to be studied and to give techniques to answer the questions. Altogether these have led to the formulation of the general life history problem.

None of the statements listed in Table 1 nor the models connected to them can be falsified or tested properly (see e.g. Rose 1983; Thompson and Stewart 1981). This has a simple reason. The same words for important concepts have quite different operational meanings to theoreticians, model makers, experimentalists and field biologists (Juhász Nagy 1970). This weak methodology often causes confusion, endless debates and inefficient research both in ecology and evolutionary biology (Pásztor 1986). Therefore, the main purpose of standardization of the terminology is to improve the actual research (Endler 1986).

What are life history traits? Model makers treat them as fitness components. They are seen as demographic variables in field studies of animal populations. Botanists measuring allocation patterns use "life history traits", the relation of which is obscure both to the fitness components and to the demographic variables (Thompson and Stewart 1981). But life history traits even as fitness components are interpreted in different ways with different consequences. Population geneticists, who are interested in the explanation of the maintenance of genetic variation ascribe differences in the fitness components of genotypes and advise making intrapopulational studies. For quantitative geneticists, life history traits are "major fitness components" (Falconer 1981, p. 302); they are treated as metric characters. Evolutionary ecologists using optimization methods try to interpret both intra- and interpopulation comparisons (Reznick 1985) though they are sometimes strongly advised to be cautious about predicting anything (Oster and Wilson 1978). All these approaches may be appropriate in their own place but their relevance to each other is not really known.

Since Fisher's basic work (1930) the intrinsic rate of growth, r, or its discrete analogue λ, defined by the Euler-Lotka equation:

$$1 = \int_0^\infty e^{-rx} l(x) m(x) dx$$

$$1 = \sum_{x=0}^\infty \lambda^{-rx} l(x) m(x)$$

have been accepted as the most immediate measure of the absolute fitness of a phenotype. These measures also work in the case of density-dependent population growth and selection (Charlesworth 1973, 1980; Sibly and Calow 1983). When population growth is density-dependent all genotypes have negative intrinsic rates of growth ($r < 0$ or $\lambda < 1$) at equilibrium, expect the one which maintains the equilibrium density. As each optimization criterion should be attributed to population growth rates, demographic optimization models play a crucial role at least among other optimization models.

In this chapter life history traits are demographic variables and considered as fitness components. Differences between the average values of demographic variables of populations at different localities can be interpreted with the help of our general model (Meszéna and Pásztor 1986). These populational averages are not supposed to be equal to the optimal strategies but the differences between these average values can be interpreted by analyzing the differences between the optimal strategies in different environments.

3 A Systematic Approach to Life History Optimization

In optimization theory the differences between populations at different localities are interpreted as the result of adaptation to different environmental conditions. From a technical point of view it means that the changes of the optimal strategy with changing environment has to be calculated. A life history strategy is represented by the fitness component vector c. All the possible strategies constitute the fitness component set C in a given environment. C can be represented in an n-dimensional space. In Fig. 1 a plausible case is shown where an increasing number of offspring per parent reduces both juvenile and parental survival. The optimal strategy maximizes the adaptive function, i.e. in the case of demographic optimization models, that strategy is the optimal one which has the maximal r or λ in a given environment. If the values of the considered fitness components depend on the environment, the optimal strategy also depends on the environment. The optimal strategy can change by changing the adaptive function or the C set. However, the adaptive function can change only by changing at least one of the fitness components, i.e. changing the C set.

As those environmental factors which do not change the shape of the fitness component set cannot change the optimal strategy, first we have to determine how the environment can change the C set. The fitness components can be independent of or dependent on each other. If a set of some fitness components cannot be separated into independent groups, it is said that a trade-off exists between the involved fitness components. Environmental factors may alter the C set by changing the independent, demographic variables and/or the shape of the trade-off functions directly. For example, higher predation rate may decrease the survival rate of each genotype. However, the environmental changes may influence the C set indirectly as well, by changing some phenotypic traits or the population density. Deteriorating environmental conditions can reduce fertility by reducing the individual growth rate of each genotype (e.g. Mills and Eloranta 1985). In a population where population size is limited in the breeding season, for example, juvenile survival can be density-dependent in the breeding season. In an environment where mortality is increased in the non-breeding season (e.g. because of severe winter conditions) the relatively lower density and reduced competition improves juvenile survival in these populations in the breeding season.

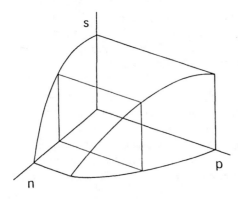

Fig. 1. Three-dimensional fitness component set. There is a trade-off between fertility (n) and young survival (s) and between fertility and parental survival (p), but young and parental survival are independent of each other

In general, the admissible set of the fitness component vectors (C) is determined by the phenotypic traits (t_i), the environmental conditions (E_j) and the critical population densities (N_k). Critical population densities are those which are limiting in certain periods, e.g. the density of breeding pairs in spring. (This concept is analogous to that of the critical age groups; see e.g. in Michod 1979.)

How do the optimal life history strategies differ in different environments? In order to solve this problem we must:

1. Specify the trade-off structure (separate the fitness components into independent groups);
2. Specify the adaptive function (simplify the Euler-Lotka equation if possible);
3. Give the environment- and density-dependent boundaries of the C set;
4. Measure the difference between the environments of the compared populations in each particular case.

The present form of our model does not consider the environmental modification of the phenotypic traits creating trade-offs between the fitness components explicitly.

Just to illustrate the effectivity of this general approach it is applied to a classic problem of the theory of life history evolution. What is the optimal fertility under different circumstances (Table 1)?

1. Fitness Structure. The two main effects of increasing number of offspring per parent (n) are the reduction of both juvenile (s) and parental (p) survival. We suppose that s and p depend on fertility (n) but they are independent of each other and other fitness components (Fig. 1). This simplification means that phenomena such as variation in parental care are not considered.

2. Adaptive Function. If a female of a certain phenotype produces n female offspring whose survival rate is s until the next census, the number of its female offspring is ns. The expected number of this phenotype at the next census is ns+p if p is the probability of the survival of the parent until the next census. As n, s, p are supposed to be independent of the other fitness components, the individuals are counted when their survival becomes independent of their family size. The fitness is proportional to ns+p if the parents and their offspring are indistinguishable at the time of the next reproduction. Under these restrictions maximization of ns+p is equivalent to the maximization of r or λ.

3. The C Set. In the simplest possible case the environment does not change the trade-off structure itself and only one component is changed directly. For example, nest predation may decrease only juvenile survival. Environmental changes are characterized by the E variable. If E increases, one or more fitness components decrease. In order to specify the density dependence of the C set, the type of density regulation had to be specified. In the simplest conceivable case there is only one factor that limits population growth; at equilibrium it has the same effect in each generation. In a seasonal environment this limiting factor can operate in the breeding season ("Lack-type regulation"), i.e. n, s or p can be density-dependent; or in the non-breeding season ("Birch-type"), i.e. survival in the non-breeding season (l) is density-dependent; or in both seasons (compound type).

4. Characterization of the Environment. The difference between the environments of the compared populations has to be expressed as their effects on the shape of the C set of the populations. The Malthusian parameter is:

$$\lambda = l\,(ns + p)$$

in the seasonal environments introduced above. Our adaptive function is:

$$\lambda' = ns + p\,.$$

The primary difference between the environments can produce some change in l or λ'. As a consequence population density may change and it can cause secondary effects. Considering two types of population regulation, four main situations can be distinguished (Table 2).

In which cases will an increased mortality be balanced by an increase in the optimal fertility? As the value of n, s, p depends on the environment and critical population density their optimal values will be environment- and density-dependent as well. So by the help of the optimality conditions the effect of small environmental and density changes on the optimum can be analyzed. (The treatment given follows the one which was given in more detail by Meszéna and Pásztor (1986.)

The optimality conditions are:

$$\left(\frac{\partial \lambda'}{\partial n}\right)_{N,E} = 0\,; \qquad \frac{\partial^2 \lambda'}{\partial n^2} < 0\,. \tag{1}$$

The population number does not change, therefore

$$\lambda = 1\,. \tag{2}$$

Table 2. Conditions for increasing optimal fertility with increasing mortality (i.e. conditions for $dn/dE > 0$)[a]

Density dependence	Decrease in the interbreeding season (1)	Decrease in the breeding season (λ')
1 Birch-type regulation		$snS_{En} + pP_{En} < pP_n\,(P_E{-}S_E)$
λ' Lack-type regulation	$snS_{Nn} + pP_{Nn} > pP_n\,(P_N{-}S_N)$	$\dfrac{\lambda'_{En}}{\lambda'_E} < \dfrac{\lambda'_{Nn}}{\lambda'_N}$

[a] Notations: n = number of offspring; s = survival of the offspring from birth until the end of the effect of the trade-off; p = survival of the parents from the beginning of the breeding season until the end of the effect of trade-off; 1 = survival in the interbreeding season (independent of n, s, p and age); R = ns + p; E = environmental variable (the demographic variables are decreasing functions of E); N = population density at the beginning of the breeding season (in case of Lack-type regulation) or population density at the beginning of the interbreeding season (in case of Birch-type regulation).
These quantities (which are positive in most cases) measure the relative disadvantage caused by

$S_E = -\partial \ln s/\partial E$	$P_E = -\partial \ln p/\partial E$	Harshening environment
$S_N = -\partial \ln s/\partial N$	$P_N = -\partial \ln p/\partial N$	Crowding
$S_n = -\partial \ln s/\partial n$	$P_n = -\partial \ln p/\partial n$	Increasing fertility
$S_{En} = -\partial^2 \ln s/\partial E\partial n$	$P_{En} = -\partial^2 \ln p/\partial E\partial n$	
$S_{Nn} = -\partial^2 \ln s/\partial N\partial n$	$P_{Nn} = -\partial^2 \ln p/\partial N\partial n$	
$\lambda'_E = -\partial \lambda'/\partial E$	$\lambda'_N = -\partial \lambda'/\partial N$	
$\lambda'_{En} = -\partial^2 \lambda'/\partial E\partial n$	$\lambda'_{Nn} = -\partial^2 \lambda'/\partial N\partial n$	

The first equation implicitly defines n_0 as a function of N,E. Applying the implicit function theorem we get:

$$\frac{\partial^2 \lambda'}{\partial n^2}\bigg|_{n_0} dn_0 + \frac{\partial^2 \lambda'}{\partial n \partial E}\bigg|_{n_0} dE + \frac{\partial^2 \lambda'}{\partial n \partial N}\bigg|_{n_0} dN = 0 , \tag{3}$$

but λ has to remain constant:

$$\frac{\partial \lambda}{\partial n} dn_0 + \frac{\partial \lambda}{\partial E} dE + \frac{\partial \lambda}{\partial N} dN = 0 . \tag{4}$$

Expressing dN from Eq. (4) and replacing to Eq. (3) after rearrangement, it can be written

$$\frac{dn_0}{dE} = -\frac{1}{\dfrac{\partial^2 \lambda'}{\partial n^2}} \left(\frac{\partial^2 \lambda'}{\partial n \partial E} - \frac{\partial^2 \lambda'}{\partial n \partial N} \frac{\dfrac{\partial \lambda'}{\partial E}}{\dfrac{\partial \lambda'}{\partial N}} \right) . \tag{5}$$

Optimal fertility increases with increasing mortality, i.e.

$$\frac{dn_0}{dE} > 0 \quad \text{if} \quad \frac{\dfrac{\partial^2 \lambda'}{\partial n \partial E}}{\dfrac{\partial \lambda'}{\partial E}} < \frac{\dfrac{\partial^2 \lambda'}{\partial n \partial N}}{\dfrac{\partial \lambda'}{\partial N}} . \tag{6}$$

If λ' is density-dependent (Lack-type) and the environment changes only l, then from Eq. (3)

$$\frac{dn_0}{dE} > 0 \quad \text{if} \quad \frac{\partial^2 \lambda'}{\partial n \partial N} < 0 . \tag{7}$$

If l is density-dependent (Birch-type) and the environment changes λ', then

$$\frac{dn_0}{dE} > 0 \quad \text{if} \quad \frac{\partial^2 \lambda'}{\partial n \partial E} > 0 . \tag{8}$$

As $\lambda' = ns + p$ (after some simple algebra) one can arrive at the results summarized in Table 2.

For easier evaluation we had to introduce some new notations. S_E, P_E, S_N, P_N, are the relative environment and density dependence of juvenile and parental survival rates respectively. They measure the magnitudes of the direct and indirect environmental effects (Fig. 2). S_n, P_n are the relative (specific) fertility dependence of juvenile and parental survival rates which measure the relative disadvantage (cost) of increasing fertility in young and parental survival, i.e. they characterize the strength of the trade-off. Direction and magnitude of the change in optimal fertility depends on the magnitudes of the direct and indirect environmental effects (S_E, P_E, S_N, P_N), the strength of the trade-off (S_n, P_n), the relative importance of the two parts of the adaptive function (ns and p) and degree of the environmental effect with different fertilities (S_{En}, P_{En}, etc.).

Without giving full details, the main conclusion of our model for populations with Lack-type regulation is as follows:

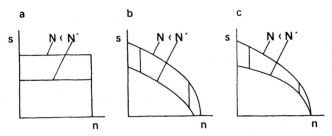

Fig. 2a–c. Fitness component sets. **a** The two fitness components (n = fertility, s = young survival) are independent of each other ($ds/dn = 0$). The boundary of the C set decreases with increasing population density ($ds/dN < 0$). **b** There is a trade-off between n and s. At higher densities the boundary of the C set (absolute disadvantage of higher fertilities) is reduced with the same value ($\partial^2 s/\partial n \partial N = 0$). **c** At higher densities the boundary of the C set (relative disadvantage of higher fertilities) is reduced proportionally with young survival ($\partial^2 \ln s/\partial n \partial N = 0$ i.e. $S_{Nn} = 0$)

$$sn S_{Nn} + p P_{Nn} > p P_n (P_N - S_N) .$$

It is a very plausible idea and a basic assumption of the r,K-strategy is that the relative disadvantage increases with increasing population density in large families (e.g. Mac-Arthur and Wilson 1967), i.e. S_{Nn} and P_{Nn} are positive. If this is true, then the left-hand side of the inequality is positive. The right-hand side will be positive only if parental survival is more sensitive to increasing density than young survival. This is rather unlikely. Consequently, in a population with Lack-type density regulation optimal fertility will increase in most cases because of increased mortality in the interbreeding season. It seems to be a very general mechanism and robust result.

In both type of populations increased juvenile survival (e.g. caused by lower predation rate or decreased density because of increased mortality in the interbreeding season or both) may lead to higher optimal fertility, even if $S_{Nn} < 0$ or $S_{En} > 0$, i.e. even if the relative disadvantage of higher fertility slightly increases with decreasing density or improving environmental conditions. If P_n is high, then improved juvenile survival upgrades the value of the offspring compared to that of the parents and optimal fertility will increase. Improved parental survival works against the increase of the optimal fertility when P_n is high, since in this case the value of the parents is upgraded compared to that of the offspring.

These results are not really surprising, though the effects of age-specific mortality factors and density dependence were discussed explicitly only in the reproductive effort model (Michod 1979; Goodman 1979). If we interpret s and p in the conventional way and use the concept of effective fertility, then our result will be very similar to that of Schaffer and Gadgil (1974), i.e. optimal reproductive effort will decrease with crowding if juvenile survival is more sensitive to increased density. Discussion of Cole's principle (Charnov and Schaffer 1973; Goodman 1974), analysis of the effect of fluctuating environment (Schaffer 1974) and the analysis of the n-p trade-off by Sibly and Calow (1983) have already shown that age specificity of mortality (i.e. if young or parents are affected) has an important role in determining the optimal strategy. What we might see more clearly now is the series of assumptions that must be tested in each particular situation.

4 Some Consequences

There are some popular explanations of fertility trends in life history literature. It may be useful to discuss some of them according to Stearns's widely cited review (1976, Table 5). The trend in clutch size (see below) is the "increased number of progeny of a constant size". The constant size assumption means that the optimal value of a phenotypic trait (size) with an effect on the n-s-p trade-off is not supposed to vary.

Increased food availability leading to an increase in total resources available for reproduction favours increasing clutch size (Lack 1954).

Lack's hypothesis on global clutch size variation seems to be self-evident. If parents have more time to collect food (total resources available for reproduction are increased), then more young can be reared and optimal clutch size will increase. However, this hypothesis is self-evident only if food availability is determined by the physiological capacity of the birds and the amount of time they can spend foraging per day and it is independent of population density.

However, in life history theory food availability is usually thought to be determined by population density, i.e. population number related to the amount of the limiting resource. Increased foraging time may then lead to increased population density instead of increased optimal fertility. If the environments of the compared populations differ only in the level of limitation (e.g. in the amount of the limiting resource), then in the λ'/λ' part of Table 2 the inequality becomes an equation and optimal fertility will not change (Meszéna and Pásztor 1986). Gadgil and Bossert (1910), who even introduced a new concept ("degree of satisfaction: the fraction of the resources actually available to the individual that is maximally utilizable") for separating population and individual levels of limitation, arrived at the same conclusion. A more special but population genetics model had similar results (de Jong 1984).

In populations with Lack-type density regulation the effect of increased day length might not be so simple; the increase in foraging time might increase the limiting resource level, but might also influence the mode of density dependence. Therefore, both density and optimal fertility may change depending on the actual mechanisms.

These considerations narrow down the applicability of the original hypothesis for explaining global clutch size variation. First, it should be tested whether food availability for clutches are determined by population density or not.

A decrease in the resources required for competition, predator avoidance, maintenance and migration leads to larger clutches (Skutch 1949, 1967; Cody 1966).

Competition, predation, requirements of maintenance and migration can be considered as sources of mortality in a local population which may differ in different environments. Our table can be used after specifying the age specificity of the effects. As already discussed, an increase in juvenile survival will increase the optimal fertility where strong n-p trade-off exists. If n-p trade-off is weak, then the selectivity of these mortalities will be decisive. For example, if bigger clutches had higher predation risk ($S_{En} > 0$; where E is the intensity of predation), then lower predation rate will go together with higher optimal clutch size. All these effects will be different if parental survival affected. The hypothesis is plausible in the case of competition and predator avoidance, if it is mainly survival of the young that varies.

Higher clutch size is favoured by r-selection:
1. *Increased density-independent mortality relative to density-dependent mortality (MacArthur and Wilson 1967);*
2. *More variable or unpredictable environment (Murphy 1968);*
3. *Decrease in adult survival: decrease in competition (Pianka 1970).*

What is the relation of our approach to r- and K-selection? In our approach both density-dependent and density-independent selection can be described. As was proven several times (e.g. in Sibly and Calow 1983) the maximization criteria are the same in both cases. Our model is related to density-dependent selection when the C-set (i.e. the set of the admissible strategies) is density-dependent. The interesting questions are: what are the conditions (e.g. type of population regulation, etc.) in which an "r-strategy" such as high fertility is optimal, which environmental changes lead in this direction?

1. As it was discussed above, it seems to be a very general and plausible mechanism that an increase in interbreeding mortality is compensated by increased fertility if the critical population density is that of the parental population at the beginning of the breeding season and juvenile survival is density-dependent. In this case decreased density-dependent mortality actually corresponds to higher optimal fertility. But the increase in the interbreeding mortality is not necessarily density-independent. In populations with compound-type density regulation winter mortality may increase because the limitation becomes more severe (as it is supposed, e.g. in Ricklefs 1980) and optimal fertility can increase as in the case of Lack-type regulation.

2. The effect of variable and unpredictable environments cannot be discussed on the basis of our model.

3. If the decrease of adult survival means decrease of adult survival in the non-breeding season, then this hypothesis is very plausible in populations with Lack-type regulation, as was shown before. If competition is decreased because of decreased adult survival in the breeding season, then the situation is described by the λ'/λ' part of Table 2. Without further specifications such self-evident generalization is not possible.

Without using the concept of reproductive effort the problem of "partitioning reproductive effort into progeny of various sizes" is a problem of finding the optimal size of the offspring in addition to the optimal values for the life history traits. In each hypothesis it is supposed that both fertility and young survival depend on the size of the offspring. The causal systems listed by Stearns give those ecological conditions which are generally supposed to modify the survival rate of young with different sizes (size-selective predation, competition, etc.). There are only few explicit models of such situations but with interesting results (e.g. Christiansen and Fenchel 1979; Stearns 1984; Sibly et al. 1985).

These considerations should have some lessons for those who would like to investigate differences between demographic variables of different populations as differences between optimal strategies either in nature or in the laboratory.

All the life history variables have to be considered which can be responsible for the empirical differences to be explained. The independent and dependent life history variables should be separated. The experimental system itself can often influence the trade-off structure. This should be taken into consideration. Our results and the existence of such trends as the global clutch size variation confirm that density dependence might play an important role in many variations. Consequently, it is necessary to specify and test the mode of population regulation and its relation to the considered life history traits as well. Direct and indirect effects of the environment should be separated.

It is certainly true that such general models like ours are not appropriate for empirical investigations without further specification. However, the presented approach provides a chance to build up a program which tests the *existence of relations*: e.g. whether a trade-off exists or not, is density dependence important or not, which age-specific rates change, etc. Experimental analysis of the environment and density dependence of trade-off structures would be extremely important to construct more specific models.

5 Relations to Some Evolutionary Concepts in Fashion

A widely accepted but rarely fulfilled demand of current evolutionary theory is that "organisms must be analysed as integrated wholes" (Gould and Lewontin 1979). What is the contribution of life history theory to this requirement?

Similarly to other evolutionists, evolutionary ecologists propose adaptive stories (sensu Gould and Lewontin 1979) when they want to explain the existence of trade-offs between life history traits. Furthermore, they sometimes call these trade-offs constraints (e.g. Ricklefs 1983). On the other hand, life history theory and the optimization approach in general are often blamed that "trade-offs among competing selective demands exert the only brake upon perfection" (Gould and Lewontin 1979) or that design constraints cannot be treated properly within the scope of the theory (Stearns 1977, 1982).

In the adaptive stories of life history theory trade-offs between life history traits are supposed to be determined by some particular phenotypic traits (e.g. size, developmental time, foraging behaviour, etc.) in a particular ecological situation (e.g. competition, predation, harshening conditions, etc.). The variation of these phenotypic traits can be limited or biased just as any variation of other phenotypic traits by developmental constraints. Developmental constraints (sensu Maynard Smith et al. 1985) on life history evolution are those features of the developmental systems which lead to bias on the production of life history strategies under specified environmental conditions. In order to evaluate the effects of developmental constraints on the optimal life history strategies, their effects on the boundaries of the C set must be described under specified environmental conditions in a given ecological situation. The effect of the so-called design constraints can be determined by analyzing their effects on the trade-off structure as well.

One point in the fitness component space belongs to one genotype in a given environment. The C set contains many genotypes. The boundaries of the C set are determined by some phenotypic traits and the environment. The expression of the genes may be different under different environmental conditions, i.e., various phenotypic traits and fitness components may belong to the same genotypes in different environments. Changing the effective environmental variable and measuring the change in the phenotypes and the related fitness components, we get something that is usually called a reaction norm. The new boundaries of the C set at the new value of the environmental variable are determined by the reaction norms of all possible genotypes besides the direct effects of the environment. In our approach the problem of phenotypic plasticity can be treated in this way.

Trade-offs between life history traits are not equal to trade-offs among competing needs (such as needs of reproduction and maintenance) and it is just their structure that should be analyzed in order to consider organisms as "integrated wholes". Life history traits can be defined operationally but "competing needs" cannot, at present. Competing needs must be resolved into phenotypic traits, the effect of which on the trade-off structure depends on the environment, in order to make the concept operational. If unfolding the structure of fitness itself is the focus (Christiansen 1984), then the differences between population genetics and optimization approaches do not seem to be so large. Some population genetics and optimization models with an analogous fitness structure have similar results (Prout 1980; de Jong 1984; Eva Kisdi, in preparation).

Acknowledgements. We had very inspiring discussions on the presented material in the population biology group of our university whose existence is due to Gábor Vida. Géza Meszéna was very helpful in problem solving. F.B. Christiansen gave useful comments. Eva Ludvig prepared the drawings and corrected my English. Ágnes Major, Jolán Mohay and Éva Kisdi, by taking over my teaching duties, made it possible for me to work on this paper. I am very grateful to all of them.

References

Baldwin JD, Dingle H (1986) Geographic variation in the effects of life-history traits in the large milkweed bug *Oncopeltus fasciatus*. Oecologia 69:64–71

Charlesworth B (1973) Selection in populations with overlapping generations. IV. Natural selection and life histories. Am Nat 107:303–311

Charlesworth B (1980) Evolution in age structured populations. Cambridge University Press, Cambridge, UK

Charnov EL, Krebs JR (1973) On clutch size and fitness. Ibis 116:217–219

Charnov EL, Schaffer WM (1973) Life history consequences of natural selection: Cole's result revisited. Am Nat 107:791–793

Cheverud JM, Rutledge JJ, Atchley WR (1983) Quantitative genetics of development: genetic correlations of age-specific trait values and the evolution of ontogeny. Evolution 37:895–905

Christiansen FB (1984) The definition and measurement of fitness. In: Schorrocks B (ed) Evolutionary ecology. 23rd Symp Br Ecol Soc, Leeds 1982. Blackwell, Oxford London

Christiansen FB, Fenchel TM (1979) Evolution of marine invertebrate reproductive patterns. Theor Popul Biol 16:267–282

Cody ML (1966) A general theory of clutch size. Evolution 20:174–184

de Jong G (1984) Selection and numbers in models of life histories. In: Wöhrmann K, Loeschcke V (eds) Population biology and evolution. Springer, Berlin Heidelberg New York, pp 87–102

Endler JA (1986) Natural selection in the wild. Princeton Univ Press

Falconer DS (1981) Introduction to quantitative genetics. Longman, London New York

Fisher RA (1930) The genetical theory of natural selection. Clarendon, Oxford

Gadgil M, Bossert WM (1970) Life historical consequences of natural selection. Am Nat 104:1–24

Goodman D (1974) Natural selection and a cost-ceiling on reproductive effort. Am Nat 108:247–268

Goodman D (1979) Regulating reproductive effort in a changing environment. Am Nat 113:735–748

Goodman D (1982) Optimal life-histories, optimal notation and the value of reproductive value. Am Nat 108:247–268

Gould SJ, Lewontin RC (1979) The spandlers of San Marco and the panglossian paradigm: a critic of the adaptationist programme. Proc R Soc London Ser B 205:581–598

Harper JL (1974) A Darwinian approach to plant ecology. J Ecol 55:247–270

Juhász Nagy P (1970) Egy operativ ökológia hiánya és szükséglete. MTA Biol Oszt Közl 12:441–464

Lack D (1947) The significance of clutch size. I. Ibis 89:302–352

Lack D (1954) The natural regulation of animal numbers. Oxford Univ Press

Levins R (1962) Theory of fitness in a heterogenous environment I. The fitness set and adaptive function. Am Nat 96:361–373

Levins R (1968) Evolution in changing environment. Princeton Univ Press

MacArthur RH (1962) Some generalized theorems of natural selection. Proc Natl Acad Sci USA 48:1893–1897

MacArthur RH, Wilson EO (1967) The theory of island biogeography. Princeton Univ Press

Maynard Smith J, Burian R, Kauffman S, Alberch P, Campbell J, Goodwin B, Lande R, Raup D, Wolpert L (1985) Developmental constraints and evolution. Q Rev Biol 60:265–287

Meszéna G, Pásztor E (1986) Denzitásfüggö életmenet stratégiák 1. A termékenység és a fiatalkori halálozás denzitásfüggésének vizsgálata. Abstr Bot 10:97–116

Michod RE (1979) Evolution of life-histories in response to age-specific mortality factors. Am Nat 113:531–550

Mills CA, Eloranta A (1985) Reproductive strategies in the stone loach Noemacheilus barbatulus. Oikos 44:341–349

Murphy GI (1968) Pattern in life history and the environment. Am Nat 102:390–404

Noordwijk AJ van (1984) Quantitative genetics in natural populations of birds, illustrated with examples from the great tit, Parus major. In: Wöhrmann K, Loeschcke V (eds) Population biology and evolution. Springer, Berlin Heidelberg New York, pp 67–83

Oster GF, Wilson EO (1978) Caste and ecology of social insects. Princeton Univ Press

Pásztor E (1986) Levél a populációbiológiáról. Módszer-Elmélet-Tudomány Világosság 27:92–100

Pianka ER (1970) On r- and K-selection. Am Nat 104:592–597

Pritts MP, Hanckok JF (1983) Seasonal and life time allocation patterns in the woody goldenrod Solidago pauciflosculosa Michaux. (Compositae) Am J Bot 70:216–221

Prout T (1980) Some relationships between density-independent selection and density-dependent population growth. Evol Biol 13:1–68

Reznick D (1985) Cost of reproduction: an evaluation of the empirical evidence. Oikos 44:257–267

Ricklefs RE (1980) Geographical variation in clutch size among passerine birds: Ashmole's hypothesis Auk 97:38–49

Ricklefs RE (1983) Comparative avian demography. In: Johnston RF (ed) Current ornithology, vol 1. Plenum, New York Lodon, pp 1–32

Rose MA (1983) Theories of life-history evolution. Am Zool 23:15–23

Schaffer WM (1974) Optimal reproductive effort in fluctuating environments. Am Nat 108:783–790

Schaffer WM (1983) The application of optimal control theory to the general life history problem. Am Nat 121:418–431

Schaffer WM, Gadgil MD (1974) Selection for optimal life histories in plants. In: Cody ML, Diamond J (eds) The ecology and evolution of communities. Harvard Univ Press, Cambridge, Mass, pp 142–157

Sibly R, Calow P (1983) An integrated approach to life-cycle evolution using selective landscapes. J Theor Biol 102:527–547

Sibly R, Calow P, Nichols N (1985) Are patterns of growth adaptive? J Theor Biol 112:553–574

Skutch AF (1949) Do tropical birds rear as many young as they can nourish? Ibis 91:430–455

Skutch AF (1967) Adaptive limitation of the reproductive rate of birds. Ibis 109:579–599

Stearns SC (1976) Life-history tactics: a review of the ideas. Q Rev Biol 51:3–47

Stearns SC (1977) The evolution of life history traits: a critic of the theory and a review of the data. Annu Rev Ecol Syst 8:145–171

Stearns SC (1982) The role of development in the evolution of life histories. In: Bonner JT (ed) Evolution and development. Springer, Berlin Heidelberg New York, pp 237–259

Stearns SC (1984) How much of the phenotype is necessary to understand evolution at the level of the gene? In: Wöhrmann K, Loeschcke V (eds) Population biology and evolution. Springer, Berlin Heidelberg New York, pp 31–49

Taylor HM, Goureey RS, Lawrence CF, Kaplan RS (1974) Natural selection of life history attributes: an analytical approach. Theor Popul Biol 5:104–122

Thompson K, Stewart AJA (1981) The measurement and meaning of reproductive effort in plants. Am Nat 117:205–211

Tinkle DW (1969) The concept of reproductive effort and its relation to the evolution of life histories in lizards. Am Nat 103:501–516

Williams GC (1966) Adaptation and natural selection. Princeton Univ Press

Theory of Phenotypic Evolution: Genetic or Non-Genetic Models?

R. F. HOEKSTRA [1]

1 Introduction

Many studies in evolutionary biology can be characterized as attempts to provide functional explanations of phenotypes. A particular phenotype (such as a morphological structure, a life history, a behaviour or a reproductive system) is interpreted as the result of various relevant evolutionary forces and constraints imposed upon the system, due for example to its history or to physical limitations. To a large extent this work is guided by results obtained from the analysis of theoretical models.

If we limit (for a moment) our attention to the last 40 or 50 years, we may observe two things with respect to these models. First, theoretical population genetics has become, rather quickly after its inception, the main formalism in which these models are framed; and secondly, there is a growing tendency in recent years to use non-genetic models, for example based on dynamic programming or game theory (Oster and Wilson 1978; Maynard Smith 1978a). This tendency has, quite naturally, generated the question whether population genetics is necessary for a proper modelling of phenotypic evolution (e.g. Maynard Smith 1982; Stearns 1984).

It is my aim in this chapter to consider some aspects of this question, which have perhaps received less attention than they deserve in the recent literature.

2 Historical Perspective

At first sight, the claim that population genetics is necessary for proper functional explanations of phenotypes becomes hilarious, if one realizes that Darwin has provided many interesting evolutionary explanations without a knowledge of (population) genetics. It is, however, instructive, to have a short look at the developments in the field of evolutionary biology from Darwin's time to the establishment of population genetics as the basic descriptive formalism for the process of (micro-)evolution. A thorough historical account has been given by Provine (1971). In a very condensed from this history is as follows. Very soon after the publication of the *Origin*, disagreement

---

1 Vakgroep Genetica, Biologisch Centrum, Kerklaan 30, NL-9751 NN Haren, The Netherlands

Population Genetics and Evolution
G. de Jong (ed.)
© Springer-Verlag Berlin Heidelberg 1988

arose among evolutionists concerning the nature of the variation upon which natural selection could act. Darwin believed that evolution proceeded by natural selection acting upon small individual heritable differences (what we now call "continuous variation"). But many biologists, notably T.H. Huxley, criticized this aspect of Darwin's theory, and held discontinuous variation (so-called saltations) to be important for evolution. Darwin was convinced that these saltations (or "sports") were too rare to be the primary source of variation, while Huxley was impressed by the fact that saltations appeared to be of the same order as observed differences between species, living and fossil. This controversy increased after the rediscovery of Mendel's work around 1900. During the first decade of this century there was a sharp conflict between the Mendelians and the biometricians. The former favoured the idea of discontinuous evolution by large mutational leaps, while the latter advocated gradual evolution by natural selection on continuous variation, but denied the relevance of Mendelian heredity. But during the following decade there was a gradual decline in the antagonism between Mendelism and Darwinism, and after another decade Mendelian heredity and Darwinian selection were quantitatively synthesized into population genetics.

The important thing to notice in this course of events is that these controversies were about the *mechanism* of evolution (and not about the general idea of evolution as such); and that theoretical population genetics has emerged quite naturally as a formalism to describe the effects of evolutionary forces on the genetic structure of populations – in other words, to describe the mechanism of (micro-)evolution.

Viewed in this way, population genetics is indeed essential for a complete understanding of evolution. It does not follow, however, that the more modest undertaking of finding a functional explanation of a particular phenotype should also be based on population genetics. For answering this last question it is useful to consider briefly the status of functional explanations.

3 Functional Explanations

Philosophers have written a great deal about the special status of functional explanations in biology. I have neither the intention nor the competence to review this subject here, but a few remarks will be relevant for the theme of this chapter. At first sight functional explanations in biology seem to explain certain characteristics of organisms in terms of their *consequences* instead of their *causes*. This has led to the idea that functional explanations are logically different from causal explanations, and therefore form a special category of explanation. However, it appears that functional explanations can be reconstructed as causal explanations, which shows that they are not logically different (for extensive discussions, see Ruse 1973 and Hull 1974). Thus, the functional explanation "Phenotype x of (individuals of) species z has as function y" is equivalent to the following two statements:

1. 'y is caused by x' (proximate part);
2. 'y (and therefore also x) has adaptive value for (individuals of) z (ultimate part).

For example, if x is a particular foraging behaviour in species z, and y stands for maximizing the energy intake per unit time spent foraging, then a complete functional explanation should be as follows. First, to demonstrate the proximate part, one should demonstrate that this particular foraging behaviour indeed maximizes the energy intake (compared with alternative foraging strategies). Secondly, to demonstrate the ultimate part it is necessary to show that natural selection can produce this foraging behaviour under certain specified selection forces.

Generally, evolutionary biologists are much more careful in demonstrating the proximate part than the ultimate part. In non-genetic models only the proximate part of the functional explanation is demonstrated, and the ultimate part is nearly taken for granted. The only aspect of the ultimate part which receives attention in these studies, is specification of the relevant selection forces, since they often play a role in the formulation of the optimization criterion. An adavantage of population genetics models is that demonstration of the ultimate component of a functional explanation is implicit in the methodology.

In the next section I will give a short overview of various models that attempt to explain the phenomenon of anisogamy, as an illustration of the fact that neglect of the ultimate component of a functional explanation leads to an unsatisfactory explanation.

4 Functional Explanations of Anisogamy

Anisogamy is morphological sexual differentiation in gametes, resulting in two types of gametes: relatively large and often immobile (female) gametes, and small motile (male) gametes. Generally, higher organisms are anisogamous, but among the Algae and Fungi isogamy is not uncommon, especially in the more primitive groups.

4.1 First Qualitative Explanation

The first explanation of anisogamy was advanced in the last part of the 19th century by a number of German evolutionists. Among them were Weismann (1886) and Hertwig (1906). They considered isogamy to be the primitive condition, from which anisogamy has been evolved due to two conflicting selective forces. The first is selection for increasing efficiency of finding a mating partner, and the second is selection for provisioning the zygote with a sufficient amount of reserve, necessary for development. In a situation where the gametes are released into the surrounding water (as in the Algae), the first selection force will favour the production of a large number of small motile gametes. On the other hand, the second selection force will favour the production of large gametes. This conflict is resolved by a differentiation of two types of gametes: small motile gametes specializing in mating efficiency, and large gametes specializing in provisioning of the zygote.

This explanation can be characterized as follows: it is qualitative, and it lacks an indication of a possible evolutionary mechanism of the system.

4.2 Quantitative Explanation: Proximate Part

The next attempt to explain anisogamy was undertaken by to Kalmus (1932). Like Weismann, he assumed two conflicting selection forces, namely a drive for numerical gamete productivity and a drive towards larger gametes for zygote provisioning. But unlike Weismann, he used a mathematical model in order to make more accurate quantitative statements about this problem. From the analysis of this model he concluded that anisogamous populations will be favoured before isogamous populations if the degree of anisogamy is sufficiently high (at least a six-fold difference in gamete size between micro- and macrogametes). If this condition is fulfilled, anisogamy will evolve.

This explanation uses quantitative reasoning, but, as in the previous explanation, the problem of how anisogamy could have evolved is not considered. From the result that in (highly) anisogamous populations more viable zygotes are produced than in isogamous populations, it is simply *assumed* that therefore anisogamy will evolve. (Although not relevant for the present discussion, note that Kalmus' explanation of anisogamy requires the effective operation of group selection).

4.3 Quantitative Explanation: Ultimate Part

After another interval of some 30 years, the evolution of anisogamy is considered again in a publication of Kalmus and Smith (1960). This work is remarkable from the point of view adopted here. Kalmus and Smith referred to Kalmus' 1932 explanation of anisogamy as still being valid, but stated that a theory explaining the *origin* of anisogamy is still lacking. They then provided a verbal suggestion of a *genetic* model of the origin of anisogamy, as follows. Suppose that in a diploid population the genotype *aa* specifies the production of microgametes. In an isogamous *aa* population a dominant mutation *A* causes the production of (fewer) macrogametes, thus introducing anisogamy. This gene *A* is expected to increase in frequency because of the selective advantage of the bigger zygotes *Aa* over *aa*; however, *A* will not reach fixation because at high frequencies *A* gametes will have difficulties in finding a mating partner, and then *a* will be at an advantage.

Thus, Kalmus and Smith arrived at the following position. There is the established Kalmus (1932) theory to explain the existence of anisogamy; and now there is also the genetic theory sketched above to explain the *origin* of anisogamy. This position is rather curious, since their genetic theory explains in principle not only the origin, but also the maintenance of anisogamy, once evolved. Therefore, their adherence to the Kalmus (1932) theory is difficult to understand. However, if we consider this work in the light of what has been said before about functional explanations as consisting of a proximate and an ultimate part, the standpoint of Kalmus and Smith becomes clearer. Kalmus' 1932 theory can be considered as a demonstration that anisogamy is under certain conditions optimal (in some sense), and forms therefore the proximate part of the explanation. Apparently, Kalmus and Smith (1960) felt that an evolutionary mechanism was lacking in this explanation, and they suggested such a mechanism in their (qualitative) genetic model, thus providing the ultimate part of the explanation.

4.4 Recent Explanations

The next theoretical study of anisogamy I will discuss here, is that of Parker et al. (1972). These authors stated explicitly that Kalmus' 1932 theory lacks a specification of a possible evolutionary mechanism, and they presented their model as an alternative to the model of Kalmus and Smith (1960). The same two conflicting selection forces which have been assumed in all the previous explanations to be responsible for the evolution of anisogamy also form the basis of the model of Parker et al. (1972): selection for increasing numbers of gametes, and selection for a large zygote size. Their model is also a genetic model, and much better elaborated (using computer simulations) than the Kalmus and Smith (1960) model. Thus, the functional explanation of anisogamy provided by Parker et al. (1972) is more complete than the previous explanations, since both the proximate and the ultimate part are analyzed quantitatively.

Further developments in the theorization of anisogamy evolution comprise partly generalizations of the model of Parker et al. (1972) (Bell 1978; Charlesworth 1978; Hoekstra 1980), partly extensions of certain aspects of the model (Maynard Smith 1978b, 1982; Wiese et al. 1979; Wiese 1981; Schuster and Sigmund 1982; Cox and Sethian 1985) and partly alternative explanations (Cosmides and Tooby 1981; Hoekstra 1984). This work will not be discussed here, however, since my objective is not to review theories of anisogamy, but to use developments in this field to illustrate certain weaknesses in the explanations.

If we consider the successive explanations of anisogamy proposed by Weismann (1886), Kalmus (1932), Kalmus and Smith (1960) and Parker et al. (1972), the following may be observed.

1. There is a tendency towards increasing degree of precision, reflected in transitions from qualitative to quantitative models. Although it was not explicitly stated, presumably Kalmus (1932) came up with a mathematical model in order to improve the older qualitative theory of Weismann.

2. The older theories (before Kalmus and Smith 1960) lack the consideration of a possible evolutionary scenario. This is considered a serious shortcoming by Kalmus and Smith (1960) and also by Parker et al. (1972), and for this reason they formulate population genetics models.

3. The relevant selection forces are the same in all four explanations. This clearly shows that merely a specification of plausible selection forces is not sufficient for a functional explanation to be acceptable. It should be demonstrated that the character under study actually can be produced by the specified selection forces, and this requires in principle a population genetics model.

The foregoing discussion might give the impression that with the model of Parker et al. (1972) we have an entirely satisfactory theory of the evolution of anisogamy. For we have seen that their model satisfies the criteria of a proper functional explanation. However, this view would be much too optimistic. The problem is that hardly anything is known about anisogamy at the genetic and physiological level. And clearly, a complete understanding of a character should include knowledge on the underlying genetics, physiology, etc. In the next section this point will be elaborated.

5 Heuristic Function of the Models

The situation described above for anisogamy is characteristic for a large part of evolutionary biology. Of the many characters or processes studied the actual proximate causation is largely unknown. This implies that the functional explanations advanced are necessarily based on highly idealized and possibly unrealistic assumptions. Consequently, these explanations are still in a *heuristic* (or *exploratory*) stage. To be sure, this is a very important function, and heuristic models are excellent tools for guiding further research, but obviously they cannot be considered to be final explanations.

This heuristic function of many models of phenotypic evolution may have bearing on the question whether population genetics models are necessary, or desirable. It was argued above that a complete functional explanation should contain a population genetics analysis concerning the evolution of the character under study; however, if in a particular study a model is used mainly for the purpose of exploring interesting aspects of the problem, it is perhaps less important to use a population genetics approach. In fact, I believe that there are rather strong arguments to use non-genetic models in many such situations. In the first place, non-genetic models are often easier to handle mathematically than genetic models. Especially for models whose main function is heuristic, simplicity and transparency are very important characteristics. Secondly, it has been shown that some non-genetic models are (qualitatively) equivalent to some (simple) genetic models (Lloyd 1977; Eshel 1982; Lessard 1984). I will not go into details here with respect to these important works, but clearly they provide a justification in some cases for the use of non-genetic instead of genetic models.

However, heuristic models are necessarily a first stage in the process of understanding phenotypic evolution, and one would like to go further. This requires of course the incorporation into the models of more realism, genetic, physiological, or otherwise. In the next section I will discuss an example where more realism is put into the models, with the consequence that an initial functional explanation based on highly simplified models has to be modified.

6 Incorporating More Realistic Genetics: Vegetative Incompatibility in Fungi

In many species there exist superimposed upon the basic sexual differentiation various mechanisms to restrict (intraspecific) zygote formation. The fungi in particular show a great variety of these so-called incompatibility systems. Roughly, there are two types of incompatibility:

1. Homogenic incompatibility, characterized by the fact that matings between *identical* genotypes (with respect to one or a few loci) are not possible. This system is very common in fungi and in higher plants.

2. Heterogenic incompatibility, characterized by the fact that matings between *non-identical* genotypes (with respect to one or a few incompatibility loci) are prevented. This system is less common than homogenic incompatibility, but is known to occur in various species (Esser and Blaich 1973), especially in a number of ascomycete fungi, such as *Neurospora* and *Podospora*. Heterogenic compatibility in *Podospora anserina* has been investigated very thoroughly by the group of Esser (see Esser 1974a)

and the group of Bernet (Bernet 1965; Boucherie and Bernet 1980). As a result many genetic and physiological details of this phenomenon in *Podospora anserina* are known. A short and simplified description of this complex system is as follows. Heterogenic incompatibility in this species can be caused by so-called allelic mechanisms and by non-allelic mechanisms. In the allelic form, hyphal fusions leading to heterokaryosis are prevented due to the formation of a barrage zone between individuals which are not identical on the relevant loci (up to five of these loci are known). The sexual compatibility between these individuals is not affected by this type of heterogenic incompatibility. The non-allelic incompatibility mechanism involves the interaction of specific alleles of two separate loci. There are various independent systems of this type known, altogether involving up to nine loci. They cause a vegetative as well as a sexual incompatibility. The latter takes the form of a non-reciprocal incompatibility, i.e. one of the reciprocal crosses does not lead to zygote formation. Two of these non-allelic mechanisms may interact to prevent any zygote formation. In a sample of 19 geographical races only 7% of all possible pairwise combinations proved to be capable of unrestricted exchange of genetic material, and in 45% there was no exchange at all. In the remaining combinations there was only partial fertility.

Two adaptive explanations have been offered for the occurrence of heterogenic incompatibility in fungi. Esser (1974b) suggests that this system has evolved due to its effects of restricting outbreeding and promoting speciation. Caten (1972) proposed that heterogenic incompatibility might serve as a defense mechanism against cytoplasmic invasion of harmful viruses or mutant-suppressive mitochondria. A similar explanation has been advanced by Hartl et al. (1975) with respect to vegetative incompatibility in *Neurospora crassa*, namely the prevention of a kind of exploitation of heterokaryons by nuclei that would be non-adaptive in homokaryons.

In a study by M. Nauta and myself (to be published elsewhere), the population genetics of non-allelic heterogenic incompatibility in *Podospora anserina* has been analyzed. Due to the complexity of the genetics, the model is intractable analytically, and we had to resort to computer simulation. The results were very briefly as follows. Interactions between partly incompatible races (some 50% of all possible interactions between races) will inevitably result in the formation of recombinant genotypes which are compatible with *both* parental genotypes. These recombinant genotypes will increase in frequency as long as the two parental strains are in contact. This finding implies that the heterogenic incompatibility system in *Podospora* is basically unstable. The only explanation we can find for the occurrence of this form of incompatibility, is that contact between different geographical races must be extremely rare, and that the system has evolved as an accidental by-product of random genetic divergence.

This example was meant as an illustration of the fact that population genetics may be necessary to make the step from a simple heuristic explanation towards a more complete explanation where realistic details are incorporated. And that it is important to try to make this step, since the original explanation may appear to be too simplistic or even wrong.

It is only fair to say that this example is biased in the sense that the added realism is *genetic* realism; clearly, when biochemical or physiological realism is to be incorporated, the model has to be adapted in other respects. But nevertheless, whenever detailed knowledge of the genetics of a particular character is to be obtained, population genetics will be necessary to adapt the functional explanation of this character to the new level of factual knowledge.

7 Conclusions

With respect to the modelling of phenotypic evolution three different situations have been considered in this chapter, and the role of population genetics in these three cases has been evaluated.

1. When an attempt is made to formulate a functional explanation of some phenotypic character, population genetics is a necessary ingredient to demonstrate the ultimate (i.e. evolutionary) part of the explanation.

2. When a model is used mainly for heuristic purposes, population genetics is not necessarily the best formalism to use. On the contrary, often population genetics leads to unnecessary mathematical complications, and other formalisms, e.g. game theory, can be better used.

3. Heuristic explanations are only a first step in the process of understanding a phenotype, and it is important to try to incorporate more realistic features into the models. When this is done, a population genetics formulation of (part of) the model is again indispensable.

References

Bell G (1978) The evolution of anisogamy. J Theor Biol 73:247–270
Bernet J (1965) Mode d'action des gènes de barrage et relation entre l'incompatibilité cellulaire et l'incompatibilité sexuelle chez le *Podospora anserina*. Ann Sci Nat Bot 6:611–768
Boucherie H, Bernet J (1980) Protoplasmic incompatibility in *Podospora anserina*. A possible function for incompatibility genes. Genetics 96:399–411
Caten CE (1972) Vegetative incompatibility and cytoplasmic infection in fungi. J Gen Microbiol 72:221–229
Charlesworth B (1978) The population genetics of anisogamy. J Theor Biol 73:347–357
Cosmides LM, Tooby J (1981) Cytoplasmic inheritance and intragenomic conflict. J Theor Biol 89:83–129
Cox PA, Sethian JA (1985) Gamete motion, search, and the evolution of anisogamy, oogamy, and chemotaxis. Am Nat 125:74–101
Eshel I (1982) Evolutionarily stable strategies and viability selection in Mendlian populations. Theor Pop Biol 22:204–217
Esser K (1974a) *Podospora anserina*. In: King RC (ed) Handbook of genetics. Plenum, New York, pp 531–581
Esser K (1974b) Breeding systems and evolution. In: Carlile MJ, Skehel JJ (eds) Proc 24th Symp Soc Gen Microbiol. Cambridge Univ Press, Cambridge, pp 87–104
Esser K, Blaich R (1973) Heterogenic incompatibility in plants and animals. Adv Genet 17:107–152
Hartl DL, Dempster ER, Brown SW (1975) Adaptive significance of vegetative incompatibility in *Neurospora crassa*. Genetics 81:553–569
Hertwig O (1906) Allgemeine Biologie, 2. Aufl. Fischer, Jena
Hoekstra RF (1980) Why do organisms produce gametes of only two different sizes? Some theoretical aspects of the evolution of anisogamy. J Theor Biol 87:785–793

Hoekstra RF (1984) Evolution of gamete motility differences II. Interaction with the evolution of anisogamy. J Theor Biol 107:71–83

Hull DL (1974) Philosophy of biological science. Prentice-Hall, Englewood Cliffs, NJ

Kalmus H (1932) Über den Erhaltungswert der phänotypischen (morphologischen) Anisogamie und die Entstehung der ersten Geschlechtsunterschiede. Biol Zentralbl 52:716–726

Kalmus H, Smith CAB (1960) Evolutionary origin of sexual differentiation and the sex ratio. Nature (London) 186:1004–1006

Lessard S (1984) Evolutionary dynamics in frequency-dependent two-phenotype models. Theor Pop Biol 25:210–234

Lloyd DG (1977) Genetic and phenotypic models of natural selection. J Theor Biol 69:543–560

Maynard Smith J (1978a) Optimization theory in evolution. Annu Rev Ecol Syst 9:31–56

Maynard Smith J (1978b) The evolution of sex. Cambridge Univ Press, Cambridge

Maynard Smith J (1982) Evolution and the theory of games. Cambridge Univ Press, Cambridge

Oster G, Wilson EO (1978) Caste and ecology in the social insects. Princeton Univ Press

Parker GA, Baker RR, Smith VGF (1972) The origin and evolution of gamete dimorphism and the male-female phenomenon. J Theor Biol 36:529–553

Provine WB (1971) The origins of theoretical population genetics. Univ Press, Chicago London

Ruse M (1973) The philosophy of biology. Hutchinson, London

Schuster P, Sigmund K (1982) A note on the evolution of sexual dimorphism. J Theor Biol 94:107–110

Stearns SC (1984) Models in evolutionary ecology. In: Wöhrmann K, Loeschcke V (eds) Population biology and evolution. Springer, Berlin Heidelberg New York, pp 261–265

Weismann A (1886) Die Bedeutung der sexuellen Fortpflanzung. Fischer, Jena

Wiese L (1981) On the evolution of anisogamy from isogamous monoecy and on the origin of sex. J Theor Biol 89:573–580

Wiese L, Wiese W, Edwards DA (1979) Inducible anisogamy and the evolution of oogamy from isogamy. Ann Bot (London) 44:131–139

Empirical Analysis of Sex Allocation in Ants: From Descriptive Surveys to Population Genetics

J. J. BOOMSMA [1]

1 Introduction

Fisher (1958) showed that equal parental investment in offspring of both sexes can be explained as the only evolutionary stable outcome of frequency-dependent natural selection. By implication, this means that a numerical sex ratio of unity is expected only in populations or species where parental investment per average male offspring equals that per average female offspring. As soon as the sex-specific cost per individual diverges, the same process of frequency-dependent selection is expected to shift the numerical sex ratio until a new equilibrium of equal cumulative investment in both sexes is reached. During the 1960s W.D. Hamilton made two crucial additions to this framework of sex-ratio theory. Firstly, that altruistic tendencies to invest in sibs or in their offspring are likely to evolve if the cost imposed by the loss of gene copies in own offspring is less than the gain of gene copies in the offspring of sibs (Hamilton 1964). Accordingly, each individual is supposed to maximize its inclusive fitness by allocating investments to its own offspring and/or to offspring of sibs depending on the respective degrees of relatedness (proportion of genes identical by descent) and on the cost/benefit ratio of the different acts of investment. Secondly, that Fisher's argument depended on the assumption of random mating (population-wide competition for mates) and that female-biased sex investments are selected when mating occurs in small sub-populations before dispersal of the inseminated females (Hamilton 1967). This shift was shown to be due to local mate competition: If related males compete for access to a limited number of females, the marginal reproductive success of genes in a male decreases as the local number of males increases, thus rendering a selectional advantage to any sex-ratio mutant investing more in female offspring (Hamilton 1967; Grafen 1984; Harvey 1985).

Fisher (1958) underlined his theoretical argument about equal investments in the sexes with empirical data on human populations, showing that the male biased sex ratio at birth and during parental investment is likely to have been affected by a larger mortality of male children, but that differential mortality after the period of parental

1 Department of Population Biology and Evolution, University of Utrecht, Padualaan 8, NL-3584 CH Utrecht, The Netherlands

Population Genetics and Evolution
G. de Jong (ed.)
© Springer-Verlag Berlin Heidelberg 1988

investment does not affect the relative investment in the sexes. According to this reasoning, the average male is cheaper per individual born, but more expensive per individual raised to maturity. Similar combinations of male-biased sex ratios at birth and increased male mortality during the period of parental care were shown to occur in some other mammals (Clutton-Brock and Albon 1982). For other taxa, the prediction that the numerical ratio of sons to daughters is the inverse ratio of their cost is not very extensively supported by empirical data, apart from the general notion that roughly equal numbers of males and females are produced in most animals (Charnov 1982). References to adult numerical sex ratios, however, implicitly assume equality of cost of a male and female and neglect possible effects of differential mortality after the period of parental care. Additional difficulties arise if investments are derived from the body size of adults, since metabolic rate is usually not directly proportional to body weight. Instead, power functions with slopes of ca. 0.75 in homeotherms (Kleiber 1961) and 0.75-1.00 in insects (Keister and Buck 1964) were shown to give the best empirical fits for data across species. Consequently, smaller individuals are generally supposed to need a larger proportion of their assimilated energy for maintenance. Even if sexual dimorphism is substantial at the end of the parental care period, this need not imply that the smaller sex is cheaper to produce. Males and females of the European sparrowhawk *(Accipiter nisus)*, for instance, appeared to have a numerical sex ratio of unity, to receive equal amounts of food (investment) per individual as nestlings but yet to differ by a factor of two in body weight at the time of fledging, without suffering any differential mortality during the period of parental care (Newton and Marquiss 1979). It should be realized therefore that many sexual dimorphisms may not reflect cost differences for parents and that the X-Y sex determination system as occurring in many diploid taxa might have inhibited selectional changes away from a 1:1 numerical sex ratio (Maynard Smith 1984), and thus also the evolution of deviating male/female costs.

In the haplodiploid Hymenoptera, however, males originate from unfertilized eggs. This enables ovipositing females to manipulate the sex of each offspring, by allowing differential fertilization of eggs with sperm stored in their spermatheca. Haplodiploids can thus optimize their primary sex ratio behaviorally, without suffering the fitness losses occurring in diploids after abortion or neglect of the less preferred sex (Maynard Smith 1980). In many Hymenopteran families, a high degree and frequency of sexual dimorphism is found as well, making these insects extremely suitable for applying sex allocation theory. Parasitic wasps have repeatedly been shown to adjust their primary sex ratio in response to the size of the hosts. The mechanism comes down to laying relatively fewer fertilized eggs in smaller hosts, because daughters presumably lose more fitness by being small adults than sons (see Charnov 1982 for a review). Evidence for Fisher's principle of numerical overproduction of the cheaper sex was found in solitary ground nesting bees and wasps (Trivers and Hare 1976; Charnov 1982), but most consistently in empirical data sets of social Hymenoptera (Trivers and Hare 1976; Nonacs 1986): the correlation between male/female cost ratio (weight units) and numerical sex ratio in Nonacs' data set on monogynous ants appears to be -0.66 ($P < 0.001$), when both parameters are expressed as angular transformed frequencies.

The purpose of this chapter is to review some aspects of the empirical research on sex allocation in social Hymenoptera. I concentrate on ants as being a homogeneously eusocial family of perennial insects (Formicidae), whose sex-ratio data have been explained over the past decade both by the asymmetry in genetic relatedness (AGR) and by local mate competition (LMC). A summary of empirical arguments is presented against LMC as a major explanatory factor for sex allocation in monogynous ants, but simultaneously the descriptive sex-ratio data used so far in AGR interpretations are shown to be inaccurate. Ultimately, the approach of direct estimation of genetic relatedness and worker reproduction in ant populations is advocated as being the most promising way of testing evolutionary theories on sex allocation in social Hymenoptera.

2 The Case of Haplodiploid Social Insects: Sex Ratios in Monogynous Ants

In an influential work, Trivers and Hare (1976) combined arguments from Fisher (1958) and Hamilton (1964) to show that workers of social Hymenoptera serve their inclusive fitness interests better by raising full sibs than own offspring, but only if they capitalize on the asymmetry in their relatedness towards full sisters ($r = 3/4$) and brothers ($r = 1/4$) and oppose the interest of their mother who is equally related to sons and daughters ($r = 1/2$). It was claimed by Trivers and Hare that both the recurring independent evolution of eusociality in the Hymenoptera, and the occurrence of female-biased sex investments in this group were due to kin selection operating on the effectively controlling worker caste. Leaving aside the controversial issue of the origin of insect sociality which has recently been reviewed by various authors (e.g. Anderson 1984 and references cited), it is obvious that Trivers and Hare's theoretical arguments on conflicting optimal investments for workers and queens have been confirmed repeatedly by later workers, using various techniques of modelling and simulation (e.g. Charnov 1978; Oster and Wilson 1978; Craig 1980; Uyenoyama and Bengtsson 1981; Pamilo 1982). However, the interpretation of empirical data by Trivers and Hare has been criticized. Particularly, their conclusion that the calculated mean of 3:1 dry-weight investment in favor of gynes would imply complete worker control over sexual investments, effective monoandry and effective worker sterility throughout monogynous ants has been under immediate attack. As argued by Alexander and Sherman (1977), the quality of Trivers and Hare's data set was probably affected by errors and biases due to sample size (concerning both the number of specimens and the number of colonies involved), inappropriate pooling of data and unjustifiable regression analysis. Their main argument, however, concerned LMC, which they advocated as a more powerful explanation of female-biased sex investments in social Hymenoptera than AGR.

3 The Pros and Cons of Local Mate Competition in Ants

The main argument of Alexander and Sherman (1977) against worker control and AGR was based on the large variance in sex investments across the species analyzed by Trivers and Hare, with the additional consideration that AGR could not explain any investment ratio estimate higher than 3:1. However, their examples in favor of LMC in social Hymenoptera mostly referred to highly specialized or polygynous ant species which did not form the core of the reasoning of Trivers and Hare. In fact, only one of the 21 monogynous, non-slave-making ant species listed by Trivers and Hare was argued by Alexander and Sherman to have potential sib mating and even this case was supported only by circumstantial evidence. Recently, Nonacs (1986) added 12 literature records to the list of sex-ratio data for monogynous ants. He argued that commonly observed traits like (1) a large numerical male bias, (2) a high proportion of sexual specialist colonies and (3) the fact that ant males seem to be generally unable to inseminate more than one gyne, are at variance with LMC as an explanatory mechanism of female-biased sex investments across monogynous ants. Surprisingly, however, the discussion of the AGR/LMC controversy has so far not explicitly dealt with the most fundamental assumptions of LMC, i.e. whether mating precedes dispersal and whether the size of breeding (sub-)populations is generally small enough to reach a sex-investment shift as large as that expected under AGR and worker control. Recent population genetics models on sex-ratio selection in subdivided haplodiploid populations show that under LMC an equilibrium sex investment of 0.6 $\left(\dfrac{F}{F+M} \right)$ is expected only if breeding patches contain less than seven females (= colonies in monogynous ants), whereas a skewness of 0.7 would require mating groups of three or less (Bulmer 1986). Even smaller figures apply to models where mating occurs after dispersal and only females disperse (Wilson and Colwell 1981; Bulmer 1986).

To evaluate the practical value of these models for the explanation of female-biased sex investments in monogynous ants, two questions need to be answered:

1. What empirical data are available on small mating groups, limited dispersal and timing of mating relative to dispersal?
2. To what extent are such data likely to affect the interpretation of literature data on sex allocation in ants, especially those for monogynous ants summarized by Trivers and Hare (1976) and Nonacs (1986)?

Mating in ants occurs normally either before or during dispersal of the gynes. According to Hölldobler and Bartz (1985), females either mate before dispersal by releasing male-attracting pheromones while still on the nest of origin ("female-calling syndrome"), or fly off towards aggregations of males which hover around on or above specific mating sites ("male-aggregation syndrome"). Mating swarms in the latter category involve such large amounts of synchronously flying ant sexuals that LMC should be negligible. The former mating pattern, which is largely restricted to primitive and/ or parasitic ants, often implies a rather limited dispersal of gynes. However, as males may be attracted from considerable distances, breeding groups are not likely to be very small unless intranidal mating occurs in high frequencies. For Europe, the number of species with supposedly regular intranidal mating is in the order of 10% of the total ant fauna (Bernard 1968; Collingwood 1979; Kutter 1977) but none of these species belong to the category of monogynous ants whose sex ratio data formed the core of the 3:1 investment argument of Trivers and Hare and Nonacs. It is further important to note that the few references of Alexander and Sherman (1977) on sib mating in monogynous ants are basically dealing with the possibility of a limited amount of inbreeding in populations which otherwise have normal nuptial flights. After tracing the original references of Trivers and Hare, Alexander and Sherman and Nonacs, and adding data from Collingwood (1979), 79% of the monogynous ants listed by Nonacs (1986) appeared to be known to have a mating flight. For most of the remaining species, the data were collected from rather dense populations and no suggestions for population subdivision or inbreeding are given.

We can conclude that the mating behavior of the large majority of ant species does not satisfy the conditions needed for LMC to have a major effect. Even small mating swarms of monogynous ants will normally combine enough unrelated sexuals to restrict selectional shifts towards a more gyne-biased sex investment to an order of magnitude of a few percent. This is much less relevant than the degree of female bias expected from AGR and worker control and, therefore, it seems very unlikely that the data sets of Trivers and Hare (1976) and Nonacs (1986) can primarily be interpreted from a LMC point of view. Having reached this conclusion, however, we still have to face Alexander and Sherman's objections against the AGR interpretation, as summarized at the beginning of this section.

4 Patterns of Variance and Covariance in Colony-Level Sex Allocation

Variation in sex investment among individual families is not or hardly counterselected. In the extreme case of an infinitely large population with a population sex ratio at equilibrium, every sex allocation is equally fit, and even a combination of all son and all daughter families is evolutionarily stable (Charnov 1982). If these conditions are not met, selection on the sex ratio is supposed to be relatively weak and the sex ratio of individual clutches could show wide scatter due to random influences (Oster and Wilson 1978). Only in extremely subdivided populations is selection likely to reduce such variation substantially (Nonacs 1986). In ants, a large variance of sex investment among colonies of the same population seems to occur in most if not all species, but the variation is not random. In a number of species the variance of sex investment decreases with increasing levels of total sexual production but the mean of the relative investment in gynes increases with productivity, often not significantly within populations but significantly after pooling the data (Fig. 1). A similar pattern of increasing mean gyne bias with total production was shown to occur in many other ant species by Nonacs (1986). Because of this correlation, however, the calculated cumulative investment $\Sigma F/(\Sigma F + \Sigma M)$, reflecting the mating flight composition, will not give the same estimate as the mean investment per colony $\Sigma\{F/(F+M)\}/n$. Pooled or weighted estimates are mainly affected by the contribution of highly productive colonies and will therefore be systematically more female-biased than unweighted average values per colony. In case of a representative sample from a stable population, weighted estimates of sexual investment give the best approximation in terms of average lifetime reproductive success. However, if highly productive and large colonies are overrepresented in a sample, weighted estimates of sexual investment tend to overestimate the average degree of female bias, and unweighted means may approach the true mean of the population(s) more closely.

Both Trivers and Hare (1976) and Nonacs (1986) used literature records on sexual investment, which were often based on very small and not necessarily representative samples of populations of monogynous ants. As small samples are most likely to contain an excess of larger colonies, their figures may have overestimated the degree of

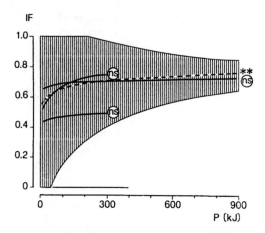

Fig. 1. Relationship between the mean and variation of relative investment in gynes $IF = F/(F+M)$ and total sexual investment $P = (F+M)$ per colony, as found in 201 colonies of the monogynous ant *Lasius niger* occurring in three coastal dune populations on the Frisian island of Schiermonnikoog. F and M are numbers of sexuals multiplied by the average cost per individual. For this plot, both F and M were expressed in kJ after estimations of the cumulative amount of energy invested per gyne and male (Boomsma and Isaaks 1985). The *shaded* area represents the total variation present (cf. Boomsma et al. 1982), containing the investments of 98.5% of the 198 colonies included in the analysis. Three apparently orphaned and exclusively male-producing colonies (Van der Have et al. 1988) are represented by the *horizontal line* at $IF = 0$. The positive correlations between IF and P were not significant (*ns*) within populations, but significant after pooling the data (**: $P < 0.01$). Comparable patterns of mean and variation of IF as function of P occur in at least two other monogynous ant species (Boomsma, unpublished)

gyne bias in a number of species. Furthermore, their estimates are based on weight data which do not include costs of respiration. According to the arguments given above, such partial estimates are expected to induce a relative overestimation of the cost of the larger sex, in relation to the specific degree of sexual dimorphism. For instance, in the extremely dimorphic ant *Lasius niger*, the individual gyne/male dry weight ratio at the onset of the mating flight was found to be 16.6, whereas the cumulative energetic investment per individual was estimated to differ only by a factor 7.7 (Boomsma and Isaaks 1985). In accordance with these expectations, 29% of the variation in the (weight) investments across monogynous ants reviewed by Nonacs (1986) could be explained after introducing sexual dimorphism (f/m weight ratio) and sample size (number of colonies) as covariates in multiple regression (both significant at $P < 0.05$). As full details on this are given in Boomsma (1987 and in prep.) it suffices to restate the conclusions formulated in those works: Assuming that weight estimations are unbiased only for species without sexual dimorphism, the literature records can be corrected for the effect of this apparently artifactual covariate. After doing so, the mean relative investment in gynes across species appeared to be 0.65 (1.8:1), with a 95% confidence interval of 0.54–0.74 (1.2:1 to 2.8:1). Also, the records for separate species now covered fairly accurately the whole range of theoretical equilibrium sex in investments given by Trivers and Hare (1976) and Oster and Wilson (1978). It follows that the logically correct objections of Alexander and Sherman (1977) against the explanatory power of AGR in Trivers and Hare's data set are likely to be no longer relevant. The recalculated pattern of sexual investment across monogynous ants seems largely understandable in terms of worker control and AGR, if variation in mating frequency and male parentage is allowed for. It is obvious, however, that data sets across species can only indicate general trends of sex allocation in field populations of ants. More adequate empirical tests of the AGR concept have recently been made by estimating directly the genetic relatedness between workers and the sexuals produced in their colonies.

5 The Population Genetics Approach

Both multiple mating (Page 1986) and worker reproduction (Bourke, 1988) have recently been shown to occur regularly in monogynous ants. Either of them has the effect of considerably decreasing the 3:1 optimum investment for workers applying to monoandry and worker sterility. For controlling workers, the expected mean ratio of investment in gynes in a population is given by:

$$\frac{\bar{F}}{\bar{M}} = \frac{\bar{r}_{f,f}}{r_{m,f}} \cdot \frac{1+p}{2}$$

(Oster and Wilson 1978), where $r_{f,f}$ and $r_{m,f}$ are the relatedness of workers towards other females and males respectively, and where p is the proportion of queen-produced males. Accordingly, the expected ratio of investment can be calculated for any population of ants, if estimations on worker reproduction and relatedness are available. As only a limited number of ant species is known to have either totally sterile workers or obligatory male-producing workers (Brian 1983), the degree of worker reproduction must usually be estimated separately in each population. Electrophoretic analysis of polymorphic enzyme loci has proven to be a useful technique to estimate not only the level of worker reproduction, but also the average relatedness of workers towards the sexuals raised in their colonies. In colonies where all workers are heterozygous for a particular enzyme locus, male parentage can usually be directly deduced (Crozier 1974; Pamilo and Rosengren 1983; Ward 1983a; Van der Have et al. 1988). The same data sets also permit the estimation of relatedness by two separate calculation procedures: the regression coefficient of relatedness (Pamilo and Crozier 1982; Pamilo 1984) and the nested ANOVA method of Wilson (1981). A summary of such estimations, drawn from the literature is presented in Fig. 2, separately for monogynous and polygynous ant species. In general, workers appear to be equally, but only slightly related to gynes and males in polygynous species, whereas a significant asymmetry occurs in monogynous ants. The variance around the mean relatedness among females in monogynous ants is substantial and appeared to be related to different levels of worker reproduction and multiple mating in *Lasius niger* (Van der Have et al. 1988). If data on both the realized sexual investment and the investment ratio expected for worker control and AGR are available for several conspecific populations, the significance of AGR can be directly tested from the principal component axis of the correlation between IF_{exp} and IF_{obs} (Fig. 3). The statistical significance of the axis indicates the importance of worker control relative to queen control and/or random

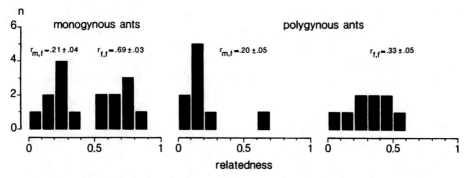

Fig. 2. Frequency (*n*) of estimates on relatedness of ant workers to the gynes ($r_{f,f}$) and males ($r_{m,f}$) produced in their colonies, as derived from allozyme analysis. The data are presented separately for monogynous *(Rhytidoponera, Lasius)* and polygynous *(Rhytidoponera, Formica)* ants. Means ± SE of relatedness in the different categories are given as well. The data included in this figure were taken from: Pamilo and Rosengren (1983), Pamilo and Varvio-Aho (1979), Van der Have et al. (1988) and Ward (1983a). Other references for polygynous ants, which were not included in this plot, as they presented estimations for $r_{f,f}$ only, showed an even lower mean for $r_{f,f}$ than the figure given in the plot (Craig and Crozier 1979; Pamilo 1981; Ward and Taylor 1981; Pearson 1983; Crozier et al. 1984; Pamilo and Rosengren 1984)

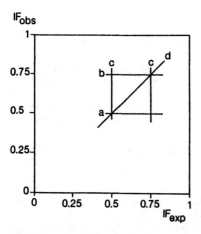

Fig. 3. Graphic design to test the significance of AGR for explaining sex investment variation in conspecific populations of monogynous ants. Alternative mechanisms of sex allocation are expected to give different relationships between IF_{exp} and IF_{obs}. *Line a:* queen control; ratio of investment is 1:1 ($IF = 0.5$) independent of worker reproduction and mating structure. *Line b:* worker control, but colonies are subdivided into full-sib worker nests, which invest only in full sisters and brothers $\overline{F/M} = 3$; $IF = 0.75$). *Line c:* relatedness structure fixed; the observed variation in IF is due to ecological factors, affecting the outcome of the worker-queen conflict. *Line d:* IF_{exp} is variable due to variation in mating structure and worker reproduction; worker control throughout, but without discrimination between full-sibs and half-sibs; IF_{obs} is entirely predicted by IF_{exp}

ecological effects, whereas the position of the axis relative to the expected worker control diagonal gives additional information about the amount of kin recognition involved. For *Lasius niger*, the correlation between IF_{exp} and IF_{obs} appeared to be significant at $P < 0.01$ and the position of the line was slightly below the diagonal, indicating no discrimination between sibs and half sibs in colonies having a multiply mated queen (data from Van der Have et al. 1988). Also, the results of Ward (1982a,b) indicate a prevalence of worker control in queenright colonies of *Rhytidoponera* ants, as both $r_{f,f}$ and IF_{obs} were close to 0.75. In summary, although the number of population genetics studies is still rather limited, the data reviewed in Fig. 2 seem to fit the AGR expectations quite well. Much further empirical work is needed, however, to assess if AGR is indeed the predominant explanation for the sex-investment strategies observed throughout monogynous ants, and to evaluate whether ecological covariates are additionally relevant in determining the optimal investments of separate populations.

Acknowledgement. This work benefited from discussions with Alan Grafen during my stay in the Oxford Zoology Department. I also thank Tom van der Have for comments on the final version of the manuscript. I am supported by a C. and C. Huygens stipendium of the Netherlands Organization for the Advancement of Pure Research (Z.W.O.).

References

Alexander RD, Sherman PW (1977) Local mate competition and parental investment in social insects. Science 196:494–500

Anderson M (1984) The evolution of eusociality. Annu Rev Ecol Syst 15:165–189

Bernard F (1968) Les Fourmis d'Europe occidentale et septentrionale. Faune Eur Bass Med 3: 1–411

Boomsma JJ (1987) The empirical analysis of sex allocation in monogynous ants. In: Eder J, Rembold H (eds) Chemistry and biology of social insects. Proc 10th Int Congr IUSSI, München, pp 355–357

Boomsma JJ (submitted) Sex investment ratios in ants: has female bias been systematically overestimated?

Boomsma JJ, Isaaks JA (1985) Energy investment and respiration in queens and males of *Lasius niger* (Hymenoptera: Formicidae). Behav Ecol Sociobiol 18:19–27

Boomsma JJ, Lee GA van der, Have TM van der (1982) On the production ecology of *Lasius niger* (Hymenoptera: Formicidae) in successive coastal dune valleys. J Anim Ecol 51:975–991

Bourke AFG (1988) Worker reproduction in the higher eusocial Hymenoptera. Q Rev Biol, in press

Brian MV (1983) Social insects: ecology and behavioural biology. Chapman & Hall, London New York

Bulmer MG (1986) Sex ratio theory in geographically structured populations. Heredity 56:69–73

Charnov EL (1978) Sex ratio selection in eusocial Hymenoptera. Am Nat 112:317–326

Charnov EL (1982) The theory of sex allocation. Monographs in population biology, vol. 18. Princeton Univ Press

Clutton-Brock TH, Albon SD (1982) Parental investment in male and female offspring in mammals. In: Current problems in sociobiology. Cambridge Univ Press, pp 223–248

Collingwood CA (1979) The Formicidae (Hymenoptera) of Fennoscandia and Denmark. Fauna Ent Scand, vol 8. Scand Science Press, Klampenborg

Craig R (1980) Sex investment ratios in social Hymenoptera. Am Nat 116:311–323

Craig R, Crozier RH (1979) Relatedness in the polygynous ant *Myrmecia pilosula*. Evolution 33: 335–341

Crozier RH (1974) Allozyme analysis of reproductive strategy in the ant *Aphaenogaster rudis*. Isozyme Bull 7:18

Crozier RH, Pamilo P, Crozier YC (1984) Relatedness and microgeographic genetic variation in *Rhytidoponera mayri*, an Australian arid-zone ant. Behav Ecol Sociobiol 15:143–150

Fisher RA (1958) The genetical theory of natural selection. Dover, New York, 2nd edn

Grafen A (1984) Natural selection, kin selection and group selection. In: Krebs JR, Davies NB (eds) Behavioural ecology: An evolutionary approach 2nd edn. Blackwell, Oxford, pp 62–84

Hamilton WD (1964) The genetical evolution of social behavior I, II. J Theor Biol 7:1–52

Hamilton WD (1967) Extraordinary sex ratios. Science 156:477–488

Harvey PH (1985) Intrademic group selection and the sex ratio. In: Sibley RM, Smith RH (eds) Behavioural ecology: Ecological consequences of adaptive behaviour. Blackwell, Oxford, pp 59–73

Have TM van der, Boomsma JJ, Menken SBJ (1988) Sex investment ratios and relatedness in the monogynous ant *Lasius niger* (L.). Evolution, 160–172

Hölldobler B, Bartz S (1985) Sociobiology of reproduction in ants. In: Hölldobler B, Lindauer M (eds) Experimental behavioral ecology and sociobiology. Fischer, Stuttgart, pp 237–257

Keister M, Buck J (1964) Respiration: some exogenous and endogenous effects of rate of respiration. In: Rockstein M (ed) Physiology of insecta III. Academic Press, London New York, pp 617–658

Kleiber M (1961) The fire of life: an introduction to animal energetics. Wiley & Sons, New York

Kutter H (1977) Hymenoptera Formicidae. Insecta helvetica 6. Fotorotar, Zürich

Maynard Smith J (1980) A new theory of sexual investment. Behav Ecol Sociobiol 7:247–251

Maynard Smith J (1984) The ecology of sex. In: Krebs JR, Davies NB (eds) Behavioural ecology: An evolutionary approach 2 nd edn. Blackwell, Oxford, pp 159–179

Newton I, Marquiss M (1979) Sex ratio among nestlings of the European sparrowhawk. Am Nat 113:309–315

Nonacs P (1986) Ant reproductive strategies and sex allocation theory. Q Rev Biol 61:1–21

Oster GF, Wilson EO (1978) Caste and ecology in the social insects. Monographs in population biology, vol 12. Princeton Univ Press

Page RE, Jr (1986) Sperm utilization in social insects. Annu Rev Entomol 31:297–320

Pamilo P (1981) Genetic organization of *Formica sanguinea* populations. Behav Ecol Sociobiol 9:45–50

Pamilo P (1982) Genetic evolution of sex ratios in eusocial Hymenoptera: allele frequency simulations. Am Nat 119:638–656

Pamilo P (1984) Genotypic correlation and regression in social groups: multiple alleles, multiple loci and subdivided populations. Genetics 107:307–320

Pamilo P, Crozier RH (1982) Measuring genetic relatedness in natural populations: methodology. Theor Popul Biol 21:171–193

Pamilo P, Rosengren R (1983) Sex ratio strategies in *Formica* ants. Oikos 40:24–35

Pamilo P, Rosengren R (1984) Evolution of nesting strategies of ants: genetic evidence from different population types of *Formica* ants. Biol J Linn Soc 21:331–348

Pamilo P, Varvio-Aho S-L (1979) Genetic structure of nests in the ant *Formica sanguinea*. Behav Ecol Sociobiol 6:91–98

Pearson B (1983) Intra-colonial relatedness amongst workers in a population of nests of the polygynous ant, *Myrmica rubra* Latreille. Behav Ecol Sociobiol 12:1–4

Trivers RL, Hare H (1976) Haplodiploidy and the evolution of the social insects. Science 191:249–263

Uyenoyama MK, Bengtsson BO (1981) Towards a genetic theory for the evolution of the sex ratio II. Haplodiploid and diploid models with sibling and parental control of the brood sex ratio and brood size. Theor Popul Biol 20:57–79

Ward PS (1983a) Genetic relatedness and colony organization in a species complex of Ponerine ants I. Phenotypic and genotypic composition of colonies. Behav Ecol Sociobiol 12:285–299

Ward PS (1983b) Genetic relatedness and colony organization in a species complex of Ponerine ants II. Patterns of sex ratio investment. Behav Ecol Sociobiol 12:301–307

Ward PS, Taylor RW (1981) Allozyme variation, colony structure and genetic relatedness in the primitive ant *Nothomyrmecia macrops* Clark (Hymenoptera: Formicidae). J Austr Ent Soc 20:177–183

Wilson DS, Colwell RK (1981) The evolution of sex ratio in structured demes. Evolution 35:882–897

Wilson J (1981) Estimating the degree of polyandry in natural populations. Evolution 35:664–673

Fitness and Mode of Inheritance

H.-R. Gregorius[1]

Abstract

It is briefly demonstrated with the help of well-known selection models that the principle of survival of the fittest critically depends on the measure of reproductive success applied. The common measure, Darwinian fitness (number of successful gametes), has only limited value for maintaining this principle, since it does not explicitly reflect the effects of the mating system and the mode of inheritance on the persistence of a type. Herewith, mating system and mode of inheritance should not be viewed separately, since the realization of the latter may depend on the former. It is pointed out that, in essence, the problem of persistence (survival) of a type becomes relevant only when this type is present at low frequencies and that, therefore, the appropriateness of any fitness measure should be evaluated at extreme frequencies.

A new fitness measure, called the "amount of self-replication", is proposed, which counts only those offspring appearing in the Darwinian fitness that show the same type as its parent. The use of this measure is demonstrated for a general biallelic selection model including arbitrary viabilities, fertilities or fecundities, mating systems and forms of segregation distortion. The central result is that an allele becomes established if the rare homozygote or heterozygote exceeds the prevalent homozygote in amount of self-replication. This leads to the formulation of a new overdominance principle: If the amount of self-replication of the heterozygote exceeds that of either homozygote as the latter becomes frequent, then both alleles remain in the population. Hence, based on this fitness measure, the principle of survival of the fittest can be maintained. Tentative arguments are given for why the results might serve as clues to the extension to more complex modes of inheritance. It is pointed out that reproductive success (fitness) and evolutionary success should be properly distinguished, and that, depending on the objective in mind, fitness measures may serve to specify causes or effects of evolution.

---

1 Abteilung für Forstgenetik, Georg-August-Universität Göttingen, Büsgenweg 2, D-3400 Göttingen-Weende, FRG

Population Genetics and Evolution
G. de Jong (ed.)
© Springer-Verlag Berlin Heidelberg 1988

1 Introduction

In sexually reproducing populations the measurement of fitness basically refers to *pairs* of individuals rather than to single individuals. This is inevitable, since in the last analysis fitness must be directly related to reproductive success, which requires mating and thus pairs of individuals. Nevertheless, our concepts of fitness are still largely based on the reproductive performance of single individuals, as is evident from the definition of Darwinian fitness as the average contribution of a type (genotype) to the gene pool of the population. In this sense the fitness of an individual is identical to its number of successful gametes (Gregorius 1982a, 1984a), and it is the primary variable on which ESS analyses rest.

Moreover, in the classical selection theory of population genetics, allelic fitnesses play a central role in that they directly determine the change in allelic frequencies. Thus, in this view individuals are individual genes, and their fitnesses have again to be specified in terms of the number of successful gametes (Gregorius 1982a). The efficacy of allelic fitnesses in understanding evolutionary processes is, however, limited to the special case where there is sexual symmetry in the viabilities or fertilities, and where the reproductive system works in a way that results in random fusion of the gametes. In essence, the classical selection theory allows only for situations producing Hardy-Weinberg frequencies among the zygotes, so that the primary determinants of evolution, namely changes in genotypic frequencies, can be traced with the help of allelic frequencies. In a sense, this procedure aims at reducing the complications of diploidy (or even higher degrees of ploidy) and the associated Mendelian inheritance to asexual reproduction at the cost of frequency-dependent (allelic) fitnesses. As soon as the genotypic frequencies among the zygotes do not comply with Hardy-Weinberg frequencies, or with any other genotypic structure that cannot be reconstructed from the underlying gene frequencies alone, the concept of allelic fitness becomes inadequate. This follows immediately from the results obtained for standard models of constant amounts of either self-fertilization or positive assortative mating (Crow and Kimura 1970, Sects. 3.8 and 4.6), where, due to identical allelic fitnesses, the allelic frequencies do not change over the generations but the genotypic frequencies do. In extreme cases, this process may lead to reproductive isolation among genetically defined parts of a population and may thus provide conditions for speciation. Hence, *evolution is basically a change in genotypic rather than gene frequencies.*

In general, maximization of "individual" fitness in the Darwinian sense is neither a necessary nor a sufficient condition for the persistence of a type in sexually reproducing populations. The reason is that Darwinian fitness does not explicitly reflect the effects of the mode of inheritance and the mating system on the persistence of a type. This can be demonstrated with the help of two well-known biallelic viability selection models, in both of which Darwinian fitness is constant (frequency-independent) and proportional to viability, as follows:

1. Segregation Distortion and Random Mating. This model demonstrates the effect of the mode of inheritance. For appropriately chosen viabilities of the three genotypes, fixation of the type with the lowest fitness occurs if segregation distortion in favour of its allele is sufficiently large. An example is provided by $v_{11} = \frac{1}{3}$, $v_{12} = \frac{2}{3}$, $v_{22} = 1$

and $\delta_1 > \frac{3}{4}$, where v_{ij} is the viability of genotype A_iA_j and δ_1 is the proportion of A_1-alleles among the successful gametes of the heterozygote A_1A_2. In this case global fixation of A_1A_1 occurs.

2. Self-Fertilization and Regular Segregation. Kimura and Ohta (1971, p. 190ff.) showed that if the heterozygote exceeds both homozygotes in viability and the homozygotes are not equally viable, then a stable polymorphism existing for small amounts of self-fertilization vanishes for a sufficiently large amount of self-fertilization. Hence, due to a change in mating system the fittest type (the heterozygote) disappears.

These examples raise the question as to whether fitness concepts based on the reproductive performance of individuals can have any predictive value at all. Clearly, if there were such a concept, it ought to be a considerable modification of Darwinian fitness, including aspects of the mode of inheritance and the mating system. In the present chapter a measure will be considered which minimally reflects the net effect of these aspects. It pursues the Darwinian idea, namely that the reproductive success of an individual determines the representation of its type in the next generation, by counting only those offspring that are of the same type as its producer. This extends the self-centered (or asexual) nature of Darwinian fitness to sexual reproduction in that it deducts the loss in identical self-replication due to the mode of inheritance and the mating system which determines the realization of this mode.

The thus extended (actually reduced) version of Darwinian fitness does not, of course, account for the possibility that a type be produced by parents both of which differ from their offspring, as is the case with a heterozygote resulting from mating between the corresponding homozygotes. Hence, a deficit of a type in self-replication may in some sense be compensated by the "genetic altruism" of other types. The predictive power of the amount of self-replication of a type may, therefore, again be expected to be limited. However, this should not be considered as a disadvantage, since, unlike Darwinian fitness, the amount of self-replication is a highly specific variable. Thus, its limits for predicting persistence of a type are likely to be better defined, which in turn allows more specific inferences to be made on possible alternatives (such as "genetic altruism") for understanding evolutionary principles.

At the outset, studies of persistence are generally concerned with finding conditions for the establishment of a new type. In its most basic form, this corresponds to the establishment of a new allele at a gene locus, which will also be the topic of the present chapter. Particular emphasis will be put on the role played by the amount of self-replication, in which all net effects of survival, fertility or fecundity, mating system and mode of inheritance via general segregation distortion are summarized.

2 Fitnesses

In an attempt to arrive at a generally valid representation of transition equations for genotypic frequencies, the present author (Gregorius 1984a) concluded that the following variables are of fundamental significance:

$I_{i,j}(k)$: = Relative frequency of k-type zygotes among those produced by fertilizations between i-type and j-type individuals *(inheritance operator)*; $I_{i,j}(k) =$ $I_{j,i}(k)$ and $\Sigma_k I_{i,j}(k) = 1$.

$w_{i:j}$: = Average number of successful gametes produced by a j-type individual that result from fertilizations by i-type individuals $(i \neq j)$.

$w_{i:i}$: = One-half the average number of successful gametes produced by an i-type individual that result from fertilizations by i-type individuals.

The $w_{i:j}$'s are termed *fractional fitnesses*, and they are related to the Darwinian fitnesses w_i of an i-type individual by

$$w_i = 2 \cdot w_{i:i} + \sum_{j, j \neq i} w_{j:i} \,,$$

which is the average number of successful gametes of an i-type individual. If P_i denotes the relative frequency of i-type individuals at the stage of census for which the fitnesses are determined, then the fractional fitness must obey the relationship

$$w_{i:j} \cdot P_j = w_{j:i} \cdot P_i \,. \tag{1}$$

Moreover, the population fitness $\bar{w}$ (average number of zygotes per member of the population) is given by

$$\bar{w} = \sum_{i \leqslant j} w_{i:j} \cdot P_j = \frac{1}{2} \sum_i w_i \cdot P_i \,.$$

The basic transition equation for the relative frequencies of the types then reads:

$$P'_k = \sum_{i \leqslant j} I_{i,j}(k) \cdot w_{i:j} \cdot P_j / \bar{w} \,. \tag{2}$$

This transition equation cannot be simplified to a form in which Darwinian fitnesses replace the fractional fitnesses without imposing very restrictive assumptions. The previous statement (supplied with examples) that Darwinian fitnesses are very unreliable predictors for the persistence of a type now becomes particularly evident. Further evidence comes from the fact that the inheritance operator may have very different forms, even for the same types and associated fractional fitnesses.

The point of interest is now to find out whether an "individual" concept of fitness, such as w_i, which, however, summarizes the fitness fractions and particular components of the mode (operator) of inheritance, can provide more insight into the complex nature of Eq. (2). For this purpose the following variables will be studied:

r_i: = $\Sigma_j w_{j:i} \cdot I_{j,i}(i)$ or the average number of offspring (zygotes) of an i-type individual which are again of type i (amount of *self-replication* of an i-type individual).

Note that $r_i \leqslant w_i - w_{i:i} \leqslant w_i$, so that r_i can again be considered as a fraction of the Darwinian fitness.

The fractional fitnesses of an individual represent its net reproductive success, which results from the joint effects of survival, fertility or fecundity, mating success, etc. However, they also depend on the frequencies. For example, when type 2 becomes very rare while type 1 remains frequent, then $w_{2:1}$ approaches zero while $w_{1:2}$ need not. If both types become rare, the behaviour of $w_{1:2}$ and $w_{2:1}$ is less obvious. However, biological realism suggests that $w_{1:2}$ and $w_{2:1}$ should simultaneously approach zero simply because their chances to mate with each other decline. This is also in accordance with requirements of continuity *(natura non saltat)*.

The situation is different for $w_{i:i}$ as P_i approaches zero. In bisexual populations with the opportunity for individuals to reproduce by self-fertilization, $w_{i:i}$ may remain positive. If self-fertilization is not relevant, then the encounters of individuals or gametes preceding mating or fertilization are subject to some degree of randomness, and $w_{i:i}$ may be expected to approach zero as P_i does. These considerations become relevant when problems of the establishment of new (and thus rare) types are to be studied, as will be done in the next section. Most basically, this situation arises with a mutant gene which has to compete with its corresponding wild-type allele. The situation can then be conceived of as selection at a biallelic locus with alleles A and B, for instance. Hence, there are three types A, B and H, where for convenience A and B denote the homozygotes for the two alleles and H denotes the heterozygote.

In the biallelic case, the inheritance operator is characterized by $I_{A,A}(A) = I_{B,B}(B) = I_{A,B}(H) = 1$ and $I_{A,A}(B) = I_{A,A}(H) = I_{B,B}(A) = I_{B,B}(H) = I_{A,B}(A) = I_{A,B}(B) = I_{A,H}(B) = I_{B,H}(A) = 0$. For the remaining cases Mendelian segregation implies $I_{H,H}(H) = \frac{1}{4}$, for example. However, since no assumptions on regularity in segregation shall be applied, all we require is that $I_{X,Y}(A) + I_{X,Y}(B) + I_{X,Y}(H) = 1$, where X and Y are any of the types A, B or H. It is thus even allowed that some of the I's are frequency-dependent, as may occur in models of pollen-tube growth (Steiner and Ross, personal communication). The amounts of self-replication now attain the form:

$$r_A = w_{A:A} + w_{H:A} \cdot I_{A,H}(A), \quad r_B = w_{B:B} + w_{H:B} \cdot I_{B,H}(B),$$

$$r_H = w_{A:H} \cdot I_{A,H}(H) + w_{B:H} \cdot I_{B,H}(H) + w_{H:H} \cdot I_{H,H}(H).$$

If the A-allele is very rare, so that its frequency $p_A = P_A + \frac{1}{2} P_H$ is close to zero, then $w_{H:A} = w_{A:H} = w_{A:B} = w_{H:B} = 0$, and therefore $r_A = w_{A:A}$, $r_B = w_{B:B}$, $r_H = w_{B:H} I_{B,H}(H) + w_{H:H} I_{H,H}(H)$ and $w_A = 2 w_{A:A} + w_{B:A}$, $w_B = 2 w_{B:B}$ and $w_H = 2 w_{H:H} + w_{B:H}$. Hence, $r_B = \frac{1}{2} w_B$ in this case.

3 Conditions for Establishment

Rearranging the general transition Eq. (2) with the help of Eq. (1) such that the fractional fitnesses appearing in the amount of self-replication of a type are pooled together, one obtains for the biallelic case,

$$P'_A = \frac{r_A \cdot P_A + w_{H:H} \cdot I_{H,H}(A) \cdot P_H}{\overline{w}}, \tag{3a}$$

$$P'_H = \frac{r_H \cdot P_H + w_{B:A} \cdot P_A}{\overline{w}}, \tag{3b}$$

$$P'_B = \frac{r_B \cdot P_B + w_{H:H} \cdot I_{H,H}(B) \cdot P_H}{\overline{w}}, \tag{3c}$$

and thus for the frequency $p_A = P_A + \frac{1}{2} P_H$ of the A-allele,

$$p'_A = \frac{\left[r_A + \frac{1}{2} w_{B:A} \right] \cdot P_A + \left[\frac{1}{2} r_H + w_{H:H} \cdot I_{H,H}(A) \right] \cdot P_H}{\overline{w}}. \tag{3d}$$

In the Appendix (Sect. 5) it is shown that p_A'/p_A becomes equal to Θ_A as p_A approaches zero, where

$$\Theta_A = \frac{1}{r_B} \cdot \left[\frac{1}{2}(r_A + r_H) + \sqrt{\frac{1}{4}(r_A - r_H)^2 + w_{B:A} \cdot w_{H:H} \cdot I_{H,H}(A)} \right].$$

Hence, Θ_A is the multiplication rate of the A-allele when it is very rare, so that $\Theta_A > 1$ means initial increase in frequency and thus establishment, and $\Theta_A < 1$ means initial decreases in frequency and thus the impossibility to become established. Clearly, all quantities appearing in Θ_A refer to the situation where the A-allele is very rare.

Since $\Theta_A \geqslant \left[\frac{1}{2}(r_A + r_H) + \frac{1}{2}\left| r_A - r_H \right| \right]/r_B = \max\left\{r_A, r_H\right\}/r_B$, one arrives at the central result:

A sufficient condition for the A-allele to become established is that the newly arising homozygote or heterozygote exceed the prevailing homozygote in the amount of self-replication, i.e. $\max\{r_A, r_H\} > r_B$ for small p_A.

For arbitrary selection regimes with Mendelian segregation it is known that a similar relationship with respect to Darwinian fitness ($\max\{w_A, w_H\} > w_B$) is necessary for the establishment of the A-allele (Gregorius 1984b). However, this relationship is by no means sufficient. For example, if mating takes place after random encounter of individuals or gametes (which does *not* necessarily imply random mating), then $w_{A:A} = w_{H:H} = 0$ for very low frequency of the A-allele, as was previously argued, and $\Theta_A = \max\{r_A, r_H\}/r_B = r_H/r_B$, since in this case $r_A = 0$. Hence, the A-allele becomes established for $\Theta_A > 1$, i.e. for $r_B < r_H$. However, it cannot become established for $\max\{w_A, w_H\} > w_B$ if $w_H < w_B < w_A$. Only in a few cases of non-random encounter and non-random mating may the underdominance implicit in $w_H < w_B < w_A$ allow for the establishment of the A-allele (see Gregorius 1984b). This shows that if there is a principle of the "survival of the fittest", it cannot be based on the measurement of Darwinian fitnesses. Measurements based on the amount of self-replication are more likely to be suitable for the formulation of such a principle, since for this measure and for the presently considered case of two alleles survival of the fittest indeed holds under all circumstances.

If $w_{B:A} \cdot w_{H:H} \cdot I_{H,H}(A) = 0$, then $\max\{r_A, r_H\} < r_B$ is sufficient to prevent establishment of the A-allele. This product may equal zero for a variety of reasons, some of which were already mentioned (random encounter, random mating). Another example is provided by a case of gametophytic incompatibility, in which pollen carrying the A-allele cannot grow on the stigmata of heterozygous plants, in which case $I_{H,H}(A) = 0$. This may happen even when heterozygous plants self-fertilize.

In general terms, the product $w_{B:A} \cdot w_{H:H} \cdot I_{H,H}(A)$ may be conceived of as representing the effects of the mutual "genetic altruism" of the rare types (A and H) on their establishment, since the mating type A × B produces only heterozygotes, which differ from both parents, and in $w_{H:H} \cdot I_{H,H}(A)$ only those A-offspring from the mating H × H are counted which again differ from their parents. The altruism is, in fact, mutual among the rare types, since heterozygotes supply A-homozygotes via $w_{H:H} \cdot I_{H,H}(A)$, and A-homozygotes supply heterozygotes via $w_{B:A}$. Hence, it is not very surprising that sufficiently large amounts of mutual genetic altruism among the rare types may guarantee establishment where self-replication fails, i.e. where

$\max\{r_A, r_H\} \leqslant r_B$. The exact amount can be obtained by setting $\Theta_A = 1$ and solving this equation for the second summand under the root, which gives $w_{B:A} \cdot w_{H:H} \cdot I_{H,H}(A) > (r_B - r_A)(r_B - r_H)$. If the inequality holds in the reverse direction, then the A-allele cannot become established. In summary:

If $\max\{r_A, r_H\} \leqslant r_B$, then the A-allele becomes established if

$$w_{B:A} \cdot w_{H:H} \cdot I_{H,H}(A) > (r_B - r_A)(r_B - r_H),$$

and it cannot become established if the sign of inequality is reversed. The rate of establishment or extinction is given by Θ_A.

While $w_{B:A}$ is independent of all three amounts of self-replication, $w_{H:H}$ is not, since it appears in r_H as $w_{H:H} \cdot I_{H,H}(H)$. Thus, even for a small amount of self-replication of the rare types, there is still a chance for the A-allele to become established provided $w_{H:H} \cdot I_{H,H}(A) > 0$ and the A-homozygote produces abundant heterozygotes by matings with the prevalent homozygote. This underlines the importance of the production of heterozygotes by matings of the rare with the frequent homozygote for the establishment of an allele. However, this effect vanishes as matings among the heterozygotes decline with their frequency.

3.1 Protection of a Polymorphism

The above statements, being derived for the establishment of the A-allele, apply analogously to the B-allele. This leads to the formulation of *a new overdominance principle* which holds for arbitrary selection regimes, mating systems and modes of inheritance at a biallelic locus and thus vastly generalizes the old overdominance principle based on Darwinian fitness:

If the amount of self-replication of the heterozygote exceeds that of either homozygote as the latter becomes frequent, then both alleles remain in the population and the biallelic polymorphism is protected.

It ought to be emphasized that this superiority of the heterozygote need not be realized over the whole range of genotypic frequencies (as it almost never is) in order to protect the polymorphism; it is sufficient that the superiority show only at extreme frequencies. The principle thus merely rephrases and makes more precise the common idea that a single type (the heterozygote) may protect a polymorphism if it is sufficiently "fit" under the relevant conditions.

In all other situations, i.e. where $r_H \leqslant r_B$ for large frequency of the B-homozygote or $r_H \leqslant r_A$ for large frequency of the A-homozygote, protectedness of the polymorphism cannot be achieved by superiority of a single type. In these situations, either a frequency-dependent change in ranking between the amounts of self-replication of the two homozygotes is required ($r_A > r_B$ for large P_B and $r_A < r_B$ for large P_A), or even more complex relationships among all three genotypes must hold.

However, since probably in the majority of biologically relevant cases random encounter of individuals or gametes plays a significant role, the above principle of heterozygote superiority can be expected to be the most frequently applicable explanation for protected biallelic polymorphisms.

4 Conclusions

The present results generalize those recently obtained for viability-fertility selection and random fusion of gametes (Ziehe and Gregorius 1985) and for viability-fecundity selection and random mating (Gregorius and Ziehe 1986). In both types of models, Mendelian segregation and frequency independence of the selection parameters was assumed. This led to constancy of the ratios of certain fractional fitnesses on which the conclusions were based. As is now evident, relaxation of these restrictions does not invalidate the principle that superiority in the amount of self-replication of the heterozygote over the frequent homozygote protects biallelic polymorphisms. This new overdominance principle differs from the old (Darwinian) one fundamentally in that it is reduced to the situations which are essential for the maintenance of a poly-morphism: extreme frequencies. Only when alleles become rare is the polymorphism endangered.

In contrast, the overdominance principle based on Darwinian fitness assumes global (i.e. over the whole range of frequencies) superiority of the heterozygote over both homozygotes. Nevertheless, it may easily fail if, for example, self-fertilization be-comes effective, as was pointed out for the model of Kimura and Ohta (1971). This model may serve to briefly demonstrate the new overdominance principle. It allows for viability selection, a constant degree of self-fertilization and random cross-fertili-zation. The fractional fitnesses are given in Gregorius (1984a) and, when applied to the amounts of self-replication, yield for regular Mendelians segregation:

$$r_A = v_A \left(s + (1 - s) \left(1 - \frac{v_B P_B}{\overline{v}} \right) \right), \quad r_B = v_B \left(s + (1 - s) \left(1 - \frac{v_A P_A}{\overline{v}} \right) \right),$$

$$r_H = (1 - s) v_H \left(1 - \frac{\frac{1}{2} v_H P_H}{\overline{v}} \right) + \frac{1}{2} s \cdot v_H,$$

$$w_A = 2 v_A, \quad w_B = 2 v_B, \quad w_H = 2 v_H,$$

where the v's are the viabilities, $\overline{v}$ is the average viability and s is the proportion of self-fertilization. It is seen that the amounts r of self-replication are generally frequency-dependent while the Darwinian fitnesses are not. Only for complete selfing (s = 1) does the frequency dependence of the r's vanish.

The r-values relevant for protection are those for $p_A = 0$, in which case $r_A = r'_A = v_A \cdot s$, $r_B = r'_B = v_B$, $r_H = r'_H = \left(1 - \frac{1}{2} s \right) v_H$, and for $p_B = 1 - p_A = 0$, in which case $r_A = r''_A = v_A$, $r_B = r''_B = s \cdot v_B$, $r_H = r''_H = \left(1 - \frac{1}{2} s \right) v_H$. Hence, $r'_H = r''_H$, and the over-dominance principle requires $\left(1 - \frac{1}{2} s \right) v_H > \max\{v_A, v_B\}$, which implies protected-ness of the biallelic polymorphism. Overdominance in Darwinian fitness implies $v_H > \max\{v_A, v_B\}$, which is necessary but not sufficient for protection of the polymor-phism. The difference between the principles clearly lies in the fact that the r-values reflect the effect of the mating system via s, while the w-values do not. Thus, simple superiority of the heterozygote may not yield a stable polymorphism, but $\left(1 - \frac{1}{2} s \right)^{-1}$-fold superiority always does.

The amount of self-replication must not be confused with the ratio of the frequency of a type in two successive generations P_i'/P_i. This was discussed as a measure of "individual" fitness by Prout (1965), and Michod (1984) used the normalization $1 - P_i'/P_i$, which he called the "per capita rate of increase". Clearly, this fitness measure differs from the amount of self-replication in that it additionally counts individuals which are not offspring of the type in question, thus this consideration also distinguishes it from Darwinian fitness. Michod (1984) demonstrated that P_i'/P_i (or $1 - P_i'/P_i$) has almost no predictive significance for the maintenance of a polymorphism even in the basic classical viability selection model. Both w_i and P_i'/P_i are concepts of fitness which more or less implicitly assume genetic identity between parents and offspring, and they are thus, in a strict sense, limited to asexual reproduction. To some degree, the amount of self-replication r_i may also be conceived of as measuring an asexual component of fitness, however, this component is mimicked by sexuality. Considering the fact that among the three measure w_i, P_i'/P_i and r_i, the last reflects the smallest portion of the reproductive success of a type, it is the more remarkable that it is the only one which provides simple sufficient conditions for the establishment of an allele and the protectedness of a biallelic polymorphism under a rather wide range of selection regimes including all types of segregation distortion and mating systems. The reason for this is probably that it explicitly separates the sum total of all those effects of the genetic system which allow an individual to reproduce its own type from those which do not. This separation, in turn, provides a useful basis for assessing the significance of the interaction of both categories of effects, as it shows up in the representation of Θ, the initial multiplication rate of a rare allele.

The present analysis is only a first step towards the formulation of more generally valid rules for the selective maintenance of polymorphisms. The extendability to multiple alleles and multiple loci as well as extranuclear inheritance has yet to be investigated. However, even for more complex modes of inheritance, the establishment or protectedness of a single gene, whether nuclear or extranuclear, remains a fundamental issue for explaining the existence of polymorphisms. Highly integrated systems are more likely to evolve by the successive integration of the genetic units of transmission, i.e. the genes, which would then be tantamount to the establishment of an allele against the sum total of "wild types". In this sense the present analysis may provide a useful clue to further generalization.

The most important statement implicit in the above considerations is probably that one has to distinguish properly between *reproductive success* and *evolutionary success*. Reproductive success is measured by various concepts of individual fitness, while evolutionary success is, most basically, determined by the conditions for establishment or maintenance as summarized in Θ. Thus, the principle of survival of the fittest may simply be understood as an attempt to link evolutionary success to reproductive success, or vice versa, and it is shown that this attempt critically depends on the concept of fitness applied. Yet, predictability of evolutionary success need not be the sole criterion for evaluating a fitness measure. On the contrary, any particular measure of fitness can be considered as a trait, the expressions of which are the outcome of an evolutionary or adaptive process. This is the case, for example, in all models of sexually asymmetric viability, fertility or fecundity selection, in which Darwinian fitness is frequency-dependent (Gregorius 1984a). In these models, the dynamics of the frequencies of the genotypes are not a function of their Darwinian fitnesses only, so that the fitness of a genotype may evolve in a non-self-adjusting manner. Hence, fitness may also be a character of interest in itself such as height, flower colour, biomass, bristle number, etc., and it is only this point of view which makes studies such as on the evolution of overdominance in fitness (Ross 1980) meaningful. In other words, depending on the objective in mind, fitness measures may serve to specify causes or effects of evolution.

5 Appendix

In order to analyze the establishment of the A-allele, consider the variable $x_A := p_A/p_A$. Then $1 - x_A = \frac{1}{2} p_H/p_A$. The transition equation for x_A follows from division of Eqs. (3a) by (3d):

$$x_A' = \frac{r_A \cdot x_A + 2 w_{H:H} \cdot I_{H,H}(A) \cdot (1 - x_A)}{\left[r_A + \frac{1}{2} w_{B:A} \right] \cdot x_A + [r_H + 2 w_{H:H} \cdot I_{H:H}(A)] \cdot (1 - x_A)}.$$

Dividing the numerator and denominator by $\bar{w}$, which leaves the equation unchanged, the denominator becomes equal to p_A'/p_A. Hence, we are concerned with a transition equation of the form

$$x' = \frac{a \cdot x + b \cdot (1 - x)}{c \cdot x + d \cdot (1 - x)} = g(x),$$

where $0 \leqslant x \leqslant 1, 0 \leqslant a \leqslant c$ and $0 \leqslant b \leqslant d$, such that $0 \leqslant x' \leqslant 1$.

These are exactly the prerequisites for the analysis of protectedness introduced by Gregorius (1982b). Under the assumption that the parameters in the transition equation for x_A can be continuously extended to the boundary $p_A = 0$, which applies here, it is shown that as p_A approaches zero, p_A'/p_A with one exception uniquely converges to

$$\Theta_A = \frac{1}{2}(a + d - b) + \sqrt{\frac{1}{4}(a + b - d)^2 + b(c - a)},$$

where $a = r_A/r_B$, $b = 2 w_{H:H} I_{H,H}(A)/r_B$, $c = (r_A + \frac{1}{2} w_{B:A})/r_B$, $d = [r_H + 2 w_{H:H} \cdot I_{H,H}(A)]/r_B$, of which all quantities are evaluated at the boundary $p_A = 0$. The division by r_B results from the fact that for $p_A = 0$, $\bar{w} = \frac{1}{2} w_B = r_B$. Since $c - a = \frac{1}{2} w_{B:A}/r_B$ and $d - b = r_H/r_B$, we obtain

$$\Theta_A = \frac{1}{r_B} \cdot \left[\frac{1}{2}(r_A + r_H) + \sqrt{\frac{1}{4}(r_A - r_H)^2 + w_{B:A} \cdot w_{H:H} \cdot I_{H,H}(A)} \right].$$

Thus, the A-allele becomes established or does not become established according to whether $\Theta_A > 1$ or $\Theta_A < 1$ respectively.

The exception is $d = b > 0$, $c > 0$ and $a = 0$, i.e. $r_A = r_H = 0$, $w_{B:A} > 0$, $w_{H:H} \cdot I_{H,H}(A) > 0$, and thus $w_{B:H} I_{B,H}(H) = I_{H,H}(H) = 0$. In this case, $g^2(x) = x$ for all x, so that the dynamics consists only of cycles of length 2. The multiplication rate of the A-allele (p_A'/p_A) for small p_A and $x_A = x$ is $\mu(x) = c \cdot x + d \cdot (1 - x)$. Consequently, $\mu(x) \cdot \mu(g(x)) = c \cdot d = w_{B:A} \cdot w_{H:H} \cdot I_{H,H}(A)/r_B^2 = \Theta_A^2$. Hence, even in this case, $\Theta_A > 1$ and $\Theta_A < 1$ are still the conditions for the establishment of the A-allele.

References

Crow JF, Kimura M (1970) An introduction to population genetics theory. Harper & Row, London New York

Gregorius H-R (1982a) The relationship between genic and genotypic fitnesses in diploid populations. Evol Theor 6:143–162

Gregorius H-R (1982b) Selection in populations of effectively infinite size. II. Protectedness of a biallelic polymorphism. J Theor Biol 96:689–705

Gregorius H-R (1984a) Fractional fitnesses in exclusively sexually reproducing populations. J Theor Biol 111:205–229

Gregorius H-R (1984b) Allele protectedness in frequency dependent biallelic selection models with separated generations. J Theor Biol 111:425–446

Gregorius H-R, Ziehe M (1986) The significance of over- and underdominance for the maintenance of genetic polymorphisms. II. Overdominance and instability with random mating. J Theor Biol 118:115–125

Kimura M, Ohta T (1971) Theoretical aspects of population genetics. Princeton Univ Press

Michod RE (1984) Constraints on adaptation, with special reference to social behavior. In: Price PW, Slobodchikoff CN, Gaud WS (eds) A new ecology: novel approaches to interactive systems. Wiley & Sons, New York

Prout T (1965) The estimation of fitness from genotypic frequencies. Evolution 9:546–551

Ross MD (1980) The evolution and decay of overdominance during the evolution of gynodioecy, subdioecy, and dioecy. Am Nat 116:607–620

Ziehe M, Gregorius H-R (1985) The significance of over- and underdominance for the maintenance of genetic polymorphisms. I. Underdominance and stability. J Theor Biol 117:493–504

Quantitative Genetics and Evolution

The Maintenance of Genetic Variation: A Functional Analytic Approach to Quantitative Genetic Models

R. Bürger[1]

1 Introduction

If one tries to understand phenomena of quantitative genetics in terms of classical population genetics one is faced with several problems. On the one hand, the genetic structure underlying quantitative traits is usually very complex and not well understood. On the other hand, models that have been treated mathematically on the basis of classical population genetics usually depend on a number of special assumptions like a very limited number of alleles or loci, special types of fitness functions, particular mutation schemes, etc. However, for describing dynamic and equilibrium properties of metric traits it is not always necessary to have a detailed knowledge of the dynamics of gene frequencies. Instead, it will often be sufficient to have appropriate information on mean values of characters, on their pattern of variation and covariation and perhaps on higher moments of the character distribution. To this aim it would be useful to have models and techniques that are less complicated to handle than the corresponding multilocus systems.

It is the purpose of the present chapter to present an approach that helps to resolve some of the above mentioned difficulties for the well-known problem whether genetic variation can be maintained through a balance between mutation and selection. This approach is based on mathematical methods from functional analysis that have long been used in theoretical physics and, only recently, also in the theory of age-structured populations (e.g. Webb 1985). Before describing the general model, the methods of analysis (without mathematical details) and the results I will shortly review what has been done so far in connection with the problem of the maintenance of genetic variation.

Crow and Kimura (1964) introduced a model that took into account mutations as a source for genetic variability in quantitative traits. This model supposes that new alleles are produced continuously through mutation and that the expression of a metric character is determined by a continuum of possible alleles. Additionally, it is assumed that the trait is subject to stabilizing selection. The dynamics of the model is given by an integro-differential equation. Kimura (1965) approximated this equation by a reac-

---

1 Institut für Mathematik, Universität Wien, Strudlhofgasse 4, A-1090 Wien, Austria

Population Genetics and Evolution
G. de Jong (ed.)
© Springer-Verlag Berlin Heidelberg 1988

tion-diffusion equation, demonstrated that the equilibrium distribution of allelic effects is Gaussian with an average effect at the optimum and calculated the genetic variance at equilibrium.

Latter (1970) investigated a corresponding discrete-time model. Lande (1975) extended this to many linked loci and proposed that large amounts of genetic variation in polygenic characters can be maintained by mutation, even when there is strong stabilizing selection. This hypothesis, however, was questioned by Turelli (1984). He reviewed and extended several analyses of models concerning the maintenance of variability by mutation and performed extensive computer simulations of one- and multilocus models allowing also for various levels of linkage. In particular, he simulated a discrete-time, one-locus haploid model with 21 alleles, which he intended as an approximation to Kimura's continuum-of-alleles model.

Turelli (1984) found that unless selection is very weak or the mutation rate very high the equilibrium variance is considerably lower than expected from Kimura's analysis. Similarly, his multilocus diploid simulations show that Lande overestimated the equilibrium variance considerably. Another interesting result of Turelli's simulations is that, as asserted by Lande (1975) and Fleming (1979), linkage has very little effect on the amount of additive genetic variance maintained. Moreover, it turned out that the results from the one-locus haploid model can be adequately extrapolated to multilocus models. Turelli also investigated analytically a model proposed by Kingman (1977, 1978), the so-called house-of-cards model. In this model it is assumed that the probability of mutations from one gene to another depends solely on the target gene, for example because only mutants of the wild type are counted, which is reasonable if the wild type predominates, or because the effects of new mutations are generally much greater than the existing genetic variance. Turelli obtained that, in contrast to Kimura's approximate analysis, this model gives values for the equilibrium variance that are close to his simulations of the Kimura model for a wide range of parameters.

The reason that Kimura and Lande overestimated the equilibrium variance is that the diffusion approximation of Kimura is inadequate. In particular, the equilibrium distribution of allelic effects in the population is not Gaussian for the exact model (cf. Fleming 1979; Nagylaki 1984; Turelli 1984; Bürger 1986). The first exact analysis of Kimura's one-locus haploid model was performed by the present author (Bürger 1986). There it has been shown, as already conjectured by Fleming (1979) and Nagylaki (1984), that in Kimura's model a unique equilibrium in fact exists and that it is globally stable. Furthermore, estimates of the equilibrium variance have been derived analytically that are in best accordance with Turelli's simulation results.

Different analyses, based on multilocus models with two or three alleles per locus have been published by Latter (1960), Bulmer (1972, 1980), Barton (1986) and Turelli (1984) respectively. All these authors, except Barton, obtained an equilibrium variance much lower than the values of Kimura and Lande. Barton showed that a deviation of the mean from the optimum leads to values of the variance which may be of the same order as that based on Kimura's and Lande's Gaussian approximation.

To sum up, the situation is rather intricate since some authors used models with a continuum of alleles (Kimura, Lande, Fleming, Nagylaki, Bürger), others with two alleles (Latter, Bulmer, Barton) or with three and even 21 alleles (Turelli). They used in part different mutation schemes (e.g. Gaussian model or "house-of-cards"). Some of them performed approximate analyses, some computer simulations. The only mathematically exact analysis seems to be that by the present author. In the sequel I will present a general model that makes it possible to derive estimates for the equilibrium variance that can be maintained through a mutation-selection balance. These estimates are independent of the number of alleles per locus and of the particular mutation scheme. To this aim I will make use of the techniques and results developed in Bürger (1986, 1987).

2 A General Model

Although the results of the present chapter are aimed at diploid populations, I will first develop a model for an infinite, asexually reproducing population with overlapping generations. Individuals will be characterized by their type x. x is a vector in some locally compact subset M of n-dimensional Euclidean space $\mathbb{R}^n$. In most applications n = 1 will be chosen. Instead of providing an exact definition, I list the subsequent typical examples for the state space M.

1. Consider the classical haploid one-locus model with n alleles. Then each allele represents a type x and M = $\{1,2,...,n\}$ will be the natural state space.
2. If one deals with Ohta and Kimura's ladder model, in which an infinite sequence of possible alleles at one locus is considered M = $\{...,-2,-1,0,1,2,...\}$ will be the appropriate choice for the state space.
3. If M = $\mathbb{R}$ is assumed each x $\in$ M may be considered as an average allelic effect on a quantitative trait or as a genotypic value. This leads to so-called continuum-of-alleles models. These are treated below in greater detail.
4. Some further sensible choices would be M = $[0,1]^n$, M = $[0, \infty]^n$ or M = $\mathbb{R}^n$. Here, each x $\in$ M can be considered as a vector of measurements of quantitative characters. Since there is a variety of possible choices of M the scale of measurement can be chosen quite arbitrarily..

Throughout, p(x,t) denotes the (normalized) density of type x in the population at time t. m(x) denotes Malthusian fitness of type x. Mutation is assumed to occur from type y to type x with probability density u(x,y), i.e. $\int_{M\setminus\{y\}} u(x,y)dx = 1$ for all y. (Note that $\int_{M\setminus\{y\}}$ reduces to $\Sigma_{x \neq y}$ if M is a finite or discrete set and equals $\int_M$ if dx is the usual measure.) For simplicity it is also assumed that mutation occurs at the same rate μ for all types.

The dynamics of type densities is then given by

$$\partial p(x,t)/\partial t = [m(x)-\bar{m}] \, p(x,t) + \mu[\int_{M\setminus\{x\}} u(x,y) \, p(y,t)dy - p(x,t)] \,, \tag{1}$$

where $\bar{m}(t) = \int_M m(x) \, p(x,t)dx$ is the mean fitness of the population. The first term on the right-hand side of Eq. (1) describes the change due to selection and the second term that due to mutation. For populations with discrete generations the dynamics is given by

$$p_{t+1}(x) = (1 - \mu) \, p_t'(x) + \mu \int p_t'(y) \, u(x,y)dy \,, \tag{2}$$

where $p_t'(x) = w(x)p_t(x)/\int w(y)p_t(y)dy$ is the density function for allelic effects after selection and $w(x)$ denotes relative fitness [cf. Turelli 1984, Eq. (2.7), for example]. Malthusian and relative fitness are related through $m(x) = \ln w(x)$.

Among others, the following special cases have been dealt with in the literature:

(a) The classical model, i.e. $M = \{1,2,...,n\}$. After changing notations $p(x,t) \to p_i(t)$, $m(x) \to m_i$ and $u(x,y) \to u_{ij}$ Eq. (1) can be rewritten as

$$\dot{p}_i = (m_i - \bar{m})p_i + \mu \Sigma_{i \neq j} u_{ij} p_j - \mu p_i, \quad i = 1,2,...,n . \tag{1'}$$

This equation as well as the corresponding difference equation has been analyzed, e.g. by Crow and Kimura (1970), Moran (1976) and Nagylaki (1977).

(b) Kimura's continuum-of-alleles model is obtained if one chooses $M = \mathbb{R}$, $m(x) = - sx^2$ and $u(x,y) = (2 \pi\gamma^2)^{-1/2} \exp[(x-y)^2/2 \gamma^2]$ (see Kimura 1965; Bürger 1986). The corresponding discrete time equation, with Malthusian fitness replaced by relative fitness $w(x) = \exp(-x^2/2V_s)$, $s = 1/2V_s$, has been investigated by Fleming (1979), Nagylaki (1984) and Turelli (1984). More generally one can consider mutation terms of the form $u(x,y) = u_1(x-y)$, such that $\int u_1(x)dx = 1$.

(c) A continuous time version of Kingman's house-of-cards model (Kingman 1977, 1978), as investigated by Turelli (1984), is obtained by choosing $M = \mathbb{R}$, $m(x) = - sx^2$ and $u(x,y) = 2 \pi\gamma^2)^{-1/2} \exp(- x^2/2 \gamma^2)$. In this model it is assumed that the mutation probability depends only on the target gene, which is a reasonable approximation, if the effects of new mutations are, in mean, much greater than the existing genetic variance.

3 Analysis and Results

In this section I will provide answers to the following questions.

1. Does Eq. (1) exhibit a stable stationary type density? It is unique?
2. Is it possible to derive estimates for the equilibrium variance of this type of density, at least if $m(x) = - sx^2$?

The answer to question (2) is closely related to the problem whether additive genetic variation can be maintained through a balance between mutation and stabilizing selection. However, estimates for an equilibrium variance make sense only if the corresponding equilibrium is stable. Most of the investigators of this problem have concentrated on estimates for the equilibrium variance and have neither proved that an equilibrium in their model exists, nor investigated its stability properties. Only recently it has been shown that for Kimura's continuum-of-alleles model in fact an equilibrium exists which is uniquely determined and globally stable (Bürger 1986). For the classical haploid model, i.e. for the discrete time version of Eq. (1'), the same result has been shown by Moran (1976). Kingman (1977, 1978) proved also some stability results for his house-of-cards model (which he formulated in a slightly different guise).

In order to obtain analogous results for the much more general Eq. (1), methods from functional analysis are required which I will briefly outline.

$x \to p(x,t)$ determines a function $p(t)$ in the Banach space of absolutely integrable functions $L^1(M)$ [remember that $f \in L^1(M)$ if and only if $\int_M |f(x)|dx < \infty$]. Next, we introduce the following linear operators on $L^1(M)$:

$$Tf(x) = -m(x)f(x) \qquad \text{(selection operator)}$$

$$Uf(x) = \mu \int_M u(x,y)f(y)dy \qquad \text{(mutation operator)}.$$

Here, T is a multiplication operator, which corresponds to a diagonal matrix for finite M, and U is a kernel operator. (If M is finite or discrete put $u(i,i) = u_{ii} = 0$.) Then Eq. (1) can be rewritten as

$$p = \partial p/\partial t = -Tp + Up + p \cdot \int_M (Tp(x) - Up(x))dx . \qquad (3)$$

Stationary or equilibrium solutions of Eq. (3) are given by $\dot{p} = 0$, or equivalently by

$$(T-U)p = \int_M (T-U)pdx \cdot p . \qquad (4)$$

The solutions p of Eq. (4) are precisely those functions in $L^1(M)$ that satisfy

$$(T-U)p = \lambda p, \quad \text{for some real } \lambda . \qquad (4')$$

For our problem only positive solutions p of this eigenvalue problem are of interest. Eigenvalue problems of this kind have been treated in Bürger (1987). I will summarize the assumptions and results in a qualitative way.

Assumptions. (Fig. 1) (a) If M is an unbounded state space (e.g. $M = \mathbb{R}^n$) Malthusian fitness must satisfy $\lim_{|x| \to \infty} m(x) = -\infty$, i.e. selection eliminates extreme types (for bounded M this is not required!). Without loss of generality $m(x) \leqslant 0$ may be assumed by adding an appropriate constant.

(b) m(x) has no cusps. This is automatically satisfied if M is a discrete set.

(c) The mutation operator U is irreducible and $u(x,y) > 0$, if $|x-y|$ is small. Both conditions are simultaneously satisfied if, for example, $u(x,y)$ is strictly positive along the diagonal $\{(z,z), z \in M\}$.

Assumption (a) is essential to make the proof work, except in special cases like the house-of-cards model, where $\lim_{|x| \to \infty} m(x) = -C$, C some positive constant, is sufficient. I do not know whether assumption (a) is really necessary. If one only has $\lim_{|x| \to \infty} m(x) = -C$, selection becomes arbitrarily weak far away from the optimum and its effects may be swamped by mutation, if it is given by $u(x,y) = u_1(x-y)$, for example. If assumptions (b) and (c) are not fulfilled counterexamples to the results below can be constructed. The precise conditions may be found in Bürger (1987).

Result 1. Under the assumptions imposed above a uniquely determined positive equilibrium density $\bar{p}$ of Eq. (3), and hence of Eq. (1), exists. It is strictly positive and globally stable.

As already mentioned, this result has been proved for the classical model with a finite number of alleles [cf. Sect. 2(a)] by Moran (1976) and for Kimura's model [cf. Sect. 2(b)] by Bürger (1986). If the fitness function has a cusp at its maximum the equilibrium density is not an integrable function, but an atom of probability occurs. This phenomenon has already been observed by Kingman (1978). The reader may notice that uniqueness and global stability of an equilibrium as proved above is peculiar for haploid models or, equivalently, for diploid models with additive fitness. In diploid models with dominance the selection-mutation equation may become very

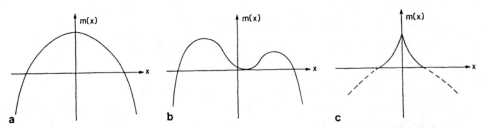

Fig. 1. a, b Two typical fitness functions on M = ℝ which fulfill the assumptions for Result 1 (see text). Note that m(x) and m(x) + c, c any real number, lead to the same differential equation. **c** A fitness function with a cusp at its optimum. In this case no integrable equilibrium density exists. Instead an atom of probability builds up

complicated. In the case of only two alleles and one locus up to three equilibria, two of which are locally stable, may exist (Bürger 1983). For models with three or more alleles at one locus the existence of stable limit cycles has been shown recently (Hofbauer 1985).

To obtain an estimate for the equilibrium variance for quadratically deviating fitness $m(x) = -sx^2$, which is assumed throughout the sequel, consider again Eqs. (4) and (4'). In Bürger (1987) it has been shown that the eigenvalue λ corresponding to the positive equilibrium solution $\bar{p}$ is strictly negative. Hence, it follows that

$$0 > \lambda = \int_M T\bar{p}(x)dx - \int_M U\bar{p}(x)dx = \int_M sx^2\bar{p}(x)dx - \mu\int_M\int_M u(x,y)\bar{p}(y)dydx$$
$$= s\int_M x^2\bar{p}(x)dx - \mu .$$

Assuming now that the mean of the equilibrium density is at zero, i.e. $\int_M x\bar{p}(x)dx = 0$, we obtain.

Result 2. μ/s is an upper bound for the equilibrium variance σ^2 maintained by the mutation-selection Eq. (1), that is

$$\sigma^2 = \int_M x^2\bar{p}(x)dx < \mu/s . \tag{5}$$

This estimate holds independently of the number of alleles (since M can be chosen quite arbitrarily) and independently of the particular form of the mutation term.

If the state space M and the mutation operator U are specified the eigenvalue λ and hence the equilibrium variance $\sigma^2 = (\lambda + \mu)/s$ can be calculated by numerical methods. Clearly, σ^2 depends also on the variance of mutational effects γ^2 [cf. Sect. 2(b)]. In Bürger (1986) the Rayleigh-Ritz technique, which had been developed in quantum mechanics, was applied to Kimura's continuum-of-alleles model. The values obtained agree precisely with Turelli's (1984) simulations of his 21-alleles model. Turelli's model was intended as an approximation to the discrete time model with a continuum of allelic effects as given by Eq. (2). By abstract arguments from spectral theory of unbounded operators it follows that, if for constant γ^2 the ratio μ/s tends to zero, the equilibrium variance approaches μ/s. If μ/s is relatively large ($\geqslant 0.01$), σ^2 is usually much smaller than μ/s, the exact value depending strongly on γ^2.

Kimura (1965) obtained from his approximate analysis that

$$\sigma^2 = \gamma(\mu/2s)^{1/2} . \tag{6}$$

This value is greater than μ/s if and only if

$$\gamma^2 > 2\,\mu/s\,. \tag{7}$$

Consequently, Kimura's Gaussian approximation is always inappropriate, if Eq. (7) is fulfilled.

There is, however, a difference between the continuous time model, as given by Eq. (1), and the discrete time model, as given by Eq. (2), that should be mentioned. Because the effects of mutations are modelled in a slightly different way, Eqs. (1) and (2) have different equilibrium solutions. Whereas for $\mu = 1$ the equilibrium solution of the discrete time model is Gaussian and hence $\sigma^2 = \sqrt{\gamma^2 V_s}$ is the equilibrium variance [recall that relative fitness is given by $w(x) = \exp(-x^2/2\,V_s)$, cf. Sect. 2(b)], this can never happen in the continuous time model. Therefore, the upper bound μ/s applies exactly only to the differential equation model, since its discrete time analogue $2\mu V_s$ may be exceeded by the variance $\sqrt{\gamma^2 \mu V_s}$ of the Gaussian approximation, which becomes exact with discrete times as $\mu \to 1$. For $\mu = 1$ one has $\sqrt{\gamma^2 \mu V_s} > 2\mu V_s$, if $\gamma^2 > 4 V_s$. This inequality is empirically probably not very realistic. However, for μ sufficiently small ($\mu < 10^{-2}$) $2\mu V_s$ appears to be an appropriate upper bound for the discrete time model as may be seen by comparison with Turelli's simulation results. Moreover, the equilibrium variance of the continuous time model with a continuum of allelic effects and Turelli's 21-alleles discrete time model agree very well (Bürger 1986).

It is interesting to notice that Turelli's approximate analysis of Kingman's house-of-cards model led to the equilibrium value

$$\sigma^2 = \mu/s\,, \tag{8}$$

which is just the upper bound given by Eq. (5). Turelli's observation that this value is close to the equilibrium variance of his 21-alleles model, if $\gamma^2 \gg 2\mu V_s$ or, for continuous time, if $\gamma^2 \gg \mu/s$, can now be supported analytically. If Eq. (7) is valid Kimura's approximation Eq. (6), is always inappropriate, but the house-of-cards model becomes relevant, since the variance of new mutations γ^2 is now much greater than the existing variance σ^2, which is always less than μ/s. Moreover, as mentioned above, for decreasing μ/s (keeping γ^2 fixed) the equilibrium variance approaches μ/s, in best accordance with Eq. (8).

In connection with Eq. (7) it should be mentioned that already Lande (1975) conjectured that the equilibrium would be nearly Gaussian if $\gamma^2 \ll \sigma^2$, or equivalently, using the Gaussian approximation, if $\gamma^2 \ll \mu V_s$. This was one of the main points of Turelli's (1984) criticism, since he claimed that $\gamma^2 \gg \sigma^2$ is much butter supported empirically. From the fact that μ/s is an upper bound for the equilibrium variance it follows that $\gamma^2 \gg \sigma^2$ is valid if $\gamma^2 \gg \mu/s$ (or $\gamma^2 \gg 2\mu V_s$). In this case the Kimura-Lande approximation cannot be valid as may be seen from inequality, Eq. (7).

To apply the above estimates for the equilibrium variance to a random mating diploid populations we may follow Kimura (1965), Lande (1975) and Turelli (1984). Assume that a quantitative trait z is determined by additive allelic effects x from N loci plus a random environmental contribution e which is genotype-independent

$$z = x + e\,, \qquad x = \sum_{i=1}^{N} (x_i + x_i')\,.$$

e is a Gaussian random variable with mean zero and variance V_e, x_i denotes the maternal and x_i' the paternal contribution from the i-th locus. To each phenotype z the Malthusian fitness value $m_p(z)$ is assigned. It follows that mean fitness of individuals with genotypic value x is given by

$$m(x) = (2 \pi V_e)^{-1/2} \int m_p(z-y) \exp(-y^2/2 V_e) dy .$$

An easy calculation shows that $m_p(z) = -sz^2$ implies $m(x) = -s(x^2 + V_e)$. This is the reason why it is sufficient to consider only the level of genotypic values, at least if one works with this special fitness function. Assuming global linkage equilibrium and independence of the additive genetic contributions under random mating, an N locus approximation for the additive genetic equilibrium variance has been given by Kimura (1965)

$$\sigma^2 = 2 \sum_{i=1}^{N} \sigma_i^2 ,$$

where σ_i^2 denotes the equilibrium variance at the i-th locus. The validity of this extrapolation, provided the σ_i^2 are correct, has been supported by extensive simulations by Turelli (1984). From the mathematical point of view this formula is not obvious. Of course, it is well known that a one-locus diploid model with additive fitness is mathematically equivalent to a one-locus haploid model. However, this does not necessarily imply that the same holds for a model with additive effects, if fitness is not linear.

Using the upper bound, Eq. (4), and assuming that the mutation rate is the same for all loci and that all loci experience the same form of selection, the additive genetic equilibrium variance of a character determined by N loci may be estimated by

$$\sigma^2 < 2 N(\mu/s) .$$

This shows that it is reasonable to investigate first asexual models and then try to extrapolate the results to diploid models. However, some caution seems to be in order, since true diploid models allowing for dominance may have a considerably more complicated dynamics, as already mentioned earlier.

4 Concluding Remarks

This chapter presents an attempt to apply deeper mathematical methods from functional analysis to models for quantitative traits, in particular to the problem whether sufficient additive genetic variation can be maintained by mutation in the face of stabilizing selection. These methods have the advantage of yielding results that are independent of assumptions on the number of alleles per locus, even on whether there are discrete or continuous allelic effects, and on the particular distribution of mutational effects.

The reformulation of models to eigenvalue problems has a long tradition in theoretical physics, where for example in order to obtain theoretical information on the states of electrons of a certain atom, eigenvalue problems of a Hamiltonian operator have to be solved.

It is clear that on the basis of the above analysis it cannot be decided whether additive genetic variation in natural populations can be maintained through a mutation-selection balance. However, a reliable upper bound for the equilibrium variance has been derived analytically. Its derivation is not based on simulations, but well supported by them. Using the results of Kimura, Lande and Turelli the haploid model can be extrapolated to a multilocus diploid model as long as additivity assumptions are imposed. I feel, like Turelli, that in order to test Lande's hypothesis it is necessary to have additional data on selection intensities and on mutation rates in metric traits. Moreover, it would be necessary to investigate the effects of pleiotropy more closely. For a more complete treatment of this topic the reader is referred to Barton (1986), Bürger (1986) and to Turelli (1984, 1986).

It may well turn out that additive genetic variability cannot be maintained at a sufficiently high level by mutation. Of other alternatives that have been proposed and analyzed I want only to mention heterogeneous or fluctuating environments and the overdominance hypothesis. The latter has been treated recently by Gillespie (1984), for example.

Acknowledgements. I thank G.P. Wagner for several helpful disucssions of this work and K. Sigmund and M. Turelli for valuable comments on the manuscript. Part of this research was done at the International Institute for Applied Systems Analysis (Laxenburg, Austria) and was supported by the Austrian "Fonds zur Förderung der wissenschaftlichen Forschung" P5994.

References

Barton NH (1986) The maintenance of polygenic variation through a balance between mutation and stabilizing selection. Genet Res Camb 47:209–216

Bulmer MG (1972) The genetic variability of polygenic characters under optimizing selection, mutation and drift. Genet Res Camb 19:17–25

Bulmer MG (1980) The mathematical theory of quantitative genetics. Oxford Univ Press, Oxford

Bürger R (1983) Dynamics of the classical genetic model for the evolution of dominance. Math Biosci 74:125–143

Bürger R (1986) On the maintenance of genetic variation: Global analysis of Kimura's continuum-of-alleles model. J Math Biol 24:341–351

Bürger R (1987) Perturbations of positive semigroups and application to population genetics. Math Zeitschrift (in press)

Crow JF, Kimura M (1964) The theory of genetic loads. Proc XIth Int Congr Genetics, pp 495–505

Crow JF, Kimura M (1970) An introduction to population genetics theory. Harper & Row, New York

Fleming WH (1979) Equilibrium distributions of continuous polygenic traits. SIAM J Appl Math 36:148–168

Gillespie JH (1984) Pleiotropic overdominance and the maintenance of genetic variation in polygenic characters. Genetics 107:321–330

Hofbauer J (1985) The selection mutation equation. J Math Biol 23:41–53

Kimura M (1965) A stochastic model concerning the maintenance of genetic variability in quantitative characters. Proc Natl Acad Sci USA 54:731–736

Kingman JFC (1977) On the properties of bilinear models for the balance between genetic mutation and selection. Math Proc Camb Philos Soc 81:443–453

Kingman JFC (1978) A simple model for the balance between selection and mutation. J Appl Prob 15:1–12

Lande R (1975) The maintenance of genetic variability by mutation in a polygenic character with linked loci. Genet Res Camb 26:221–235

Latter BDH (1960) Natural selection for an intermediate optimum. Aust J Biol Sci 13:30–35

Latter BDH (1970) Selection in finite populations with multiple alleles II. Centripetal selection, mutation and isoallelic variation. Genetics 66:165–186

Moran PAP (1976) Global stability of genetic systems governed by mutation and selection. Math Proc Camb Philos Soc 80:331–336

Nagylaki T (1977) Selection in one-and two-locus systems. Lecture Notes in Biomathematics 15. Springer, Berlin Heidelberg New York

Nagylaki T (1984) Selection on a quantitative character. In: Chakravarti A (ed) Human population genetics. Pittsburgh Symp. Van Nostrand, New York

Turelli M (1984) Heritable genetic variation via mutation-selection balance: Lerch's zeta meets the abdominal bristle. Theor Pop Biol 25:138–193

Turelli M (1986) Gaussian versus non-Gaussian genetic analyses of polygenic mutation-selection balance. In: Karlin S, Nevo E (eds) Evolutionary processes and theory. Academic Press, London New York, pp 607–628

Webb GF (1985) Theory of nonlinear age-dependent population dynamics. Dekker, New York

Quantitative Genetic Models for Parthenogenetic Species

W. GABRIEL[1]

1 Introduction

Animal and plant breeders have successfully applied quantitative genetic models (Bulmer 1980; Falconer 1981) to describe the phenotypic selection of metric characters. The extension of these concepts to problems in evolutionary ecology (Lande 1976a,b, 1982) gives general dynamic expressions for the phenotypic evolution of polygenic characters, though at the expense of making confining assumptions such as infinite population size, additive loci, and restrictions on the form of selection functions. Therefore, to connect ecology with genetics, the framework of quantitative genetics is most suitable for questions where (1) the underlying genetic systems are so complex that we cannot hope to understand them locus by locus and, for that reason, we strive to obtain models that give reasonably good approximation, and (2) where the results of the theory seem to be at least qualitatively correct in spite of the unavoidable inherent simplifying assumptions. Most of the possible errors resulting from such approximation to reality may be irrelevant in comparative studies where the emphasis is on qualitative behaviour.

The need for genetic theory in ecology is self-evident when one tries to evaluate the performance of parthenogenetic species relative to sexual ones to gain insight into the complicated reproductive strategies of the freshwater zooplankton species with long periods of parthenogenetic reproduction but sexually produced resting eggs. Here, I will demonstrate the usefulness of quantitative genetic models for investigating such problems as the consequences of obligate and cyclical parthenogenesis for the evolutionary potential of species and for the evolution of specialist versus generalist strategies.

1 Max Planck Institute for Limnology, Dept. of Physiological Ecology,
August-Thienemann-Str. 2, D-2320 Plön, FRG

Population Genetics and Evolution
G. de Jong (ed.)
© Springer-Verlag Berlin Heidelberg 1988

2 Phenotypic Evolution of a Single Character

Let us first consider a single quantitative character of a diploid organism and assume that this character is determined by n diploid loci and that its genotypic value can be written as a summation over purely additive contributions of the 2 n allelic effects. (This assumption of additivity is necessary only for cyclical parthenogenesis in order to make the decomposition of genetic variance into a hidden and expressed component tractable; for obligate parthenogenetic species the genotypic value comprises all additive and non-additive contributions.) The per generation change in the genotype distribution of a population can be estimated when the non-genetic contributions to the phenotypic variance (e.g. developmental noise), the genotypic variance and the shape of the fitness function are known. Sexual and asexual populations differ in the realized genetic variance. In the case of sexual reproduction, Lande (1976a,b) calculated an equilibrium level of genetic variance maintained by a balance between input via mutation and recombination and output by selection. Lynch and Gabriel (1983) derived analogous formulas for phenotypic evolution under parthenogenesis and found the equilibrium to be:

$$\hat{V}_g = \frac{1}{2} \left(V_m + \sqrt{V_m \left(V_m + 4 \left(V_e + V_w \right) \right)} \right) \tag{1}$$

with V_w as the variance of the fitness function assumed to be Gaussian, V_e as the variance in developmental noise, and V_m as the per generation input of genetic variance. The number of loci is implicit in V_m, where the total input of mutation is summed over loci and alleles. The corresponding formula for sexual reproduction has identical structure but V_m is always multiplied by 2 n. Thus, the number of loci appears explicitly also (rewritten from Lande 1976):

$$\hat{V}_g = \frac{1}{2} \left(2 n V_m + \sqrt{2 n V_m \left(2 n V_m + 4 \left(V_e + V_w \right) \right)} \right) . \tag{2}$$

An essential result of the models is that these equilibria (Eqs. 1 and 2) are independent of the optimal genotypic value so that a population under directional selection can approach its equilibrium level of genetic variance even if it is far from the optimal genotype. Lynch and Gabriel (1983) have shown that for most parthenogens this equilibrium level will be reached within a few hundred generations and have argued that the maintenance of this genetic variability is an important mechanism which renders rates of phenotypic evolution the same order of magnitude as bisexual organisms. For example, a change in the mean genotypic value as large as 5 phenotypic standard deviations is feasible within 100 generations. In general, higher genetic variability from sexual reproduction implies a more rapid evolution, but only if mutation rate per locus and the sensitivity of phenotypic development to environmental effects are equal for sexual and parthenogenetic reproduction. There are arguments suggesting higher mutation rates and lower environmental sensitivity in parthenogenetic species (Lynch 1985), but experimental evidence is lacking.

This study of phenotypic evolution under parthenogenesis may hold as an example of how methods of quantitative genetics can be applied to ecological questions and lead to new insights. Lynch and Gabriel (1983) have shown that the common assertion, that obligate parthenogenesis is an evolutionary dead end, is suspect.

3 Consequences of Cyclical Parthenogenesis

From the same study some unexpected consequences for cyclical parthenogenesis also follow. Under parthenogenesis there is a continuous accumulation of hidden genetic variance which can be quantified by a calculation of the covariances between allelic effects. According to Lynch and Gabriel (1983), up to 75% of this hidden genetic variance can be converted into expressed genetic variance within one generation after sex; the minimum value is 50% since covariances between all genes on different chromosomes are immediately broken up. Therefore, massive amounts of hidden genetic variance may be released when a cyclical parthenogenetic species interrupts long periods of unisexuality by a bout of recombination. Dramatic responses to selection may then occur within a few generations. The mean value of the genotype distribution may jump several phenotypic standard deviations in a single generation after sex. In the long term, the rate of evolution under cyclical parthenogenesis is essentially the same for frequencies of sexual reproduction from once per generation to once per 100 generations. Figure 1 shows the changes of genotype distributions during evolution under bisexuality (Fig. 1a) and cyclical parthenogenesis (Fig. 1b,c). For cyclical parthenogens, the genotype distribution after each bout of sexual recombination is broader than under continuous sexual reproduction due to the expression of hidden

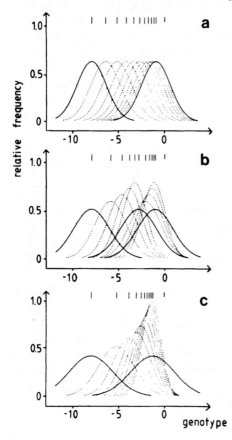

Fig. 1a–c. Comparison of the change in the genotype distribution during evolution of sexual and cyclical parthenogenetic species.
a Sexual reproduction each generation.
b Sexual reproduction every five generations.
c Sexual reproduction every ten generations.
Mean phenotype values are marked on the x-axis (optimal value at 0)

genetic variance. This may explain why resting eggs are produced sexually: if the optimal genotype of the next season is unpredictable, a broad genotype distribution in resting eggs provides the population with enough variability to track fluctuations in the environmental conditions. The nearly optimal clones can use the reproductive advantage of parthenogenesis in a two-fold way during times of more constant environment: they grow faster because they do not produce males, and they avoid the production of less fit offspring as it would necessarily happen with recombination. Therefore, cyclical parthenogenesis may guarantee high performance.

4 Muller's Ratchet

An argument often made about parthenogenesis is that deleterious mutations would accumulate by "Muller's Ratchet" (Muller 1964). Felsenstein (1974) and Maynard Smith (1978) evaluated Muller's verbal arguments with detailed genetic model studies and gave precise conditions under which Muller's ratchet can operate. If mutations reduce fitness only slightly, and if the effective population size is small, there is a chance that all individuals in the optimal class (with fewest mutations) die or fail to reproduce. Gabriel and Wagner (1988) constructed a phenotypic model as an alalogue to the studies of Felsenstein and Maynard Smith to see if Muller's arguments hold when fitness is determined by several independent quantitative characters, each with a polygenic basis. This quantitative genetic model also gives conditions for the fitness and the effective population size under which Muller's ratchet operates. Without mutations, the optimal genotype class gets lost by random drift and Muller's ratchet advances, always deleting the class with the fewest mutations. In this model each mutation decreases the fitness with high probability, but, in contrast to Felsenstein and Maynard Smith, back-mutations are not totally ignored. The surprising result is that even under a continuously operating Muller's ratchet, the mutational load of the population is limited: as soon as the population's mean fitness has decreased to a certain value, the accumulation of further mutations no longer has an effect on the population mean fitness. The main reason for the difference between the phenotypic model and Felsenstein's and Maynard Smith's models is that in the former, back-mutations become more frequent with increasing distance to the optimum due to the geometry of the multidimensional fitness function. This provides a limit for the decrease of fitness comparable to the expected mean fitness of a sexual population under random drift as approximated by Lande (1976b). The model of Gabriel and Wagner (1988) shows that Muller's ratchet may hinder the survival of the optimal genotype in a parthenogenetic species, but, nevertheless, the population can stay in a stochastic equilibrium near the optimum. This can, therefore, be interpreted as a phenotypic analogue for a transition from direct replication to stochastic replication in polynucleotides (Swetina and Schuster 1982). Under high genomic mutation rates, stochastic replication may have similar effects as recombination. In this case, it seems to be of only minor importance for performance whether reproduction is sexual or parthenogenetic.

5 Optimal Breadth of Adaptation

Another important factor for the performance of a species is its ability to adapt to changing environmental conditions. A first step towards a quantitative genetic theory for the evolution of the breadth of adaptation has been taken by Lynch and Gabriel (1987a,b) who have developed a theory of the expression of the response of individual genotypes to density-independent gradients of environmental factors. The fitness function had often been treated as fixed, and therefore, independent of evolution, but in this approach the fitness function is treated as a performance curve (tolerance curve or norm of reaction) which itself can change during the evolutionary process. The model has incorporated the following environmental variabilities: spatial variance ($V_{\varphi s}$), within generation temporal variance ($V_{\varphi tw}$), and between generation temporal variance ($V_{\varphi tb}$). Breadth ($\sqrt{g_2}$) and optimal setting (g_1) of the maximum of the performance curve of each individual are treated as quantitative characters on a scale of an environmental variable (φ) so that the fitness of an individual is determined by

$$w\,(z_1, z_2 \mid \varphi) = \frac{1}{\sqrt{2\,\pi\,z_2}} \exp\left\{-\,\frac{(z_1 - \varphi)^2}{2\,z_2}\right\},\tag{3}$$

where z_1 and z_2 are the phenotypic values corresponding to the genotypic values g_1 and g_2. The difference between genotypic and phenotypic values may, in this context, be called developmental noise. It is trivial to see that the optimal g_1 is identical with the mean value of the environment, but the question is how to predict the optimal genotypic value of breadth of adaptation dependent on given variabilities of the environment. This can be done by optimization of the expected geometric mean fitness (Lynch and Gabriel 1987b). After a scale transformation so that the mean value of the environmental state φ is 0, an approximation formula for the expected geometric mean fitness is

$$w\,(g_1, g_2) \cong \frac{1}{\sqrt{2\,\pi\,(V + V_{\varphi s})}} \exp\left\{-\,\frac{1}{2}\left[\frac{g_1^2 + V_{\varphi tb} + V_{\varphi tw}\left(1 + \dfrac{V_{\varphi s}}{V}\right)}{V + V_{\varphi s}}\right]\right\}\tag{4}$$

with

$$V = g_2\left(1 - \frac{1}{K}\right) + V_{e1}$$

$$K = \frac{g_2\,(g_2 + V_{e1})}{V_{e2}} + 2$$

and with V_{e1} and V_{e2} as the variances of the developmental noise on g_1 and g_2.

The parameters of this theory can be extracted from experimental data by a maximum likelihood procedure as demonstrated by Gabriel (1987). The theory confirms what is intuitively expected: temporal variability of the environment selects for more broadly adapted genotypes, and the within-generation heterogeneity is more important than the between generation component. But, on the other hand, the theory produces results which are at least at first counterintuitive. Spatial heterogeneity selects for more broadly adapted genotypes only when operating in conjunction with certain patterns of temporal variance: if the within-generation temporal variance is much less than the between-generation temporal variance, spatial heterogeneity can select for a *reduction* of the breadth of adaptation. The formulas of Lynch and Gabriel are results of an optimality approach for infinite populations and do not tell how fast evolution takes place and how much better sexual populations adapt than parthenogenetic or cyclic parthenogenetic populations. Below, I present some simulation results with mutation and selection on finite populations. I did not find differences in the optimal breadth of adaptation between sexual and parthenogenetic species.

6 Evolution of Generalism and Specialism

The simulations start with an effective number of parents for which the allelic contributions to the genetic values of the individuals' performance curve (as the setting of the maximum and as the variance) are stored. The first step is the production of offspring, by free recombination in the case of sexual reproduction. Each offspring carries a new mutation according to the genomic mutation rate. The phenotypic values are calculated by adding developmental noise to the genetic values. Individual fitness is calculated from these phenotypic values. The next generation of parents is created by applying viability selection, which is a function of the environmental state for each individual (see Eq. 3). From the surviving parents the new effective population ($N_e = 250$) is chosen randomly. Presented here are simulations performed with the following parameters: 20 diploid loci per character; mutation probability per genome is 0.1; average widths (square root of the variance) of mutational effect are 0.1 for g_1 and 1.0 for g_2 on a relative scale so that the developmental noise on g_1 is 1; as shown by Lynch and Gabriel (1987) the effect of developmental noise on g_2 is of minor importance, and therefore, is set at zero in this study. The mutational effects on the genotypic values g_1 are assumed to be normally distributed around its actual values, and for g_2 to be according to a beta distribution of the second kind (see Lynch and Gabriel 1987a).

Figure 2 shows the evolution of sexual and parthenogenetic populations after a drastic reduction (by a factor of 100) of the between-generation temporal variance of the environment so that a smaller breadth of adaptation is optimal (g_1 is assumed here to be at its optimum). Both sexual and parthenogenetic populations converge to the new optimal value but the sexual population is much faster due to the higher genetic variance under sexual reproduction.

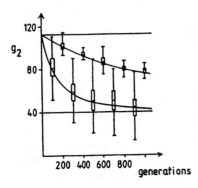

Fig. 2. Population response to a change in the optimal breadth of adaptation. *Upper curve:* evolution under parthenogenesis; *lower curve:* under sexual reproduction. The *boxes* give the quartiles, *vertical bars* give the total range of results. *Horizontal lines* are the optimal values

When these and similar simulations are performed until convergence has been obtained, it is possible to estimate the equilibrium levels of genetic variances. These values can be approximately compared with the predictions of Eqs. (1) and (2) by the following simple calculation. To reduce the problem to the estimation of genetic variance under phenotypic evolution of a single character, one allows only one character to vary. If one keeps the breadth of adaptation ($\sqrt{g_2}$) fixed, the fitness function varies with the optimal value (g_1) in such a way that $g_2 + V_{\varphi s}$ corresponds to the variance of the fitness function V_w in Eqs. (1) and (2). [With $V_{e2} = 0$, it follows from Eq. (3) that $w(g_1, g_2)$ is proportional to $\exp(-g_1^2/2(g_2 + V_{e1} + V_{\varphi s}))$; according to Lynch and Gabriel (1983) the corresponding genic fitness $w(g)$ for a single character is proportional to $\exp(-g^2/2(V_e + V_w))$.] To make these comparisons, simulations were performed for several variabilities of the environment until equilibria at the corresponding optimal g_2 values were attained. The calculations are in good agreement with simulation results for asexual populations. For sexual populations, however, the expected values are always somewhat higher (up to 40%) than the simulation results. One possible explanation of this effect may be found in the criticism of Turelli (1984) and Bürger (1986 and this Vol.). They argue that under certain circumstances the maintained genetic variance may be smaller than predicted by Lande. In the face of finite populations, this may be more important for sexual than for asexual reproduction. To get an estimate for the genetic variance of g_2 one can fix the g_1 value and see how the geometric mean fitness changes with variation of g_2. This again gives the corresponding fitness curve for application of Eqs. (1) and (2). Figure 3 shows that such curves are characteristically asymmetrical, but for the application of Eqs. (1) and (2), the fitness curve for g_2 should be Gaussian. Therefore, by calculating variances from these curves, one can get only rough estimates of V_w. However, these calculations still agree quite well for parthenogenesis, and they differ by much less than one order of magnitude for sexual reproduction. Therefore, rough estimates of expected genetic variances of the g_2 are possible by this procedure also.

The asymmetry of the expected geometric mean fitness $w(g_1, g_2)$ with respect to g_2 (see Fig. 3) may have another important implication: If one compares the fitness values at lower and higher g_2 values but with the same distance to the optimal g_2, the fitness is more reduced by lowering the breadth of adaptation than by increasing g_2.

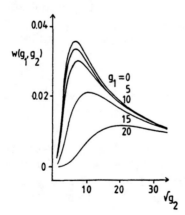

Fig. 3. The expected geometric mean fitness as a function of the realized breadth of adaptation. The curves with lower fitness are calculated for deviations of g_1 (the setting of the maximum of the performance curve) from the mean environmental state

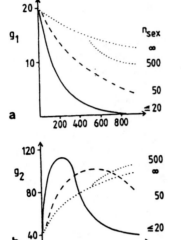

Fig. 4a,b. Evolution to a new mean environmental state (new value 0, old value 20) at various frequencies of sexual reproduction (n_{sex}). The optimal breadth of adaptation ($\sqrt{g_2}$) is identical for both environmental states. **a** Time course of the change in g_1 (setting of the maximum of the performance curve). **b** Corresponding time course of g_2 which is temporarily increased during the evolution of g_1

This higher fitness difference causes a stronger selection. Therefore, evolutionary increases of breadth of adaptation may be faster than reductions.

Figure 3 also demonstrates that the maximum of $w(g_1, g_2)$ occurs always at higher g_2 values when g_1 is not at its optimum (which is chosen to be 0). This becomes important if the environment shifts to a new mean value and the evolution of g_1 (the optimal setting of the maximum of the performance curve) is slow. Let us assume that the optimal breadth of adaptation ($\sqrt{g_2}$) is identical for the old and the new environmental mean value so that only g_1 has to be adapted. But as long as g_1 is not optimal, the fitness is higher for increased g_2 values due to the shape of $w(g_1, g_2)$. Therefore, the evolution to the new g_1 state is accompanied by a temporary increase in g_2. The breadth of adaptation returns to its optimal value when g_1 evolves near its new optimum. This is shown in Fig. 4a and b for parthenogens and cyclical parthenogens with various frequencies of sex. When sexual reproduction occurs more often than once per 25 generations, there is no difference between sexual and parthenogenetic species. At frequencies lower than every 30 generations, a difference between cyclical parthenogens and bisexuals appears. This is in contrast to the study on the evolution of a single quantitative trait where small differences appear at frequencies higher than 100. The reason for this is the small population size which reduces the hidden genetic variance stored during asexual reproduction.

7 Conclusions

The three models presented here demonstrate how concepts of quantitative genetics can be applied fruitfully to problems in evolutionary ecology; here, several qualitative and quantitative predictions are made concerning the performance of both obligate and cyclically parthenogenetic species as compared to sexually reproducing species.

The study of the evolution of a single quantitative character shows that under parthenogenesis a high degree of genetic variability can be maintained, similar in amount to that produced by sexual reproduction, by a balance between input via mutation and output via selection. Therefore, rates of evolution comparable even to sexually reproducing species are possible. Cyclical parthenogenesis has to be considered as a very powerful reproductive strategy because (1) it provides the population with a broader genetic variance after a sexual cycle than there would be under continuous sexual reproduction, (2) the long-term rates of evolution are equal under cyclical parthenogenesis and sexuality and (3) the advantage of faster population growth can be used during asexual reproduction. This model on a single quantitative character refutes the common assertion that parthenogenesis is an evolutionary dead end. A second disadvantage often associated with parthenogenesis, the accumulation of deleterious mutations by Muller's ratchet, is also found to be questionable if fitness is determined by several quantitative characters with a polygenic basis. If back-mutations are not excluded, the mutational load of parthenogenetic species is limited to a level where the mean population fitness is similar to a sexual population under random drift. A further argument conjectured against parthenogenesis is a reduced flexibility in fluctuating environments. A study of the evolution of the breadth of adaptation shows that the optimal breadth of adaptation in temporally and spatially heterogeneous environments is independent of reproductive strategy. The performance of parthenogenetic species is expected to be high even under non-constant environmental conditions. Evolution to a new optimal breadth or to a new optimal setting of the performance curve is indeed faster under sexuality, but this is important only for fast and large changes in the mean environmental state or in temporal or spatial variances. Long-term rates of evolution are identical for cyclical parthenogenesis and permanent sexual reproduction, but under the restriction that the effective population size is large compared to the frequency of sex.

These new and surprising results of the models presented may demonstrate that the application of quantitative genetic models — even with all its inherent simplifications — is a powerful and promising approach to problems in evolutionary ecology.

Acknowledgements. I thank Michael Lynch, Robert W. Sterner and Hans Georg Wolf for valuable criticism and helpful comments.

References

Bulmer MG (1980) The mathematical theory of quantitative genetics. Oxford Univ Press

Bürger R (1986) On the maintenance of genetic variation: global analysis of Kimura's continuum-of-alleles model. J Math Biol 24:341–352

Falconer DS (1981) Introduction to quantitative genetics. Longman, New York

Felsenstein J (1974) The evolutionary advantage of recombination. Genetics 78:737–756

Gabriel W (1987) The use of computer simulation to evaluate the testability of a new fitness concept. In: Möller DPF (ed) System analysis of biological processes. Vieweg, Braunschweig, pp 82–88 (Adv System Anal, vol 2)

Gabriel W, Wagner GP (1988) Parthenogenetic populations can remain stable in spite of high mutation rate and random drift. Naturwissenschaften 75 (in press)

Lande R (1976a) The maintenance of genetic variability by mutation in a polygenic character with linked loci. Genet Res 26:221–235

Lande R (1976b) Natural selection and random genetic drift in phenotypic evolution. Evolution
    30:314—334
Lande R (1982) A quantitative genetic theory of life history evolution. Ecology 63:607—615
Lynch M (1985) Spontaneous mutations for life-history characters in an obligate parthenogen.
    Evolution 39:804—818
Lynch M, Gabriel W (1983) Phenotypic evolution and parthenogenesis. Am Nat 122:745—764
Lynch M, Gabriel W (1987a) The evolution of the breadth of biochemical adaptation. In: Calow
    P (ed) Evolutionary physiological ecology. Cambridge Univ Press, pp 67—83
Lynch M, Gabriel W (1987b) Environmental tolerance. Am Nat 129:283—303
Maynard Smith J (1978) The evolution of sex. Cambridge Univ Press
Muller HJ (1964) The relation of recombination to mutational advance. Mutat Res 1:2—9
Swetina J, Schuster P (1982) Self-replication with errors: a model for polynucleotide replication.
    Biophys Chem 16:329—345
Turelli M (1984) Heritable genetic variation via mutation-selection balance: Lerch's zeta meets
    the abdominal bristle. Theor Popul Biol 25:138—193

Quantitative Genetics of Life History Evolution in a Migrant Insect

H. DINGLE[1]

1 Introduction

As he did with so much of population genetics, T. Dobzhansky provided the clearest insights into the nature of adaptations (Dobzhansky 1956). With his characteristic eloquence he stressed that adaptation does not involve simply the acquisition of single advantageous traits, but rather the development of arrays of characters that interact with one another in "complexes" or "syndromes". Frazetta (1975) called such syndromes "complex adaptations" and likened them to machines in which all parts must blend together operationally to make the machines work. For the organism the parts are morphological, developmental, physiological, and behavioral traits, the "eco-behavioral traits" of Parsons (1983a,b), and studies of the ways selection has combined such traits should produce greater understanding of adaptation than analysis of single characters alone. The truth of this assertion is amply demonstrated, for example, in studies of temperature regulation in house mice *(Mus musculus)* from both field and laboratory populations. Adaptation to cold involves size, shape, pelage, metabolism, and nest-building behavior, all of which are phenotypically and genetically correlated (Berry and Jakobson 1975; Sulzbach and Lynch 1984). The genetic architecture provides the blueprints for the machine.

For migratory species the most important complex adaptation involves the relation between migratory behavior and the schedules of reproduction and mortality (Rankin and Singer 1984; Rankin et al. 1986). Migratory behavior allows the choice of where and when to breed, and its relation to births and deaths determines the life history "strategy" (Dingle 1984, 1985). Such strategies determine fitness and for that reason are central to evolutionary theory (Bell 1980). Migrants possess the ability to colonize successfully upon arrival in a new habitat, and a large body of theory (summarized by Stearns 1976, 1977; Safriel and Ritte 1980; Parsons 1983a,b) predicts that such colonizers will display life histories characterized by the presence of correlations between migration and early and rapid reproduction ("r-strategies"). These are generally supported by empirical evidence, and an important question in evolutionary biology concerns the correspondence between such phenotypic correlations and the underly-

---

1 Department of Entomology, University of California, Davis, CA 95616, USA

Population Genetics and Evolution
G. de Jong (ed.)
© Springer-Verlag Berlin Heidelberg 1988

ing genetic architecture. This is so because the organization of the genetic variance-covariance structure will have an important bearing on the future evolutionary course of the adaptation. Traits tied together by genetic correlations may be augmented or constrained in their response to selection both within and across environments, and so may not evolve directly toward individual optima (Lande 1982; Via 1984; Via and Lande 1985).

The polygenic nature of most life history traits dictates the use of quantitative genetic techniques to assess the genetic variances and covariances contributing to the architecture of coevolved syndromes. We have used these methods in my laboratory in pursuit of an overall objective of determining the role of migration in life history evolution. We have focused on a migratory and a nonmigratory population of the milkweed bug, *Oncopeltus fasciatus*, in the belief that comparisons of within-species populations of differing phenotype and genotype may provide insight into the genetic basis of complex adaptations. Three studies of the migratory Iowa population have been published (Hegmann and Dingle 1982; Palmer 1985; Palmer and Dingle 1986), and manuscripts evaluating in detail results from the nonmigratory Puerto Rico population are in preparation (Dingle et al., in prep.; Dingle and Evans, in prep.). This chapter presents an overview and direct comparison of genetic variation and covariation among migration and life history traits in the two populations, and assesses the differences observed in terms of evolving life history adaptations.

2 Experimental Populations

The life cycles of Iowa and Puerto Rico *O. fasciatus* differ in important ways. The Iowa population is highly migratory and representative of bugs which invade the northeastern half of North America each spring and summer (Ralph 1977; Dingle 1978, 1981; Dingle et al. 1980). The insects arrive between March and July, depending on latitude, and produce from one to two generations in patches of milkweed (mostly *Asclepias syriaca*). These bugs cannot survive winter in the temperate zone, but in the autumn short days induce an adult reproductive diapause which facilitates migration. Circumstantial but reasonably compelling evidence suggests that at least a portion of the population migrates south to overwinter before returning again the following spring. The overall pattern is thus somewhat similar to that of eastern populations of the monarch butterfly, *Danaus plexippus* (Brower 1985). In contrast, the tropical Puerto Rico population breeds throughout the year. There is only local movement between patches of milkweed (primarily *A. curassavica*), and there is essentially no diapause [a very small portion of individuals show some reproductive delay under short days and relatively low temperature (Dingle 1981)].

There are also other differences between the two populations, especially in morphological and life history traits. The data for females are summarized in Table 1. Iowa bugs are significantly larger as indicated by the mean wing lengths. There is, however, also a shape component to the difference in body proportions since mean head capsule widths are the same. Under the environmental conditions presented, age at first reproduction in Iowa females (measured as the interval from adult eclosion to first egg) is delayed relative to Puerto Rico, reflecting the diapause component which is com-

Table 1. Phenotypic means estimated prior to selection in females from Iowa and Puerto Rico populations of *Oncopeltus fasciatus* reared at LD 14:10 and 27 °C

Trait	Iowa mean ± SE[a]	Puerto Rico mean ± SE
Wing length (mm)	12.74 ± 0.02	11.49 ± 0.04
Head capsule width (mm)	2.06 ± 0.01	2.06 ± 0.01
Age at first reproduction (days)	20.11 ± 0.55	10.3 ± 0.31
Clutch size (eggs)	26.0 ± 0.60	60.1 ± 2.26
Fecundity first 5 days (eggs)	157.1 ± 2.18[b]	164.4 ± 5.22[b]

[a] Iowa data from Palmer and Dingle (1986).
[b] Means for fecundity and clutch size determined from independent samples.

pletely eliminated in the Iowa bugs only at day lengths above 15 h. The clutch size difference reflects the habit in Iowa bugs of ovipositing virtually every day, while Puerto Rico females do so only every other day. These differences appear to be due to higher predation in Iowa and the packing of eggs into cocoons of leaf-rolling caterpillars in Puerto Rico (Dingle 1981). Note, however, that mean fecundity for the important (for a colonizer) first 5 days of reproduction does not differ between the populations even though the Puerto Rico bugs are smaller. The two populations, however, differ genetically with respect to the control of fecundity as revealed by hybridization studies; gene differences also contribute to the other aspects of the contrasting phenotypes (Dingle et al. 1982; Leslie and Dingle 1983). Additional phenotypic differences include various physiological responses to temperature and photoperiod (Dingle 1981).

The observed genetic and phenotypic disparities between the insects from Iowa and Puerto Rico indicate that the two populations have been evolving under different selective regimes. The question I specifically address here is whether, in spite of differences in individual traits, the pattern of genetic variance and covariance is basically the same. For example, in the Iowa population is long-distance migration simply superimposed on a genetic architecture involving early and rapid reproduction and common to both populations or are there fundamental differences in architecture between migrant and nonmigrant bugs?

3 The Genetic Architecture of Oncopeltus Life Histories

For an initial assessment of the genetic correlation structure of *O. fasciatus* life histories, we employed a half-sib design which included 1161 female offspring from 31 paternal half-sib and 104 full-sib families (Hegmann and Dingle 1982). The experiment was carried out under an LD 16:8 23 °C regimen in which Iowa bugs do not diapause (Dingle 1974). A set of 11 life history characters including wing and body lengths, age at first reproduction (adult eclosion to first egg), development time (birth to adult eclosion), interclutch interval, and size and percentage hatch of each of the first three egg clutches were measured for each female. From the resulting data set we computed heritabilities and phenotypic and genetic correlation matrices (details in Hegmann and Dingle 1982).

The results of the half-sib analysis (Hegmann and Dingle 1982) indicated sufficient genetic variation for response to selection in most of the individual traits, but the correlation structure indicated that selection on many of them would also influence the multivariate vector of means. For example, both measures of body size were positively correlated phenotypically and genetically. They were also positively correlated with clutch size, especially genetically, and negatively correlated with development time. The three traits thus share genes such that larger bugs tend to produce larger clutches and to develop faster. Of perhaps even greater interest, age at first reproduction was notable for its lack of genetic correlation with any of the other life history variables. This genetic uncoupling of age at first reproduction from the rest of the life history provides considerable flexibility in the timing of breeding since the genetic variance influencing this trait is not "tied up" by pleiotropic or linkage effects. This is likely to be of considerable importance to a migrant where the timing of reproduction can be critical (Dingle 1984).

Because the genetic architecture, revealed by the half-sib analysis, suggested the potential for considerable and potentially interesting direct and correlated responses to selection, we initiated the comparative study of the Iowa and Puerto Rico populations. Valid comparison required a common and ecologically realistic environment which did not induce diapause in the Iowa bugs. The regimen chosen was LD 14:10 27 °C because both Iowa and Puerto Rico populations experience conditions similar to this in the field for at least part of the annual cycle. In matching sets of analyses the populations were examined by first employing offspring-parent regression to estimate genetic as well as phenotypic parameters for life history characters in the common environment. Following this we subjected both populations to bidirectional selection for wing length. This trait was chosen as the target for artificial selection because (1) it shows considerable phenotypic variation over the geographic range of *O. fasciatus* (Dingle et al. 1980), (2) longer wings are associated with migratory tendency (Dingle et al. 1980), and (3) it displays high additive genetic variance and so should respond rapidly (Hegmann and Dingle 1982 and below). Full details of the protocols used in our analyses are given in Palmer and Dingle (1986); general discussions of these quantitative genetic methods can be found in Falconer (1981).

Heritability estimates from the offspring-parent regressions indicated considerable additive genetic variance for some traits in both populations but not in others (Table 2). Both wing length and head capsule width display high heritabilities across populations, as was also the case for the estimate from the half-sib analysis of the Iowa population shown for comparison. Opposite results were obtained for age at first reproduction (eclosion to first oviposition). A highly significant positive estimate of heritability was obtained for Puerto Rico while a negative and non-significant value was computed for Iowa. This Iowa result also contrasted with earlier Iowa estimates obtained at LD 16:8 23 °C (Table 2) and at LD 12:12 23 °C. The latter environment is diapause-inducing, and a high estimate of approximately 0.7 was estimated from selection (Dingle et al. 1977). This value is likely an overestimate resulting in part from maternal effects in which diapause mothers tend to produce nondiapause offspring (Groeters and Dingle 1987), but still suggests that heritability for age at first reproduction in this population under these conditions includes an element of genetic variance for critical photoperiod as opposed to the eclosion-oviposition interval per se. Whether

Table 2. Heritabilities estimated from offspring-parent regression prior to selection in females from Iowa and Puerto Rico populations of *Oncopeltus fasciatus* reared at LD 14:10 and 27 °C

Trait	Iowa heritability ± SE[a,c]	Puerto Rico heritability ± SE[c]
Wing length	0.87 ± 0.15*** 0.55 ± 0.22[b]	0.58 ± 0.15***
Head capsule width	0.71 ± 0.17***	0.61 ± 0.15***
Age at first reproduction	−0.20 ± 0.13 ns 0.25 ± 0.12[b]	0.35 ± 0.10***
Clutch size	0.14 ± 0.08[b]	0.13 ± 0.10 ns
Fecundity first 5 days	0.50 ± 0.17***	0.06 ± 0.11 ns

[a] Iowa data from Palmer and Dingle (1986).
[b] Estimated at LD 16:8 and 23 °C from half-sib analysis (Hegmann and Dingle 1982).
[c], *** $p < 0.001$.

environmental variation would produce similar variation in heritability estimates for the Puerto Rico bugs remains to be assessed, but in view of the almost complete lack of adult reproductive diapause in this population, the extremes produced by responses characteristic to critical photoperiod should be absent. Finally, of the two estimates of fecundity, clutch size displayed similar low nonsignificant heritabilities in both populations, while egg production over the first 5 days of oviposition showed a marked difference. Heritability was significantly high and positive in the Iowa sample while it was indistinguishable from zero in Puerto Rico. There thus seem to be differences in the two populations with respect to the contributions of additive genetic variance to age at first reproduction and to early in life fecundity.

Differences are also evident when the correlation structure of the two populations is examined (Table 3). The two morphological measures, wing length and head capsule width, are positively phenotypically and genetically correlated in both populations suggesting the sharing of genes acting additively to contribute to a joint phenotype

Table 3. Phenotypic (upper triangle) and genetic (lower triangle) correlations estimated prior to selection at LD 14:10, 27 °C for Iowa (upper value of each pair) and Puerto Rico (lower value) populations of *Oncopeltus fasciatus*. Iowa data from Palmer and Dingle (1986)

Trait	WL[a]	HCW[a]	AFR[a]	FEC[a]
Wing length (WL)	–	0.50 ± 0.09*** 0.50 ± 0.09***	0.06 ± 0.11 −0.05 ± 0.10	0.24 ± 0.10* −0.09 ± 0.10
Head capsule width (HCW)	0.32 ± 0.13* 0.68 ± 0.10***	–	−0.13 ± 0.11 −0.13 ± 0.10	0.13 ± 0.11 0.13 ± 0.10
Age at first reproduction	–[b] −0.15 ± 0.19	–[b] 0.09 ± 0.19	–	0.02 ± 0.11 0.13 ± 0.10
Fecundity first 5 days (FEC)	−0.57 ± 0.11*** −0.25 ± 0.46	0.62 ± 0.13*** 0.45 ± 0.38	–[b] −0.52 ± 0.37	–

[a] * Different from 0 at $p < 0.05$, two-tailed t-test; *** $p < 0.001$.
[b] Undefined because of negative heritability for AFR.

(dominance and interaction components of genetic covariance are included with environmental variance; see Falconer 1981). In the Puerto Rico population there were no significant correlations between age at first reproduction (AFR) and any of the other variables; because of the negative heritability of AFR in the analysis of the Iowa sample (Table 2), genetic correlations could not be estimated. Correlation estimates for fecundity differ between the two populations with respect to both phenotype and genotype. In Iowa bugs there is a significant positive phenotypic correlation between wing length and fecundity, whereas this is not the case in those from Puerto Rico. In Iowa there is also a positive genetic correlation between head capsule width and fecundity. A puzzling result in the Iowa sample is the highly significant negative genetic correlation estimate for wing length and fecundity. We believe, however, that this results from sampling error to which estimates of genetic correlations are notoriously prone (e.g. Falconer 1981, p. 285). This is supported by the fact that selection on wing length indicates relatively large *positive* correlations between this morphometric trait and fecundity (see below) and by the fact that the phenotypic correlation (in a controlled environment where environmental covariance should be small) is significantly positive.

The relatively high heritabilities for wing length in both populations (Table 2) predict a rapid response to artificial selection on this trait. That this is indeed the case is illustrated in Fig. 1 which displays responses to selection in females of the two samples over nine generations. There is very little difference between bugs from the two sources when selected in either direction. In all cases the change in wing length was about 10–15% that of the base population, although there is some asymmetry of response

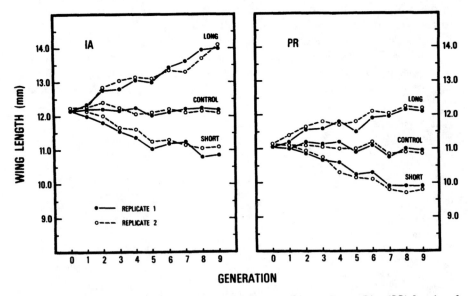

Fig. 1. Direct responses to selection on wing length in Iowa (*IA*) and Puerto Rico (*PR*) females of *Oncopeltus fasciatus*. *Points* indicate mean wing lengths of bugs in each line for each generation. Sample sizes for each mean are between 201-377 bugs; standard errors of all means are less than the diameter of the points. Iowa data from Palmer and Dingle (1986)

in favor of the long-winged lines. At least in the Puerto Rico population additive variance for the trait was not exhausted by the ninth generation as we have subsequently carried the lines to the nineteenth generation with continued response to selection (Dingle et al., in prep.).

Of much greater interest are the responses of other traits to the selection on wing length. Two responded in the same manner in both populations. First, as predicted by positive genetic correlations (Table 3), head capsule width in both replicates of both samples responded in the same direction as wing length with long-winged bugs having larger head capsules and vice versa (Palmer and Dingle 1986; Dingle et al., in prep.). Secondly, the two populations are the same with respect to age at first reproduction with neither displaying any consistent change in this trait in response to selection for wing length (Fig. 2A). Only the long-winged line of Iowa replicate I differed significantly from unselected controls, and there was qualitative inconsistency across lines and populations. These data and our earlier results (Hegmann and Dingle 1982) would seem to confirm the lack of correlation between age at first reproduction and morphometric or other life history traits.

With respect to fecundity (for the first 5 days) and flight, the two sample populations differ conspicuously (Fig. 2B and C). In the case of fecundity both replicates of Iowa bugs display high fecundity in the long-winged line and low fecundity in the short-winged line when contrasted with controls. There was no such consistency of response in the Puerto Rico bugs where there were no statistically significant differences between lines in the first replicate, and in the second, only the short-winged line differed from controls. This latter significant difference disappeared when later intervals of oviposition were considered (Dingle et al., in prep.). Our measure of flight was nonstop wing beating for 30 min or longer on a tether which has proven to be a useful index of migration (Dingle et al. 1980). In these tethered flight experiments, Iowa migrating bugs showed increased long duration flights in both long-winged lines, while the short-winged lines did not differ from controls. The nonmigrating Puerto Rico bugs differed from the Iowa bugs in two ways: their overall flight performance was very much less, in fact less than half of the performance of the lowest Iowa line in terms of flights lasting at least 30 min, and there were no detectable differences among any of the lines. The Puerto Rico bugs were retested in later generations with the same result (Dingle and Evans, in prep.). Iowa and Puerto Rico populations thus differ not only in absolute tethered flight performance, confirming earlier results (Dingle et al., 1980), but also in whether flight durations respond to selection for wing length.

4 Discussion

Our results for the Iowa population of *O. fasciatus* generally support the notion of a suite of correlated characters adapted for colonization. In these bugs the genetic correlation structure is such that genes favoring long wings also result in larger, more fecund individuals which display flight behavior characteristic of migration. Missing from the correlations is early age at first reproduction, one of the most important contributors to a high Malthusian parameter (r) and for this reason thought to be an

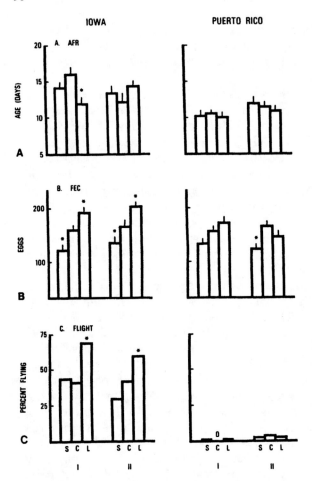

Fig. 2A–C. Correlated responses to bidirectional selection on wing length in Iowa and Puerto Rico females of *O. fasciatus*. Data are for the 6th generation of selected Puerto Rico bugs and for the 7th (*AFR, FEC*) or 9th (*FLIGHT*) generation of Iowa samples. A Age at first reproduction (*AFR*); B fecundity for the first 5 days of oviposition (*FEC*); and C percent of bugs flying for over 30 min in tethered flight tests (*FLIGHT*). *Asterisks* indicate cases in which selected lines differed significantly from controls. *S, C, L* indicate short-winged, control, and long-winged lines, respectively, for the first (*I*) and second (*II*) replicates. Standard errors are indicated where appropriate. Data for Iowa are taken from Palmer (1985) and Palmer and Dingle (1986)

important element of a colonizing syndrome (Lewontin 1965; Parsons 1983a). I have interpreted this to mean that selection in fact favors genetic (as well as phenotypic) flexibility in the timing of reproduction without the "cost" of constraint resulting from genetic correlations with other life history traits (Dingle 1984). There are certainly advantages for a migrant, or any colonizer, in this arrangement.

The correlation structure of the Puerto Rico bugs differs from Iowa in two important ways in spite of similar correlations in morphometric traits and in age at first reproduction (Fig. 2). These differences are in fecundity and flight. The latter is perhaps not surprising in view of the very short flight durations in the nonmigrant Puerto Rico bugs when contrasted to Iowa (Dingle et al. 1980). But it is not simply the case that the nonmigrants fail to show a correlation between wing length and flight; they also lack both phenotypic and genetic correlations between either wing length or head capsule width (size) and fecundity. There is thus a fundamental difference in genetic (and phenotypic) architecture between the migratory and nonmigratory populations. It is interesting to note here that in comparisons between species of *Tribolium*, it was found that correlated responses in life history traits to selection for "dispersal", as predicted on the basis of a colonizing syndrome, characterized *T. castaneum* but not *T. confusum* (Lavie and Ritte 1978; Wu 1981). The former is considered a "primary" and the latter a "secondary" colonist (Ziegler 1976). Two data points do not yield a trend, but the results from *Oncopeltus* and *Tribolium* suggest that further comparisons would be worthwhile.

Our results from *O. fasciatus* are also interesting in another context. The application of quantitative genetic theory to evolution would be considerably facilitated if it could be assumed that the pattern of genetic variances and covariances remains relatively constant across evolving populations (Arnold 1981; Lande 1982; Parsons 1983a). However, our results and those of others (Derr 1980; Giesel et al. 1982; Parsons 1983a; Service and Rose 1985) suggest that at best this assumption may be an oversimplification. Investigations of the roles of selection and genotype-environment interactions (Via and Lande 1985; Groeters and Dingle 1987) in maintaining observed across population and environment variability are clearly necessary but are still in their earliest stages.

Acknowledgments. I thank Nancy Dullum for typing the manuscript and Ken Evans for his assistance in the laboratory and for preparing the figures. The comments of Tim Prout, Gerdien de Jong, and Ken Evans were much appreciated. Supported by grants from the US National Science Foundation.

References

Arnold SJ (1981) Behavioral variation in natural populations. I. Phenotypic, genetic and environmental correlations between chemoreceptive responses to prey in the garter snake, *Thamnophis elegans*. Evolution 35:489–509

Bell G (1980) The costs of reproduction and their consequences. Am Nat 116:45–76

Berry RJ, Jakobson ME (1975) Adaptation and adaptability in wild-living house mice *(Mus musculus)*. J Zool (London) 176:391–402

Brower LP (1985) New perspectives on the migration biology of the monarch butterfly, *Danaus plexippus* L. In: Rankin MA (ed) Migration: mechanisms and adaptive significance. Contrib Mar Sci Suppl 27:748–785

Derr JA (1980) The nature of variation in life history characters of *Dysdercus bimaculatus* (Heteroptera:Pyrrhocoridae), a colonizing species. Evolution 34:548–557

Dingle H (1974) Diapause in a migrant insect, the milkweed bug *Oncopeltus fasciatus*. Oecologia (Berlin) 17:1–10

Dingle H (ed) Migration and diapause in tropical, temperate, and island milkweed bugs. In: Evolution of insect migration and diapause. Springer, Berlin Heidelberg New York, pp 254–276

Dingle H (1981) Geographical variation and behavioral flexibility in milkweed bug life histories. In: Denno RF, Dingle H (eds) Insect life history patterns: habitat and geographical variation. Springer, Berlin Heidelberg New York, pp 55–73

Dingle H (1984) Behavior, genes, and life histories: complex adaptations in uncertain environments. In: Price PW, Slobodchikoff CN, Gaud WS (eds) A new ecology: novel approaches to interactive systems. Wiley & Sons, New York, pp 169–194

Dingle H (1985) Migration and life histories. In: Rankin MA (ed) Migration: mechanisms and adaptive significance. Contrib Mar Sci Suppl 27:27–42

Dingle H, Brown CK, Hegmann JP (1977) The nature of genetic variation influencing photoperiodic diapause in a migrant insect, *Oncopeltus fasciatus*. Am Nat 111:1047–1059

Dingle H, Blakley NR, Miller ER (1980) Variation in body size and flight performance in milkweed bugs *(Oncopeltus fasciatus)*. Evolution 34:371–385

Dingle H, Blau WS, Brown CK, Hegmann JP (1982) Population crosses and the genetic structure of milkweed bug life histories. In: Dingle H, Hegmann JP (eds) Evolution and genetics of life histories. Springer, Berlin Heidelberg New York, pp 209–229

Dobzhansky T (1956) What is an adaptive trait? Am Nat 90:337–347

Falconer DS (1981) Introduction to quantitative genetics, 2nd ed. Longman, New York

Frazetta TH (1975) Complex adaptations in evolving populations. Sinauer, Sunderland, Mass

Giesel JT, Murphy PA, Manlove MN (1982) The influence of temperature on genetic interrelationships of life history traits in a population of *Drosophila melanogaster*: what tangled data sets we weave. Am Nat 119:464–479

Groeters FR, Dingle H (1987) Genetic variation and phenotypic plasticity of life history characters: response to photoperiod in two populations of the large milkweed bug *Oncopeltus fasciatus* (Hemiptera:Lygaeidae). Am Nat (in press)

Hegmann JP, Dingle H (1982) Phenotypic and genetic covariance structure in milkweed bug life history traits. In: Dingle H, Hegmann JP (eds) Evolution and genetics of life histories. Springer, Berlin Heidelberg New York, pp 177–185

Lande R (1982) A quantitative genetic theory of life history evolution. Ecology 63:607–615

Lavie B, Ritte U (1978) The relation between dispersal behavior and reproductive fitness in the flour beetle, *Tribolium castaneum*. Can J Genet Cytol 20:589–595

Leslie JF, Dingle H (1983) A genetic basis of oviposition preference in the large milkweed bug, *Oncopeltus fasciatus*. Ent Exp Appl 31:36–48

Lewontin RC (1965) Selection for colonizing ability. In: Baker HG, Stebbins GL (eds) Genetics of colonizing species. Academic Press, London New York, pp 77–94

Palmer JO (1985) Ecological genetics of wing length, flight propensity, and early fecundity in a migratory insect. In: Rankin MA (ed) Migration: mechanisms and adaptive significance. Contrib Mar Sci Suppl 27:663–673

Palmer JO, Dingle H (1986) Direct and correlated responses to selection among life history traits in milkweed bugs *(Oncopeltus fasciatus)*. Evolution 40:767–777

Parsons PA (1983a) The evolutionary biology of colonizing species. Cambridge Univ Press

Parsons PA (1983b) Ecobehavioral genetics: habitats and colonists. Annu Rev Ecol Syst 14:35–55

Ralph CP (1977) Effect of host plant density on populations of a specialized seed sucking bug *Oncopeltus fasciatus*. Ecology 58:798–809

Rankin MA, Singer MC (1984) Insect movement: mechanisms and effects. In: Huffaker CB, Rabb RL (eds) Ecological entomology. Willey & Sons, New York, pp 185–215

Rankin MA, McAnelly ML, Bodenhamer JE (1986) The oogenesis-flight syndrome revisited. In: Danthanarayana W (ed) Insect flight, dispersal and migration. Springer, Berlin Heidelberg New York Tokyo, pp 27–48

Safriel UN, Ritte U (1980) Criteria for the identification of potential colonizers. Biol J Linn Soc 13:287–297

Service PM, Rose MR (1985) Genetic variation among life history components: the effect of novel environments. Evolution 39:943–945

Stearns SC (1976) Life history tactics: a review of the ideas. Q Rev Biol 51:3–47

Stearns SC (1977) The evolution of life history traits. Annu Rev Ecol Syst 8:145–172

Sulzbach DS, Lynch CB (1984) Quantitative genetic analysis of temperature regulation in *Mus musculus*. III. Diallel analysis of correlations between traits. Evolution 38:541–552

Via S (1984) The quantitative genetics of polyphagy in an insect herbivore. II. Genetic correlations in larval performance within and among host plants. Evolution 38:896–905

Via S, Lande R (1985) Genotype-environment interaction and the evolution of phenotypic plastic-
    ity. Evolution 39:505−522
Wu A-C (1981) Life history traits correlated with emigration in flour beetle populations. Thesis,
    Univ Illinois Chicago Circle
Ziegler JR (1976) Evolution of the migration response: emigration by *Tribolium* and the influence
    of age. Evolution 30:579−592

The Evolution of Genetic Correlation and Developmental Constraints

J. M. Cheverud[1]

1 Introduction

This chapter concerns the evolutionary significance of morphological integration (Olson and Miller 1958) and its evolution in the context of developmental constraints. By morphological integration, I refer to the functional, developmental, genetic, and evolutionary coordination of morphological elements. A major argument against the comprehensive nature of neo-Darwinian evolutionary theory is that given a function-ally integrated phenotype, and by function I refer to both development and static function, it is virtually impossible to produce a viable evolutionary change by random, independent mutation of the parts. For example, if six character changes are required to adapt to a new environment, and the probability of a single, favored change is 10^{-5}, then the probability of all six changes occurring in the same organism is 10^{-30}, given no sexual reproduction (Eden 1967). Even in sexually reproducing organisms the probability of mutual adaptive evolution for all six characters may be low. In order for integrated evolution to occur, there must be some constraint on the pattern of heritable variation, and that constraint, as it evolves, should be consistent with the adaptive needs of the organism as a functionally integrated whole. Thus, only internal-ly consistent variations will be subjected to selection by the external environment. It has been suggested that modern population genetics and neo-Darwinian theory can-not explain the existence of these constraints (Gould 1980; Alberch 1980; Oster and Alberch 1982), which are said to be due primarily to the epigenetic nature of develop-ment (Waddington 1957; Frazzetta 1975; Riedl 1978).

This problem of neo-Darwinian theory is largely the result of confusion as to what is mean by random mutation among neo-Darwinians and their critics. In neo-Darwi-nian theory mutations are random with respect to the requirements of the external environment in that the environment does not direct the phenotypic effects of genetic mutations in an adaptive fashion. Mutations need not be random with respect to past evolutionary history or current morphological state. Patterns of phenotypic variation produced by genic mutation may not be uniformly distributed, although mutations are still often considered to be random and uniform at the level of DNA base pair

1 Department of Anthropology, Northwestern University, Evanston, IL 60201, USA

Population Genetics and Evolution
G. de Jong (ed.)
© Springer-Verlag Berlin Heidelberg 1988

changes (even so the frequency of purine to purine mutation is higher than the frequency of purine to pyrmidine mutation). Epigenesis, or the "mechanical" translation of genotype and environment into phenotype, can result in highly nonuniform patterns of phenotypic variation, even if mutation is uniform at the DNA level (Waddington 1957). This is not inconsistent with neo-Darwinian theory.

As a basic example, the rules for translation of genetic information (DNA or RNA sequence) into phenotypic information (protein amino acid sequence) insures that a random, uniformly distributed change in any given DNA triplet will result in a highly nonuniform change in the associated amino acid. Figure 1 provides the distribution of amino acids produced by single, random, uniformly distributed point mutations of DNA triplets producing the amino acid serine. This distribution is highly nonuniform, several amino acids (including serine itself, proline, alanine, etc.) occurring quite frequently while others (including valine, methionine, aspartate, etc.) are not represented at all. Furthermore, this distribution will depend directly on the current phenotypic state in that each amino acid will have its own particular nonuniform distribution of mutational effects. Therefore, the rules of development, or the "mechanics" of genotype-phenotype translation, constrain the pattern of phenotypic expression for mutants even when change in the DNA is random and uniform.

These rules of development are very highly burdened and cannot be easily changed in evolution because of extremely strong stabilizing selection on them (Riedl 1978). Genetically based morphological changes occurring over generations due to selection on phenotypes can only maintain their integrity if the genotype to phenotype translation follows consistent rules from one generation to the next. Although the rules of development may have been arbitrarily set relative to then current or future adaptive needs during some early period of evolution, subsequent, adaptive phenotypic evolution has proceeded under the assumption of their continuance. These rules of develop-

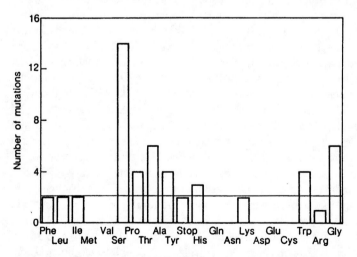

Fig. 1. Distribution of amino acids produced by single, random, uniform point mutations of DNA triplets coding for the amino acid serine. The *horizontal line* indicates a uniform frequency distribution for mutant amino acids. The *symbols* along the horizontal axis are standard abbreviations for the amino acids and special triplet codes (stop)

ment carry the burden of all subsequent phenotypic evolution (Riedl 1978). Arbitrarily changing even a single DNA triplet-amino acid association could have devastating consequences for the organism as its effects ramified through the system. However, less centrally placed developmental rules are probably less burdened and more open to evolutionary change. Indeed, at some distant time even the translation of RNA to protein was probably less burdened and itself evolved.

Epigeneticists have shown that individual developmental events and their culminative phenotypes are strongly interconnected with one another. A change in one event or character will be associated with a cascade of effects on dependent and correlated characters in a fashion determined by the rules of development. Thus, the standard answer to the problem of evolution in integrated character complexes is that the characters are correlated with one another through the epigenetic system so that change in one character will be coincident with mutually adaptive, correlated changes in functionally related characters. Instead of requiring six independent mutations, as in the example above, integrated phenotypic change of all six characters can be accomplished with one favorable mutation affecting developmental parameters.

When a single mutation at a gene locus affects may phenotypic characters, the phenotypic effect of that mutation is said to be pleiotropic. Most genes have pleiotropic, or manifold effects (Wright 1980). It has been argued independently by Riedl (1978) and Frazzetta (1975) that pleiotropy has evolved under selection for systematization whereby a developmental system which results in adaptively coordinated effects for developmentally and functionally related traits will be favored over a developmental system in which such characters are independently controlled. The pattern of pleiotropy would then reflect the patterns of developmental and functional integration. This is a theory for the evolution of epigenetic systems and makes predictions concerning their structure.

While Riedl's and Frazetta's theories of selection for the systematization of development predict that patterns of pleiotropy will mimic past and present patterns of phenotypic functional integration, it is not sufficient to predict the response of an integrated set of characters to selection, or evolutionary integration (Cheverud 1984). Since selection acts on *variation* between individuals rather than on individuals themselves, predicting response to selection requires a theory of the statistical relationship between genotypic and phenotypic *variation*, or quantitative genetic theory, rather than a theory of the "mechanical" relationship between genotype and phenotype in an individual, or epigenetic theory. However, it is the statistical effects of genetically based variation in the mechanics of development which are measured in quantitative genetics. The epigenetic system constrains and defines patterns of variation. Thus, quantitative genetic parameters measure the effects of the epigenetic system and developmental constraints on evolutionary response to selection. It is through quantitative genetic theory that neo-Darwinism accounts for developmental constraints in evolution and it should be possible to use epigenetic theory to predict the patterns of quantitative genetic variation and covariation (Cheverud 1982, 1984; Riska 1986) representing this constraint. Thus, in order to consider the role of developmental constraints in evolution we must consider the evolution of genetic variance and correlation patterns.

2 Evolution of Genetic Correlation

The evolution of genetic correlation is a subject which has only been considered in recent years (Lande 1980, 1984; Cheverud 1982). I will show that the pattern of genetic variation and correlation evolves in response to the patterns of developmental constraints and since the pattern of genetic variation affects the direction and rate of evolution in multiple character systems (Lande 1979), that it is through the genetic variance/covariance matrix that development constrains evolution.

Consider two characters, X and Y, which interact during development or in the performance of some function. Given that their interaction affects the relative fitness of individuals, the average fitness of individuals with phenotypic value X_1 will depend on their value for trait Y. This is a model of bivariate stabilizing selection. In general, with bivariate stabilizing selection, the distribution of additive genetic values will tend to approach the pattern of stabilizing selection (Lande 1980, 1984; Cheverud 1982, 1984) which in turn reflects the functional interaction of the traits. So we expect patterns of genetic correlation to partially mimic patterns of stabilizing selection.

Lande (1980) has shown that at equilibrium the genetic variance/covariance matrix at a single locus is

$$C = W^{1/2} \ (W^{-1/2} \ U \ W^{-1/2})^{1/2} \ W^{1/2} \ ,$$

where W describes the fitness surface or the pattern of stabilizing selection as experienced by the additive genetic values, and U is the mutation matrix describing the pattern of phenotypic variation produced by mutation each generation. The fitness matrix, W, represents the spread of fitness around the optimal value. Dimensions along which stabilizing selection is weak have high values while dimensions along which stabilizing selection is strong have low values. When off-diagonal values of the W matrix are nonzero, stabilizing selection will produce correlation. The mutation matrix, U, measures the pattern of variation produced by genic mutation and when off-diagonal values are nonzero, variation produced by mutation will be nonuniform.

Stabilizing selection, as used here, includes the concepts of both "internal" and "external" selection (Riedl 1978). External selection is stabilizing selection due to interaction of the phenotype with the external environment. Internal selection is stabilizing selection due to the interaction of the phenotype with other, internal characteristics of an organism. Internal selective forces relate to the need for coadaptation of traits one to another rather than to the external environment. The pattern of internal stabilizing selection is that exerted by the epigenetic system on its parts. Through its effects on patterns of genetic correlation (Cheverud 1984) internal stabilizing selection will constrain the pattern of variation available for selection by the external environment. Only developmentally (internally) consistent mutations will be offered as potential adaptations for selection by the external environment. Thus, through the fitness matrix describing stabilizing selection patterns, the effects of epigenetic systems on expressed phenotypic variability are accounted for in neo-Darwinian evolutionary theory.

Thus, we expect quantitative genetic variation to be nonuniform due to internal stabilizing selection, even if variation produced by mutation is uniform (Cheverud 1984). However, it is also quite possible that the pattern of phenotypic variation produced by genic mutation will itself have some nonuniform structure. We may find that mutations which make two characters jointly larger or smaller than average are more common than those which cause one character to be large and the other to be small. Since it is well known that mutation rates (mutator genes) and the phenotypic effects of mutations (dominance modifiers) can evolve, one might expect the pattern of phenotypic variation produced by mutation to match long-standing developmental and functional patterns of interaction. Thus, through the mutation matrix nonuniform patterns of variation produced by mutation are accounted for in neo-Darwinian evolutionary theory.

3 Simulation of Genetic Correlation Evolution by Stabilizing Selection

A computer simulation of the evolution of genetic correlations was developed in order to test the analytical models developed by Lande (1980, 1984) and explicated by Cheverud (1982, 1984) for the evolution of genetic correlation patterns. Due to difficulties in estimating the segregation variance under selection, the simulation follows individual genotypes over generations. A description of the simulation follows:

1. Input parameters:
    a) Number of loci;
    b) Number of generations (t);
    c) Population size (N);
    d) Index weights (a1, a2);
    e) Slope (m) and intercept (b) of fitness functions;
    f) Number of iterations (i);
    g) Pleiotropic allele frequencies (++) (--) (+-) (-+);
    h) Nonpleiotropic allele frequencies (+0) (0+) (-0) (0-).
2. Assign alleles to individuals and calculate trait values for X, Y, and the index (I = a1*X + a2*Y) allowing each allele a unit additive effect and summing over all loci.
3. Calculate the initial correlation between X and Y.
4. Produce the next generation with each locus being independently inherited.
    a) Randomly choose an individual;
    b) Calculate its probability of mating using the fitness function, Pr (mating) = b + m * |index| (see Fig. 2);
    c) Test the probability of mating (Pr) against a random number (RND): if Pr < RND then select another individual; if Pr > RND then choose a mate by the same procedure;
    d) Randomly choose one allele from each parent at each locus to generate the offspring genotype;
    e) Calculate offspring X, Y and index;
    f) Repeat until N offspring are produced.
5. Calculate new correlation between X and Y.
6. Repeat for t generations.
7. Repeat for i iterations.

Figure 2 gives the function describing the probability of mating and several graphs displaying its relationship to the index value. All of the simulations were run with

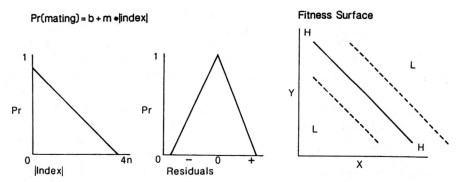

Fig. 2. The probability of mating (fitness) as a linear function of the absolute value of the index (a1*X + a2*Y). The function is graphed first against the absolute value of the index, secondly against raw index values (residuals), and finally as a fitness surface for the original variables X and Y. Note that selection is for a correlation of −1.0

both a1 and a2 equal to one and identical pleiotropic allele frequencies ($p_i = 0.25$). Nonpleiotropic alleles were not included in this set of simulations. Thus, the correlation in the original population was zero. Note that in this situation selection is entirely stabilizing and that individuals with high values of X combined with low values of Y and vice versa are favored over those with concurrent low or high values for the two variables (see Fig. 2). Thus, selection is for a correlation of minus one. The narrow-sense heritability of both X and Y is one. Thus, the phenotypic and genetic correlations and covariances are equal. Simulations were run for ten generations and repeated ten times each.

The first set of simulations varied population size (N = 25, 100) and the slope of the fitness function, or strength of selection (m = −0.1, −0.3, −0.5) with ten loci. There were no significant differences between correlations produced by the different levels of selection. Response quickly leveled off at an intermediate value (r = −0.33) after a few generations. The larger populations size showed a slightly, but not significantly, better response. There was considerable variation within each set of ten runs started with identical parameters and the extent of variation in response between repeats was much greater with small population size (N = 25), as might be expected due to genetic drift of allele frequencies.

These results were much different than those expected on the basis of the analytical models (Lande 1980, 1984). Instead of approaching r = −1.0, correlations seemed to be limited to a less extreme value. There are many differences between the analytical and simulation models which might account for the differences in results. Major differences occur largely due to the number of loci considered. Lande's (1980, 1984) models consider one locus at a time and simply sum the genetic covariances over loci, treating linkage disequilibrium as a separate source of covariance from pleiotropy. In contrast, the simulation model does not differentiate between the two sources of genetic correlation, as they would both contribute in natural situations. With small population sizes and strong selection, at least relative to the infinite population size and weak selection of the analytical models, strong linkage disequilibrium is likely to be produced. Thus, genotypes with positive pleiotropic allele combinations such as

[(++)(--)] will be favored over individuals who are homozygous for these alleles, [(++)(++)] or [(--)(--)], to the same extent as those carrying the (+-) and (-+) alleles are favored. An individual with ten (++) and ten (--) alleles will have an index value of zero and a high probability of mating, although their offspring are likely to contribute to positive correlations. This phenomenon should lead to some oscillation of correlation values as the segregation of positive pleiotropic alleles produces offspring contributing to positive correlation from parents contributing to a zero correlation. Such oscillations were noted in the simulation data. The correlation will only reach -1.0 when all of the (++) and (--) alleles are eliminated from the population. Thus, linkage disequilibrium may limit the response to correlating selection.

Another difference between the models relating to multiple alleles is in the treatment of epistatic genetic variance. In Lande's analytical models all fitness surfaces are described with respect to additive genetic values while in the simulation, as in nature, it is the phenotypic values upon which selection acts. This convention is used in the analytical models to make them tractable. While in the simulation all of the variance of X and Y is additive genetic, variance in the absolute value of the index and probability of mating is not. With two loci, for a starting population with only pleiotropic alleles each in equal frequency at each locus and m = -0.25, the total genotypic variance is 0.0502. The additive genetic variance is 0.0061, the dominance variance is 0.0120, and the epistatic variance is 0.0321. Thus, the narrow-sense heritability for the character being selected (absolute value of the index) is only 0.12. Presumably, the addition of further loci would further increase the predominance of epistatic variance as the number and kind of epistatic interactions increased. Preliminary calculations indicate that the additive genetic variance decreases as the negatively pleiotropic alleles [(+-)(-+)] increase in frequency until an intermediate value for the correlation is reached and than increases again at more extreme values. Thus, the predominance of epistatic variance, not considered in the analytical models, may limit response to correlating stabilizing selection by producing regions of very low additive genetic variance in fitness at intermediate correlation values. This is obviously a difficult theoretical problem.

A preliminary set of simulations were run with N = 100, m = -0.1 and all other input parameters held constant except for the number of loci. All simulations showed the same behavior as above, response halting at an intermediate correlation value typically between -0.3 and -0.5, except for those run with only one locus. With only one locus, the (++) and (--) alleles are quickly eliminated from the population resulting in a correlation of -1.0 after only a few generations, just as predicted by the analytical models. Further simulation showed that the rate of approach to a correlation of -1.0 was a function of the strength of selection when only one locus was considered, again as in the analytical models.

In contrast to a simple interpretation of the analytical models, the number of loci seems to greatly affect the evolutionary dynamics of correlations in response to stabilizing selection. Certainly further analytical and simulation work in this area will prove interesting. It is important to stress that these results do not in any way invalidate the analytical models, but rather point out the limits to their utility when used to interpret more natural situations. Epistatic variance in fitness and linkage disequilibrium modify expectations in unforeseen ways.

4 Conclusions

The commonly cited sources of developmental constraints on evolution, internal stabilizing selection, and nonuniform mutation have been shown to directly affect the pattern of heritable variation, as measured by the quantitative genetic variance/covariance matrix. It is through their effects on patterns of heritable variation that developmental processes constrain evolutionary change. Thus, a useful research program for considering the role of developmental constraints in evolution is to attempt to predict the pattern of genetic correlation through theories which predict the pattern of internal stabilizing selection and phenotypic patterns of mutational effects. Those studying developmental constraints and their role in evolution have concentrated primarily on the demonstration of developmental and phylogenetic patterns rather than on predicting their effects on evolutionary change. Perhaps the wider recognition that developmental constraints are accounted for in neo-Darwinian evolutionary theory by the genetic variance/covariance matrix will facilitate the joint study of development and evolution.

However, further population genetics work needs to be done on the evolution of correlation patterns themselves, especially the effects of linkage disequilibrium in small populations and the role of epistatic variance in regimes of stabilizing selection. Theoretical work in this area has just begun and hopefully future work will clarify the relationships between stabilizing selection patterns, patterns of mutational effects, and the resultant expected patterns of genetic correlation.

References

Alberch P (1980) Ontogenesis and morphological diversification. Am Zool 20:653–667

Cheverud J (1982) Phenotypic, genetic, and environmental morphological integration in the cranium. Evolution 36:499–516

Cheverud J (1984) Quantitative genetics and developmental constraints on evolution by selection. J Theor Biol 110:155–172

Eden M (1967) Inadequacies of neo-Darwinian evolution as a scientific theory. In: Moorhead P, Kaplan M (eds) Mathematical challenges to the neo-Darwinian interpretation of evolution. Wistar Inst Symp Monogr 5, Philadelphia, pp 5–19

Frazzetta T (1975) Complex adaptations in evolving populations. Sinauer, Sunderland, Mass

Gould S (1980) Is a new and general theory of evolution emerging? Paleobiology 6:119–130

Lande R (1979) Quantitative genetic analysis of multivariate evolution, applied to brain:body size allometry. Evolution 33:402–416

Lande R (1980) The genetic covariance between characters maintained by pleiotropic mutations. Genetics 94:203–215

Lande R (1984) The genetic correlation between characters maintained by selection, linkage, and inbreeding. Genet Res Camb 44:309–320

Olson E, Miller R (1958) Morphological integration. Univ Chicago Press

Oster G, Alberch P (1982) Evolution and bifurcation of developmental programs. Evolution 36:444–459

Riedl R (1978) Order in living organisms. Wiley & Sons, New York

Riska B (1986) Some models for development, growth, and morphometric correlation. Evolution 40:1303–1311

Waddington C (1957) The strategy of genes. Allen & Unwin, London

Wright S (1980) Genic and organismic selection. Evolution 34:825–843

Models of Fluctuating Selection for a Quantitative Trait

L. A. ZONTA and S. D. JAYAKAR[1]

1 Introduction

The mathematical theory of natural selection has been concerned, since its beginning in the 1920s, with searching for patterns of selection which would maintain genetic variation. One of the interesting situations analyzed is the maintenance of a one-locus polymorphism due to fluctuating selection, in other words, selection which favours different genotypes in different periods of time, e.g. seasons, years, etc. (Haldane and Jayakar 1963).

Intuitively, it seems to us that fluctuating selection (which is the rule for natural populations) would maintain more genetic and phenotypic variation than constant selection also for a polygenic quantitative trait, as has indeed been observed in laboratory populations of Drosophila (Mackay 1981). On the other hand, in a theoretical investigation using a Gaussian model, Slatkin and Lande (1976) concluded that the niche width (the phenotypic variance of the trait) would not be greater under varying selection than under constant selection. However, the Gaussian model has received severe criticism concerning its suitability in the study of the evolution of quantitative traits. A model with few loci each of which contributes towards the total variation of the trait is an alternative, reasonable model and has experimental support (Thoday 1961).

We consider it instructive to study such an oligogenic model to compare the effects of fluctuating as against constant selection in the maintenance of genetic and phenotypic variation.

2 Model

The simplest oligogenic model is one with two diallelic loci. This gives nine genotypes (ten if the cis and trans double heterozygotes are considered as different genotypes) each with its own distribution for the quantitative trait prior to selection. The popula-

---

1 The "A. Buzzati-Traverso" Department of Genetics and Microbiology, University of Pavia, Via S. Epifanio 14, I-27100 Pavia, Italy

Population Genetics and Evolution
G. de Jong (ed.)
© Springer-Verlag Berlin Heidelberg 1988

tion would consist of a mixture of these distributions in proportion to their genotypic frequencies.

The phenotypic distribution of each genotype can be defined independently, and the fitness for the trait in a given environment is also defined by a mathematical function independently of the phenotypic distributions. First, all the distributions are assumed to be Gaussian. The fitness of each genotype can be calculated by integrating the product of the corresponding phenotypic function and the fitness function. Once we have these nine (ten) genotypic fitnesses, we have reduced the situation to a two-locus, qualitative selection model.

The most general model with Gaussian functions is shown in Table 1.

A genotypic fitness is maximized with respect to ν when $\mu = \nu$, and with respect to τ when $\tau^2 = (\mu - \nu)^2 - \sigma^2$ for $\sigma^2 < (\mu - \nu)^2$, and $\tau = 0$ otherwise.

From the vast array of possible distributions, we have selected for study those in which the effects of the two loci on the phenotype are independent. Each locus can, for example, define either the mean or the variance of the trait distribution or both.

The results which we present are derived from two sets of models: (1) where both loci have additive effects on the distribution mean (multiplicative model of Karlin 1975); (2) where one locus has an additive effect on the mean and the other a multiplicative effect on the variance.

Our main interest was, of course, to find situations in which fluctuating selection maintained more chromosomes and a wider niche (population variance) as opposed to constant selection. Since models of type (1) never maintain more than one locus polymorphic, we have not investigated them in any detail. Type (2) models gave more interesting results, some aspects of which were analyzed in detail.

While for constant selection it is a straightforward matter to apply the set of criteria derived by Bodmer and Felsenstein (1967) to the numerical fitness matrix for an analysis of equilibrium stability, for fluctuating selection such an analysis is difficult. It is possible to combine the one-generation fitness matrices to obtain a two-generation fitness matrix. However, since in this case the fitness values are frequency-dependent and therefore change during the process of selection, the Bodmer-Felsenstein criteria are not applicable for a stability analysis. We have therefore resorted to a numerical

Table 1. General two-locus selection model

Loci	A		B	
Alleles	A_1, A_2		B_1, B_2	
Gametes	$A_1 B_1$	$A_1 B_2$	$A_2 B_1$	$A_2 B_2$

Genotypic functions $\equiv N (\mu_{ij}, \sigma_{ij})$ $i, j = 1,...,4$

	$A_1 B_1$	$A_1 B_2$	$A_2 B_1$	$A_2 B_2$
$A_1 B_1$	(μ_{11}, σ_{11})	(μ_{12}, σ_{12})	(μ_{13}, σ_{13})	(μ_{14}, σ_{14})
$A_1 B_2$		(μ_{22}, σ_{22})	(μ_{23}, σ_{23})	(μ_{24}, σ_{24})
$A_2 B_1$			(μ_{33}, σ_{33})	(μ_{34}, σ_{34})
$A_2 B_2$				(μ_{44}, σ_{44})

Fitness function: $W^{(k)} \equiv N (\nu_k, \tau_k)$ $k = 1,2$

Genotypic fitness: $w_{ij}^{(k)} = \exp -[(\mu_{ij}-\nu_k)^2/2 (\sigma_{ij}^2 + \tau_k^2)]/(\sigma_{ij}^2 + \tau_k^2)$

approach. The equilibrium constellation was calculated numerically by successively applying the genotypic fitness matrices, obtaining the gametic frequencies and then the panmictic genotypic frequencies. The criterion for deciding whether the process had reached an equilibrium was that the difference in each of the corresponding gametic frequencies in two consecutive generations was less than 10^{-8}.

3 Analysis

The results presented here refer to a particular case of the type (2) model. Table 2 shows the matrix of means and standard deviations of the phenotypic functions (μ_{ij}, σ_{ij}; i,j = 1,2,3) and the various selection regimes employed (ν_k, τ_k; k = 1,2). The two selection functions which alternate are in this model symmetrical around 0 ($\nu_1 + \nu_2 = 0$). In this case, a comparison of the cyclic regime with the constant regime with the fitness function having a mean 0 is the most conservative in the sense that it is the least likely to show more genetic variation being maintained by variable selection with respect to constant selection.

For every experiment, we used two values of the recombination frequency between loci (r = 0 and r = 0.5). The former case coincides with a one-locus four-allele model, and the latter with two unlinked loci. Although we realize that intermediate values are important, our experiments to investigate this are at present too incomplete to be presented.

We have also studied the effect of alternating one generation of one fitness function with several of the other.

A regular cycle of the environments is, however, an unnatural assumption. We have therefore run some trials in which the order of occurrence of the two environments was determined by chance (using a random number generator) with fixed probabilities. Because of the possibility of large fluctuations in the gametic frequencies in this experiment, we have iterated the process for 4000 generations.

Table 2. Description of the model analyzed

	Genotypic functions		
	$B_1 B_1$	$B_1 B_2$	$B_2 B_2$
$A_1 A_1$	(−2, 0.5)	(0, 0.5)	(2, 0.5)
$A_1 A_2$	(−2, 1)	(0, 1)	(2, 1)
$A_2 A_2$	(−2, 2)	(0, 2)	(2, 2)

Constant fitness regime: $F \equiv N (0, \tau)$ $\tau = 0.5, 1, 2$

Variable fitness regime: $F^{(1)} \equiv N (-\nu, \tau)$; $F^{(2)} = N (\nu, \tau)$
                                   $0 < \nu \leqslant 2.5$ $\tau = 0.5, 1, 2$

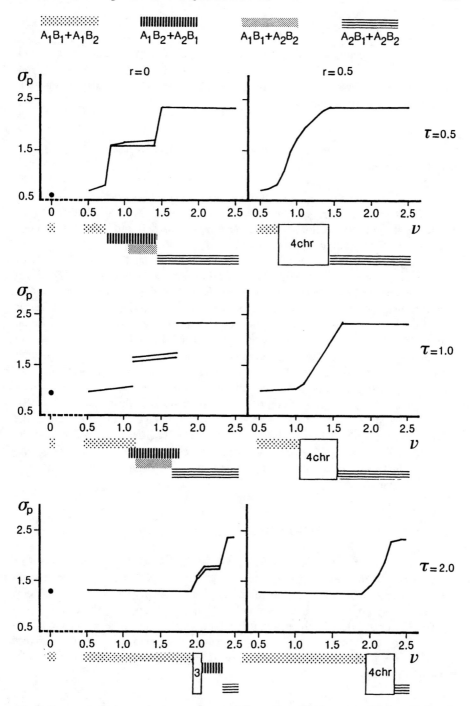

Fig. 1. Summary of the results for constant and fluctuating selection at equilibrium. The graphs show the variation of the phenotypic standard deviation with ν. Also shown is the chromosome complement present

4 Results

The results of the analyses of the model shown in Table 2 are illustrated in Fig. 1 and discussed below. The results are given in terms of equilibrium polymorphisms and phenotypic standard deviations at these equilibria. Unless otherwise stated, the equilibrium is the only stable one.

4.1 Fitness Functions Symmetric With Respect to the Mean of the B_1B_2 Heterozygotes

It is important to keep in mind that the phenotypic matrix is symmetric and the two fitness functions (f.f.s) are symmetric with respect to it. Which chromosomes are maintained at equilibrium therefore depends on which f.f. is applied first (in all our experiments the function with mean $-\nu$).

The larger the difference between the two f.f.s the wider the niche; the case of constant selection corresponds of course to no difference in the f.f.s ($\nu = 0$) and is the regime with the lowest possible niche width. Parallel to the increase in niche width, there is a change in the chromosome set maintained at equilibrium (Fig. 1). In fact, if the population maintains only chromosomes 1 (A_1B_1) and 2 (A_1B_2) the maximum niche width possible is 1.5. To increase it beyond this point, other chromosomes have to be recruited. In fact, with an increase in ν chromosome 1 (A_1B_1) is dropped and chromosome 3 (A_2B_1) is introduced. Given the symmetry of the model, however, this situation (chromosomes 2 + 3) is equivalent to retaining 1 + 4. Which of these combinations is achieved depends on the initial frequencies of the simulation process. The combination (3 + 4) of course gives the widest niche and therefore this is the polymorphism maintained beyond a certain value of ν. The transition depends on the values of τ: the higher τ, the higher is ν at the critical points of the transition. Moreover, with increasing values of τ, the region in which four chromosomes are maintained becomes narrower.

4.2 Fitness Functions Asymmetric in Their Means and/or Variances

The above results for symmetrical f.f.s are, as we have mentioned earlier, the least likely to give strong differences with respect to constant selection. We have made a few trials using asymmetric f.f.s on the same phenotypic matrix of Table 2. There is far greater polymorphism when τ is small (0.5 and 1.0); for $\tau = 2$ there is generally less polymorphism than in the symmetric situation. Further, there can be more than one equilibrium. With asymmetry in the variances of the two f.f.s the larger variance dominates the results, giving a lower degree of polymorphism, or even fixation of a single chromosome.

4.3 Unequal Number of Generations for the Two f.f.s (Regular Cycle)

We have so far investigated only strict fluctuations of the environment every generation. It is perhaps somewhat more realistic to have one or the other environment present for a longer period. In order to simulate this, we have alternated a period of one
generation for one of the environments with a higher number of generations $(2, 4, 9)$
of the other, maintaining, however, this fixed cycle. This has been done for $(\nu = 1.4,$
$\tau = 1)$. This choice was made because it was one of the experiments which showed
the most polymorphism in the $(1:1)$ regime of selection. We find no qualitative difference in the results for $(1:1)$, $(1:2)$ and $(1:4)$. The $(1:9)$ scheme fixes one of the
chromosomes. As would be expected, the $(2:1)$, $(4:1)$ and $(9:1)$ regimes give results
symmetrical to the former. The maximum niche width achieved over the whole cycle
is highest for $(1:1)$ and is successively smaller for $(2:1)$ and $(4:1)$ and is minimum for
$(9:1)$.

4.4 Unequal Probabilities of the Two f.f.s

An even more realistic regime would be to assign the environments at random with
given probabilities p and $1-p$. The polymorphisms at equilibrium are no different
from the fixed cycle as long as $0.3 \leqslant p \leqslant 0.7$ (approximately). At $p = 0.9$ chromosome 1 $(A_1 B_1)$ is fixed. At an intermediate value $(p = 0.8)$ chromosome 3 $(A_2 B_1)$ is
fixed. Here, we have the maximum niche width possible in a monomorphic situation
for this model. This result is difficult to explain.

5 Conclusions

A general conclusion of our analysis could not be better stated than in Mackay (1981):
"Is genetic variance maintained in a variable environment? The answer is: 'yes, sometimes'." However, we are also interested in the questions "when?" and, in particular,
"how does it compare with a constant environment?".

Our analysis provides answers to both questions for a very particular two-locus
model for a quantitative trait. Clearly, a two-locus model is an extremely simple one.
However, the sternopleural bristle number in *Drosophila melanogaster* gives a major
response to two loci (Thoday 1961). Our model is not therefore as much in the realm
of science fiction as it might seem.

Using three traits in *Drosophila melanogaster*, Mackay (1981) conducted laboratory
experiments to investigate the results of fluctuating selection as opposed to constant
selection. One of these characters was indeed the sternopleural bristle number; the
others were the abdominal bristle number and body weight. For the sternopleural
bristle number "the phenotypic and additive genetic variances ... are substantially
and significantly greater in populations experiencing spatial and temporal variation
than in the control populations". The other two traits did not give this result. This
could depend on the kind of genetic determination of the trait.

Our main conclusion is that fluctuating environments can under certain circumstances maintain a greater phenotypic variance than a constant environment. Whether this increase is accompanied by a greater degree of polymorphism depends on the different contributions of the genotypes to the phenotypic variance.

Comparing our results to one-locus theory, there are certain similarities in the general conclusions. Hoekstra et al. (1985), for a one-locus model of selection with two randomly alternating sets of fitness values which favoured the two different homozygotes, investigated the effects of the proximity of the heterozygote to one of the homozygotes. The maintenance of polymorphism depended both on the strength of selection and on r, the probability of one of the fitness sets. This is analogous to our results in that maintenance of polymorphism is increasingly favoured with a larger difference in the two selection regimes, and by the probabilities being closer to (1:1).

We conclude then that, given the right situation, temporally varying selection can maintain more genetic and phenotypic variability for a quantitative trait than constant selection. This is true in our two-locus model, but more importantly has also been found in laboratory populations.

References

Bodmer WF, Felsenstein J (1967) Linkage and selection: theoretical analysis of the deterministic two locus random mating model. Genetics 57:237–265

Haldane JBS, Jayakar SD (1963) Polymorphism due to selection of varying direction. J Genet 58:237–242

Hoekstra RF, Bijlsma R, Dolman AJ (1985) Polymorphism from environmental heterogeneity: models are only robust if the heterozygote is close in fitness to the favoured homozygote in each environment. Genet Res 45:299–314

Karlin S (1975) General two-locus selection models: some objectives, results and interpretation. Theor Popul Biol 7:364–398

Mackay TFC (1981) Genetic variation in varying environments. Genet Res 37:79–83

Slatkin M, Lande R (1976) Niche width in a fluctuating environment — density independent model. Am Nat 110:31–55

Thoday JM (1961) Location of polygenes. Nature (London) 191:368–370

Development and Selection

Components of Selection: An Expanded Theory of Natural Selection

J. Tuomi[1], T. Vuorisalo[2], and P. Laihonen[2]

1 Introduction

In his qualitative theory of natural selection, Darwin (1859) proved that selection is able to produce adaptive heritable changes in populations. His qualitative reasoning, however, did not satisfy the founders of theoretical population genetics. Their major task was to establish a quantitative theory which could derive the rates of evolutionary change and specify the interactions of major evolutionary forces (Haldane 1924; Fisher 1930; Wright 1931). For this purpose, the concept of fitness was introduced as a technical tool for quantifying natural selection in terms of its statistical consequences. Accordingly, selection itself was redefined as (1) the differential reproduction of genotypes (Dobzhansky 1951; Lerner 1958, 1959; Mayr 1970), or (2) the differential spread of gene alleles (Fisher 1930; Williams 1966; Dobzhansky 1970; Dawkins 1976, 1982), or even (3) the non-random changes in gene frequency (Wright 1931, 1949; Simpson 1953). As a result, natural selection is generally described as a consequence of genetic variation in fitness (Lerner 1959; Maynard Smith 1969; Lewontin 1970; Williams 1970).

Thus, classic neo-Darwinism provided not only a theory of evolutionary mechanisms, but also a conceptual system which affects our perception of evolutionary processes. Selection was described through its statistical consequences, and the founders of theoretical population genetics therefore established technical definitions of selection that were operationally workable in quantitative models. However, as will be shown below, such technical definitions can be understood as shortcuts for representing the selection process. To demonstrate this, we show that the standard neo-Darwinian definitions of selection can be derived from the same causal model of the selection process. This unifying model of natural selection integrates genetic, ontogenetic and ecological aspects of selection within a single theoretical framework.

The classic neo-Darwinian models mainly analyzed the genetic aspects of the selection process. A major challenge of modern evolutionary population biology is there-

1 Department of Theoretical Ecology, University of Lund, Ecology Building, S-223 62 Lund, Sweden
2 Department of Biology, University of Turku, SF-20500 Turku, Finland

Population Genetics and Evolution
G. de Jong (ed.)
© Springer-Verlag Berlin Heidelberg 1988

fore to develop explicit theories of the ontogenic and ecological aspects of natural selection.

2 Evolutionary Forces in a Model System

Early population genetics was based on six fundamental ideas (Lewontin 1974; Michod 1981): (1) The totality of genes in a population at a given time makes up the gene pool. (2) Gene alleles are transmitted in sexual populations from the gametic gene pool of one generation to the next through two processes: First, the relative production of gametes carrying alternative gene alleles (T_g) determines the gametic gene pool of the next generation depending on the genotypic composition of the population. Second, the combination of gametes in fertilization (T_F) determines the genotypic composition of the population as a function of gametic allele frequencies and mating patterns. (3) Consequently, genetic material is recombined in each generation, as gene alleles are brought together into genotypes which in turn are broken down in gametic production (Fig. 1). (4) When no outside force (mutation, immigration, emigration, selection or drift) is operating, gene frequencies in the gene pool remain unaltered (Castle-Hardy-Weinberg equilibrium law). (5) A change of gene frequency in the gene pool is the elementary evolutionary process, or even the evolutionary change. (6) Evolutionary forces can be expressed in terms of their effect on changes in gene frequency.

Wright (1949, p. 370) defined *selection pressure* as involving "exhaustively all systematic modes of change of gene frequency which do not involve physical transformation of the hereditary material (mutation) or introduction from without (im-

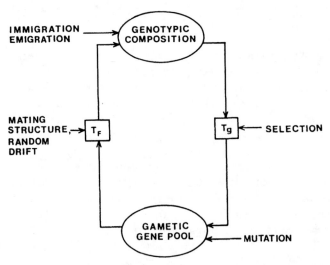

Fig. 1. A functional division of evolutionary forces in a model system where the combination of gametes in fertilization (T_F) determines the genotypic composition of the population, and the perpetuation of gene alleles (T_g) from one generation to the next determines the composition of the gametic gene pool

migration)". This definition specifies, in fact, the effects that selection and emigration may together have on the gene pool of local populations. Although their combination is mathematically justified, selection and emigration are not functionally comparable. Both emigration and immigration rates can alter the genotypic composition of the population, but they do not determine the rates at which gene alleles are perpetuated in descendants (Fig. 1).

Selection pressure may therefore be defined to include "anything tending to produce systematic, heritable change in populations between one generation and the next" (Simpson 1953, p. 138), or rather to include anything tending to produce non-random differential propagation of gene alleles from one generation to the next. Thus, selection comprises the basic evolutionary force that leads to evolutionary changes in successive generations, while immigration and emigration modify the ecological and genetic environments within which the primary forces (mutation, selection and drift) operate.

3 Fitness as a Technical Tool

The essence of selection is that individuals carrying different genotypes have differential success in leaving descendants and therefore contribute differentially to the gene pools of succeeding generations (Dobzhansky 1951). The average contribution that the carriers of a given genotype make to the gene pool, relative to the contributions of other genotypes in the same population, is a measure of the fitness of the genotype (Dobzhansky 1970). Such a *genotypic fitness value* (sensu Sober and Lewontin 1982) varies as a function of (1) the survival rates of zygotes from fertilization to maturity, (2) their age-specific survival and gametic production and (3) the frequency with which their gametes successfully participate in fertilization. Genetic variation in these components represents selection, which we can now express in terms of a single term, fitness.

Genotypic fitness provided a useful technical tool for analyzing the consequences of selection; the modeller may simply assume fixed fitness values for given genotypes and derive the corresponding selection pressures on gene frequencies (Fig. 2). If the gametic frequencies of two alleles A and a are p and q ($p + q = 1$) and the gametes are randomly combined in fertilization (T_F, Fig. 2), the frequencies of genotypes AA, Aa and aa at generation t will be p^2, $2pq$ and q^2 respectively. Now the modeller can make selection work by modifying the fitness values W_{AA}, W_{Aa} and W_{aa} of the genotypes AA, Aa and aa. First, the propagation of allele a to the next generation $t+1$ (T_g, Fig. 2) can be simulated by calculating its new gametic frequency

$$q' = q(pW_{Aa} + qW_{aa})/\overline{W} ,$$

where $\overline{W}$ is the average fitness of the population. Second, the change in its frequency

$$\Delta q = q' - q = q(pW_{Aa} + qW_{aa} - \overline{W})/\overline{W}$$

expresses selection pressure as a function of genotypic fitness values (Wright 1949; Sober and Lewontin 1982).

In this procedure, the selection process itself remains a black box (Darlington 1983), as the modeller calculates the consequences of selection from the knowledge

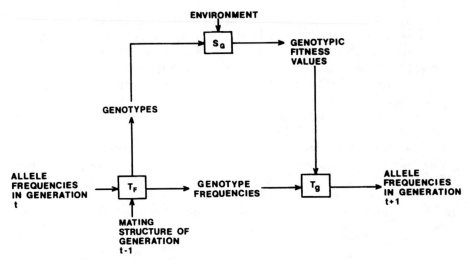

Fig. 2. A black box model of genotypic selection (S_G), in which genotypic fitness values are used as *post factum* measures of the intensity of selection (T_F and T_g as in Fig. 1)

of genotype frequencies and genotypic fitness values (Fig. 2). We therefore define *genotypic selection* (S_G, Fig. 2) as comprising those unspecified factors which actually determine the relative contribution of genotypes to the gene pool of the next generation. As fitness quantifies this relative contribution (Dobzhansky 1951, 1970; Lerner 1958, 1959), it can be understood as a technical shortcut for representing the actual selection process. Thus, fitness provides, above all, a logical tool for simplifying the functional complexity of selection into a mathematically tractable parameter (Stearns 1982). It can be used for analyzing (1) how efficiently selection changes gene frequencies in model populations, and (2) how selection can interact with other forces in producing changes in the gene pool.

4 Differential Reproduction: Selection as a Black Box

Although constant genotypic fitness values are theoretical terms unmeasurable in practice (Kojima 1971; Lewontin 1974; Michod 1981), population genetic models have proved that even relatively weak selection can effectively change gene frequencies, and that selection is "not primarily a process of elimination" but rather "a process of differential reproduction" (Simpson 1949, p. 97). Even such genetic differences which do not influence survival rates can be effective in selection if they lead to differences in reproductive rates. As such differential reproduction can lead to the establishment of new gene combinations, as soon as "selection is defined as differential reproduction, its creative aspects become evident" (Mayr 1970, p. 119; Muller 1949; but see Huxley 1963, p. xviii).

This neo-Darwinian concept of selection is derived from the genotypic model of natural selection (Fig. 2), where the relative contribution of genotypes is used as a *post factum* measure of the intensity of selection. Accordingly, selection itself "can be defined in terms of its observable consequences as the *non-random differential reproduction of genotypes*" (Lerner 1958, p. 5). This is the most common textbook definition of selection, frequently modified as "differential perpetuation of gene alleles and gene complexes" (Dobzhansky 1970, p. 97). It is even emphasized that selection *is* the differential reproduction of genotypes but *does not produce* any differences in reproductive success (e.g. Wallace and Srb 1961; Mayr 1970).

The major fallacy of this black box definition (sensu Darlington 1983) is that the selection process (S_G, Fig. 2) is replaced by its consequences. As a result, selection is understood as "a technical term in evolutionary studies" (Simpson 1953, p. 138), which provides no causal explanation for observed differences in survival or reproduction (Lerner 1958; Grene 1961; Bock 1980; Dunbar 1982). The neo-Darwinian concept of selection is thus an adequate representation of natural selection only insofar as we analyze the consequences rather than the causes of the non-random differential reproduction of genotypes. The obvious cost of this technical concept is that it gives no insight into the causal machinery of the selection process, nor does it lead to any unifying theory of natural selection.

5 Genic Selection as a Process

Classic neo-Darwinism redefined selection with reference to the perpetuation of genetic units rather than to its causal mechanisms, and genes are therefore frequently considered as the primary units of selection (Fisher 1930; Hamilton 1964; Williams 1966; Dawkins 1976, 1982). Obviously, it is the genes, and not genotypes, phenotypes or individual organisms, that increase in frequency *as a consequence of selection*; genotypes are broken down in meiosis and phenotypes are unique and will be destroyed with the death of their carriers. The differential perpetuation of genotypes can take place only in asexual populations, where individuals transmit their whole genomes to offspring.

The selection of gene alleles, or *genic selection* (Williams 1966), can most simply be modelled in terms of the rates at which gene alleles increase in frequency (Fisher 1930). These rates quantify *genic fitness values* which can be used to express selection pressure:

$$\Delta q = q(W_a - \overline{W})/\overline{W} \, ,$$

where

$$W_a = (pqW_{Aa} + q^2 W_{aa})/q = pW_{Aa} + qW_{aa}$$

is the genic fitness value for allele a (Sober and Lewontin 1982). As genic fitness values are calculated from the knowledge of genotypic fitness values, this expression of selection pressure is equivalent to the genotypic formula presented above.

In fact, genic selection describes the same causal process as the genotypic model of selection (Fig. 2); genic fitness values vary as a function of (1) the fitness values of the genotypes that make up the genetic environments of the allele, (2) the relative frequency of the genotypes into which the allele enters in fertilization and (3) the number of replicas of the allele that the genotypes are carrying (Williams 1966; Crow and Kimura 1970).

If we now apply the components of the genotypic model of natural selection, genic selection as a process (S_g) can be defined as

$$S_g = (T_F, S_G, T_g) ,$$

where genotypic selection (S_G) comprises a causal sub-process of genic selection. Thus, the major causal implication of genic selection is that neither S_G nor genotypic fitness values as such are always sufficient to predict changes in gene frequency, as other factors influencing the combination of gametes in fertilization (T_F) or the perpetuation of gene alleles (T_g) may modify the evolutionary outcomes of genotypic selection (for an example, see Templeton 1982). Thus, S_g rather than S_G alone provides the evolutionary machinery that determines the rates of spread of gene alleles.

6 Ontogeny and Organismic Selection

Although selection can be modelled as the differential spread of gene alleles, this does not mean that selection operates separately and independently on gene alleles and individual loci. Gene interactions can mask the effects of individual alleles and loci so that the "real objective of selection is the genotype as a whole" (Wright 1949, p. 381), or preferably the whole organism (Wright 1980). Mayr (1955, 1959, 1970, 1982) has repeatedly emphasized the unity of the genotype, and the fact that both genes and genotypes are selected in terms of their phenotypic products. Thus, it is necessary to realize that "natural selection favors (or discriminates against) genes or genotypes indirectly through the phenotypes (individuals) that they produce" and that "individuals (and not genes) are the target of natural selection" (Mayr 1970, p. 108 and vii).

While defining selection in terms of its consequences, Mayr (1970) as well as Wright (1949, 1980) implicitly assume that selection is the process which modifies the survival and reproduction of organisms carrying different genes and genotypes. We therefore divide genotypic selection (S_G), earlier specified in terms of its consequences (Fig. 2), into two sub-processes (Fig. 3). First, *ontogeny* comprises the sub-process (O_G) whereby the phenotypes of individual zygotes develop under the control of their genotypes and epigenetic environments, and which thus determines phenotypic differences between individual organisms. Second, *organismic selection* (S_p), in which zygotes are functionally tested throughout their life cycles, determines how phenotypic variation between individuals influences their success in survival and reproduction (Tuomi 1982).

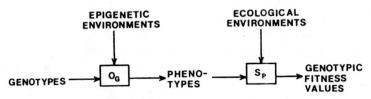

Fig. 3. Ontogeny (O_G) and organismic selection (S_p), which together determine genotypic fitness values as a function of organism-environment interactions, are the major components of genotypic selection (S_G, Fig. 2)

Table 1. Levels of natural selection (symbols as in Figs. 2 and 3)

Level	Process	Outcome
1. Organismic selection	S_P	Non-random differences in survival and/or reproductive rates of individual organisms
2. Genotypic selection	$S_G = (O_G, S_P)$	Non-random differences in the average contribution that the carriers of given genotypes make to the next generation (genotypic fitness values)
3. Genic selection	$S_g = (T_F, S_G, T_g)$	(a) Non-random differences in the rates of spread of gene alleles from one generation to the next (genic fitness values) (b) Non-random change of gene frequency from one generation to the next (selection pressure)

As now

$$S_G = (O_G, S_P)$$

genotypic selection depends on (1) how genotypic variation is mapped to the phenotypic level, and (2) how this phenotypic variation affects the survival and reproduction of organisms under given environmental conditions (Fig. 3). In other words, genotypic fitness values quantify the intensity of selection, which arises as the combined product of ontogeny and organismic selection. As the intensity of selection is thus measured by the outcomes of genotypic selection, it is important to make a clear distinction between the selection processes and their consequences (Table 1).

This unifying model of the selection process implies a division of natural selection into three interacting levels. First, *gene alleles and gene complexes* are the units (replicators, sensu Hull 1980) that carry the effects of natural selection from one generation to the next. Second, *genotypes* comprise an intermediate level, where genes are combined into functionally coherent genetic systems. Third, *phenotypes*, or rather whole organisms, form the level (interactors, sensu Hull 1980) at which the functional coherence of genetic systems is tested in phenotype-environment interactions. Consequently, it is this organismic selection which actually determines whether selection will favor the carriers of a given gene or not.

7 External and Internal Components of Organismic Selection

Organismic selection, or the struggle for existence, is a continuous process, in which each separate individual organism, as well as its individual structural, physiological and behavioural features are functionally tested in their "infinitely complex relations to other organic beings and to external nature" (Darwin 1859, p. 61; Mason and Langenheim 1961; Tuomi 1982). However, organisms are not mere passive participants in this testing process; their own phenotypic traits and behaviour can influence the frequency with which they happen to meet favourable and unfavourable extrinsic

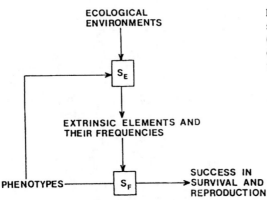

ECOLOGICAL
ENVIRONMENTS

S_E

EXTRINSIC ELEMENTS AND
THEIR FREQUENCIES

PHENOTYPES————S_F

SUCCESS IN
→SURVIVAL AND
REPRODUCTION

Fig. 4. A sub-division of organismic selection (S_P, Fig. 3) into interactions (S_E) which determine frequencies of extrinsic elements and those (S_F) which test functional capabilities of organisms

conditions. As the success of organisms in their struggle depends both on the frequency and on the functional effects of such environmental challenges (Haukioja 1982), organismic selection can be sub-divided into (1) the phenotype-environment relations (S_E) that influence the frequency of interactions with specific extrinsic elements, and (2) those relations (S_F) that influence the maintenance and reproduction functions of organisms. Therefore, organismic selection (S_P, in Fig. 3) can be redefined as

$$S_P = (S_E, S_F) \,,$$

where S_E and S_F together determine the success that organisms have in their survival and offspring production (Fig. 4).

This sub-division of organismic selection implies that there should be a relatively high selection pressure for traits which increase the frequency of favourable extrinsic conditions and decrease the probability of unfavourable conditions (S_E, Fig. 4). This is just the case in some conspicuous examples of adaptation, involved in, for example, cryptic and aposematic colouration, seed dispersal and insect pollination in plants, as well as habitat selection, food selection, predator avoidance or sexual selection in animals (e.g. Thoday 1953; Wake et al. 1983). Thus, a number of adaptations have evolved for selecting the external elements in relation to which organisms themselves are functionally tested. This explains why organisms seem to be so well adapted to their ecological niches; they select such a sub-set of the ecological environments, which is favourable for their own maintenance and reproduction functions.

The *functional selection* of organisms (S_F, Fig. 4) concerns the ability of organisms to cope with the environmental challenges that they happen to meet during a given time interval. This ability depends on (1) how well various phenotypic traits are co-adapted to support the maintenance and reproduction of organisms (internal functional selection), and (2) how well their phenotypes as a whole are adapted to the external environment (external functional selection). Consequently, phenotypic traits may increase or decrease the functional capabilities of their carriers, and these functional effects may in turn modify the success that their carriers will have in their survival and offspring production (Fig. 4). Since the maintenance and reproduction capacity of organisms essentially depends on the functional compatibility of their phenotypic traits (Bock 1980), these internal interactions may be equally or even more important driving forces of organismic selection than purely external interactions.

Organismic selection is always constrained by the internal compatibility of phenotypic traits, and therefore constraints (Gould and Lewontin 1979; Bonner 1982; Maynard Smith et al. 1985) may be considered as integrated causal elements of the selection process. If organismic selection itself can thus lead to different evolutionary pathways depending on the basic morphological and physiological features of organisms, phenotypic evolutionary changes may be relatively independent of the extrinsic environments (Wagner 1985). Accordingly, any evolutionary change in one phenotypic trait will alter the internal context for the selection of other traits, so that selection leads to the co-evolution of functionally constrained phenotypic traits (Wagner 1984). Especially irreversibility in phenotypic evolution can be expected to depend more on the internal than on the external components of natural selection.

Acknowledgements. We are grateful to Matt Ayres, Erkki Haukioja and Stephen C. Stearns for useful suggestions on earlier drafts of the manuscript, and to Terttu Laurikainen for drawing the figures. The work has been supported by The Academy of Finland.

References

Bock WJ (1980) The definition and recognition of biological adaptation. Am Zool 20:217–227

Bonner JT (ed) (1982) Evolution and development. Dahlem Konferenzen 1981. Springer, Berlin Heidelberg New York

Crow JF, Kimura M (1970) An introduction to population genetics theory. Harper & Row, New York

Darlington PJ, Jr (1983) Evolution: questions for the modern theory. Proc Natl Acad Sci USA 80:1960–1963

Darwin C (1859) On the origin of species by means of natural selection or the preservation of favoured races in the struggle for life. Murray, London

Dawkins R (1976) The selfish gene. Oxford Univ Press

Dawkins R (1982) The extended phenotype. The gene as the unit of selection. Freeman, Oxford

Dobzhansky Th (1951) Genetics and the origin of species, 3rd edn. Columbia Univ Press, New York

Dobzhansky Th (1970) Genetics of the evolutionary process. Columbia Univ Press, New York

Dunbar RIM (1982) Adaptation fitness and the evolutionary tautology. In: King's College Sociobiology Group (ed) Current problems in sociobiology. Cambridge Univ Press, pp 9–28

Fisher RA (1930) The genetical theory of natural selection. Clarendon, Oxford

Gould SJ, Lewontin RC (1979) The spandrels of San Marco and the Panglossian paradigm: a critique of the adaptationist programme. Proc R Soc London Ser B 205:581–598

Grene M (1961) Statistics and selection. Br J Philos Sci 12:25–42

Haldane JBS (1924) A mathematical theory of natural selection and artificial selection. Part 1. Trans Cambridge Philos Soc 23:14–19

Hamilton WD (1964) The genetical theory of social behavior. J Theor Biol 7:1–52

Haukioja E (1982) Are individuals really subordinated to genes? A theory of living entities. J Theor Biol 99:357–375

Hull DL (1980) Individuality and selection. Annu Rev Ecol Syst 11:311–332

Huxley J (1963) Evolution, the modern synthesis, 2nd edn. Allen Unwin, London

Kojima K (1971) Is there a constant fitness for a given genotype? No! Evolution 25:281–285

Lerner IM (1958) The genetic basis of selection. Wiley & Son, New York

Lerner IM (1959) The concept of natural selection: a centennial view. Proc Am Philos Soc 103: 173–182

Lewontin RC (1970) The units of selection. Annu Rev Ecol Syst 1:1–18

Lewontin RC (1974) The genetic basis of evolutionary change. Columbia Univ Press, New York

Mason HL, Langenheim JH (1961) Natural selection as an ecological concept. Ecology 42:158–165

Maynard Smith J (1969) The status of neo-Darwinism. In: Waddington CH (ed) Towards a theoretical biology, vol 2. Aldine, New York, pp 82–89

Maynard Smith J, Burian R, Kauffman S, Alberch P, Campbell J, Goodwin B, Lande R, Raup D, Wolpert L (1985) Developmental constraints and evolution. Q Rev Biol 60:265–287

Mayr E (1955) Integration of genotypes; synthesis. Cold Spring Harbor Symp Quant Biol 20:327–333

Mayr E (1959) Where are we? Cold Spring Harbor Symp Quant Biol 24:1–13

Mayr E (1970) Populations species and evolution. Belknap, Cambridge, Mass

Mayr E (1982) Adaptation and selection. Biol Zentralbl 101:161–174

Michod RE (1981) Positive heuristics in evolutionary biology. Br J Philos Sci 32:1–36

Muller HJ (1949) The Darwinian and modern conceptions of natural selection. Proc Am Philos Soc 93:459–470

Simpson GG (1949) The meaning of evolution. Yale Univ Press, New Haven

Simpson GG (1953) The major features of evolution. Columbian Univ Press, New York

Sober E, Lewontin RC (1982) Artifact, cause and genic selection. Philos Sci 49:157–180

Stearns SC (1982) On fitness. In: Mossakowski D, Roth G (eds) Environmental adaptation and evolution. Fischer, Stuttgart, pp 3–17

Templeton AR (1982) Adaptation and the integration of evolutionary forces. In: Milkman R (ed) Perspectives on evolution. Sinauer, Sunderland, Mass, pp 15–31

Thoday JM (1953) Components of fitness. Symp Soc Exp Biol 7:96–113

Tuomi J (1982) Evolutionary theory and life-history evolution: the role of selection and the concept of individual organism. Rep Dep Biol, Univ Turku 4/1982

Wagner GP (1984) Coevolution of functionally constrained characters: prerequisites for adaptive versatility. Biosystems 17:51–55

Wagner GP (1985) Über die populationsgenetischen Grundlagen einer Systemtheorie der Evolution. In: Ott JA, Wagner GP, Wuketits FM (eds) Evolution, Ordnung und Erkenntnis. Parey, Berlin Hamburg, pp 97–111

Wake DB, Roth G, Wake MH (1983) On the problem of stasis in organismal evolution. J Theor Biol 101:211–224

Wallace B, Srb AM (1961) Adaptation. Prentice-Hall, Englewood Cliffs, NJ

Williams G (1966) Adaptation and natural selection. Princeton Univ Press

Williams M (1970) Deducting the consequences of selection: a mathematical model. J Theor Biol 29:343–385

Wright S (1931) Evolution in Mendelian populations. Genetics 16:97–159

Wright S (1949) Adaptation and selection. In: Jepsen GL, Mayr E, Simpson GG (eds) Genetics, paleontology and evolution. Princeton Univ Press, pp 365–389

Wright S (1980) Genic and organismic selection. Evolution 34:825–843

The Genetics of Information and the Evolution of Avatars

P.-H. GOUYON[1,2] and C. GLIDDON[1,3]

1 The Problem

Evolution requires that heritable changes in organisms occur over a period of time. In his seminal work "The Origin of Species", Darwin (1859) suggested that the mechanism by which such evolutionary changes took place was natural selection which favoured individuals or groups. He stated "individuals having any advantage, however slight, over others, would have the best chance of surviving and procreating their kind." The problem which has confronted evolutionary biologists in their attempts to interpret Darwin's work is what is the precise definition of "their kind". The importance of what is transmitted (i.e. information) from one generation to the next had been noted by biologists such as Weissman (the id) and the rules of its transmission, from one generation to the next by means of sexual reproduction, had been defined by Gregor Mendel. The synthesis of these observations with the mechanism proposed by Darwin has formed the main thrust of evolutionary biology in the 20th century. Neo-Darwinism is an attempt to link the transmission of genetic information with the evolution of individuals, groups and, more recently, molecules.

Several of the major conceptual problems which have arisen in neo-Darwinism are the result of the failure to distinguish between genetic information and its material support. Recently, several biologists and philosophers have become aware that such a problem exists (see e.g. Sober 1984) and have tried to distinguish units which directly reproduce their structure (replicators, e.g. genes) from units which interact directly with their environment (interactors, Hull 1984, or vehicles, Dawkins 1984, e.g. individuals). This distinction, although useful, is still not sufficiently radical to allow discrimination between information and its material form. For example, is a gene defined by a portion of a nucleic acid molecule or by the information carried by this molecule? Or, to consider a more complex biological level, is an individual defined as a set of molecules interacting with their environment or as the information which that individual carries and by which it was produced? It would certainly appear preferable

1 Centre d'Etudes Phytosociologiques et Ecologiques, CNRS, Route de Mende, BP 5051, F-34033 Montpellier Cedex, France
2 I.N.A.P.G., 16, Rue Claude Bernard, F-75231 Paris Cedex 05, France
3 School of Plant Biology, University College of North Wales, Bangor, Gwynedd, Wales

Population Genetics and Evolution
G. de Jong (ed.)
© Springer-Verlag Berlin Heidelberg 1988

to most biologists to define such entities in terms of their material form. The problem, in this case, is that the material form is always ephemeral and never transmitted in terms of evolution; even the gene, taken as a length of nucleic acid, will vanish over time, due to the semi-conservative nature of nucleic acid replication. The confusion of information with its material support, coupled with the difficulties of dealing simultaneously with different levels of biological integration, has led inexorably to endless debate about what is the unit of selection, as has been pointed out by Brandon (1984). For instance, scientists concerned with entities which are maintained over time have felt obliged to defend the gene as the unit of selection (e.g. Williams 1966; Dawkins 1976), whereas those who took a more ecological viewpoint (e.g. Mayr 1970) have insisted that the individual, as opposed to the gene, is the unit of evolution.

It is clearly necessary to discriminate between genetic information and its material support as well as realizing that the rules which govern the transmission of genetic information are dependent on the nature of the material support. In the remainder of this chapter, *any* material support which allows the replication of genetic information will be termed an *avatar*, which is used in the Hindu religion to refer to the material forms assumed by the god Vishnu. Damuth (1985) has used the term avatar to mean the unit of selection within a community and between species. He states: „Just as populations evolve by organismic selection, communities evolve by avatar selection, and more inclusive units, the higher level analogues of the species, evolve as their component communities do. This formulation of higher level selection reveals a congruence with processes at the lower, organism-based level and suggests the most profitable direction to be taken in attempts at formal extension of selection theory." The redefinition of avatar used in this chapter is the result of pursuing such a "profitable direction".

2 The Future Synthesis

Life, the Universe and Everything can be represented, somewhat simplistically, by the scheme given in Fig. 1. Here, the information which is replicated and its corresponding avatar are presented for a number of different levels of biological integration. It is interesting to note that with the exception of the level of the gene, the biological terms describe the avatar rather than its corresponding information. The species here is represented as an avatar, because this is clearly the way in which taxonomists use the term, i.e. a set of physical characteristics of organisms. Evolutionary biologists, however, will often use species in the sense of a collection of genetic information and, in this case, the term does not describe an avatar (see e.g. Damuth 1985).

The distinction between the information and the avatar allows much clearer evolutionary questions to be phrased, since it shows that avatars do not reproduce. That is, *only information can be the target of selection but this does not imply that only the gene is the unit of selection* since information exists at all levels of biological integration. This distinction becomes particularly necessary when different types of genetic information co-occur at the same level. For instance, when studying a trait such as male sterility in plants, which is usually determined by both nuclear (equally inherited through male and female functions), and cytoplasmic (transmitted through the female

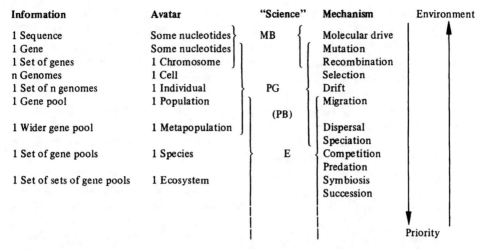

Information	Avatar	"Science"	Mechanism	Environment
1 Sequence	Some nucleotides	MB	Molecular drive	
1 Gene	Some nucleotides		Mutation	
1 Set of genes	1 Chromosome		Recombination	
n Genomes	1 Cell		Selection	
1 Set of n genomes	1 Individual	PG	Drift	
1 Gene pool	1 Population		Migration	
		(PB)		
1 Wider gene pool	1 Metapopulation		Dispersal	
			Speciation	
1 Set of gene pools	1 Species	E	Competition	
			Predation	
1 Set of sets of gene pools	1 Ecosystem		Symbiosis	
			Succession	
				Priority

Fig. 1. Life, the Universe and Everything. The hierarchical organization is incomplete, however, the separation of information from avatar is absolute. *MB* Molecular biology; *PG* population genetics; *PB* population biology; and *E* ecology. For priority and environment, see text

function only) genetic information, it is impossible to define the fitness of an individual. Consider a species where females (male-steriles) and hermaphrodites (male-fertiles) co-exist, the females producing 50% more seeds than the hermaphrodites. It is clear that the cytoplasmic genes in a female have a higher fitness than those in a hermaphrodite (females produce more ovules). However, nuclear genes in a female have a lower fitness than those in a hermaphrodite (Lewis 1941, has shown that the minimum increase in ovule production which could compensate for the loss of male function is more than two). The only reason why evolutionists have long thought that they could define the fitness of an individual in terms of numbers of gametes, irrespective of the kind of genetic information carried by those gametes, is that they have always implicitly assumed that male and female gametes were equivalent, i.e. that they were dealing with nuclear genes. This pure avatar vision of reproduction has also had a very deleterious effect on botanists who have become accustomed to measure the fitness of hermaphroditic plants by counting the number of seeds which they produce (not because they were implicitly assuming a cytoplasmic genetic determination of the studied trait but because they were forgetting that if the *individual* seems, at first sight, to be reproduced by seeds, its nuclear genome is transmitted equally through pollen).

Given that information is the target of selection and that the avatar is the means whereby it is replicated, it now becomes possible to attack the problems posed by the different levels of integration which exist within all biological systems. Selective and evolutionary events may occur, more or less independently, at different levels in the biological hierarchy. Some levels in the hierarchy are relatively easy to integrate. For example, although only recently discovered, transposons and repeated DNA sequences are already considered to be selected simultaneously by molecular processes within the nucleus and by classical selective events at the level of the individual (Charlesworth and Langley 1986). Another example in which selection occurs at four different levels in the hierarchy is the case of the T-locus in mice. In this case, selection has been suggested to be acting at the level of the gene, the chromosome, the individual and the population (see e.g. Lewontin 1962). The most general approach which considered the effects of selection acting at different levels in the hierarchy is the fundamental theorem of natural selection (Fisher 1930) in which it is shown that selection acting at the level of the individual will, given certain conditions, cause a continual increase in the fitness of the population. Whether this theorem is of general applicability or not, it illustrates the type of integrative approach which is necessary.

A peculiarity of the organization of biological hierarchies is that the avatars at a given level represent the environment of those at a less integrated (= lower) level (see Fig. 1). However difficult it may be, in order to completely explain a trait, it may be necessary to take all of the levels, together with their specific (emergent) properties, into account. Usually, a "sufficient" explanation will be provided by the study of "relevant" levels, but problems still remain concerning the objective definitions of "sufficient" and "relevant". The time constants often differ considerably at different levels in the hierarchy, i.e. there is a type of priority (see Fig. 1) with the shortest time constants being associated with the lower levels of the hierarchy. In order to become fixed at a particular level, a trait must first have been successful at all lower levels in the hierarchy and, in order to be maintained, it must not be eliminated at any higher level. It should be noted, however, that the differences in time constants previously referred to may allow the higher levels to be ignored. That is to say that time constants allow an effective uncoupling, at least for the purposes of analysis, of the different levels. For instance, the maintenance of sexual reproduction, in a given species, could be determined by forces which are peculiar to that species. For example, in aphids, the existence of eggs may be necessary for protection against frost. However, the outcome of this interindividual selection will have an effect at the level of the species. That is, those species which have lost sexual reproduction are more likely to become extinct than those which have not. This effect at the species level could well explain why the majority of extant species possess sexual reproduction. For such a trait, when examined at two different levels, there could be said to be both proximate and ultimate causes. It is interesting to note that proximate causes are quite fashionable at the moment in biology. Given the above scenario, it should not be surprising to find many different, species-specific, proximate causes for traits where there may be only one ultimate cause.

The New Synthesis, or neo-Darwinism, has two distinctive features, firstly, it attempts to study the transmission of genetic information from an evolutionary viewpoint and secondly, it emphasizes the population level. If the second feature can be broadened to include other levels of integration then one can expect neo-Darwinism to provide the final solution.

Acknowledgements. We would like to express our thanks to the "Réunions du Mardi" in Montpellier, during which most of this article was debated and to John Endler for his helpful criticism and advice.

References

Brandon RN (1984) The levels of selection. In: Brandon RN, Burian RM (eds) Genes, organisms, populations. Cambridge, Mass, pp 133–141

Charlesworth B, Langley CH (1986) The evolution of self-regulated transposition of transposable elements. Genetics 112:359–383

Damuth J (1985) Selection among "species": a formulation in terms of natural functional units. Evolution 39:1132–1146

Darwin C (1859) On the origin of species by means of natural selection, 1st edn. Murray, London

Dawkins RC (1976) The selfish gene. Oxford Univ Press

Dawkins RC (1984) Replicators and vehicles. In: Brandon RN, Burian RM (eds) Genes, organisms, populations. MIT, Cambridge, Mass, pp 161–180

Fisher RA (1930) The genetical theory of natural selection. Oliver & Boyd, Edinburgh

Hull D (1984) Units of evolution: a metaphysical essay. In: Brandon RN, Burian RM (eds) Genes, organisms, populations. MIT, Cambridge, Mass, pp 142–160

Lewis D (1941) Male sterility in natural populations of hermaphrodite plants. New Phytol 40: 56–63

Lewontin RC (1962) Interdeme selection controlling a polymorphism in the house mouse. Am Nat 96:65–78

Mayr E (1970) Populations, species, and evolution. Harvard Univ Press, Cambridge, Mass

Sober E (ed) (1984) Conceptual issues in evolutionary biology – an anthology. MIT, Cambridge, Mass

Williams GC (1966) Adaptation and natural selection. Princeton Univ Press

Sib Competition as an Element of Genotype-Environment Interaction for Body Size in the Great Tit

A. J. van Noordwijk[1]

1 Introduction

Final body size results when growth stops. In order to gain insight in the genotype-environment interactions affecting adult body size one must therefore study the growth process.

Growth has been studied in many different ways. With birds the principal ways are hand rearing on known diets, detailed descriptive field studies or studying the effects on growth patterns of manipulations carried out in field experiments or in lab studies. Using data from a descriptive field study, augmented by field experiments, I have been intrigued by apparent differences between fully grown nestlings and adults in the extent to which variation in body size is genetic (van Noordwijk 1984, 1986; van Noordwijk et al. 1987).

These results can briefly be described as showing a high heritability (60%) for body size of adults and for fully grown nestlings when feeding conditions during the breeding season were good. In contrast, no heritability for fledging size is observed when feeding conditions were poor. In the latter case there is fairly strong selection on the environmental component in the phenotype after fledging (van Noordwijk 1986). This leads to the idea that there is genetic variation for the size at which growth stops. It is further suggested that the time available for growth is limited, which leads to the observation that under poor feeding conditions the genetically determined final size is not reached. I here report some consequences of these ideas about nestling growth that were investigated with simulations based on growth curves.

The family of growth curves described by Richards (1959), of which logistic and von Bertalannfy curves are special cases, contains an error term. White and Brisbin (1980) have proposed a biologically realistic interpretation of this error term in order to get the best estimate of a growth curve. Considering the usual quantitative genetic model of a genotypic value together with an environmental deviation, it is easy to interpret the error term in the Richards growth curves as representing the environmental deviation from the genotypic value. This provides us with a way of parameterizing the environmental effects on growth in relation to genetic variation for body

1 Zoologisches Institut, Rheinsprung 9, CH-4501 Basel, Switzerland

Population Genetics and Evolution
G. de Jong (ed.)
© Springer Verlag Berlin Heidelberg 1988

size. The explicit interpretation of the environmental deviation in the growth curve allows us, in cases where the environmental conditions are known or can be predicted, to combine these environmental effects with the growth characteristics under ideal conditions, to obtain a prediction of the resulting final size.

Of particular interest are genotype-environment interactions, i.e. where different genotypes respond differently to changes in environmental conditions, or from an observational point of view, where the effect of genotype and the effect of environment on the phenotype are not additive. In some cases, additivity of genetic and environmental effects could be restored using an appropriate transformation, but in other cases this may not be possible, for example where the genetic variation does not come to expression under particular environmental conditions that nevertheless lead to phenotypic values in the normal range.

In this chapter I will first review the evidence suggesting that final size rather than growth rate has a genetic component. Next, a re-interpretation of terms in the Richards growth curves will be formulated, designed to investigate these problems. Simulation results will be discussed to evaluate the likely advances in understanding genotype-environment interactions. Special attention will be given to the role of sib competition as a process that causes deviations from optimal growth under conditions of limited food availability.

2 Growth in Nestling Great Tits

Nestling great tits fledge about 20 days after hatching. In first clutches all eggs normally hatch within 24 h, but in later broods, repeat or second clutches, hatching may be spread out over several days. Nevertheless, in all types of broods all nestlings fledge together. The three structures that are frequently measured as indices of body size grow at different rates. Adult tarsus length is already reached 11–12 days post-hatching. The fledging weight, reached by day 15 is often equal to the subsequent adult weight. Full wing length, however, is reached only after fledging (O'Connor 1977; Garnett 1981; van Noordwijk et al. 1987). Whereas tarsus length remains constant after completion of growth, weight may change considerably depending on conditions (e.g. van Balen 1967, 1980) and the wing length is constant between moulting seasons (late summer). Thus, in principle the three measures have undergone different lengths of juvenile environmental influences and they also reflect current environmental conditions differently. In the simulations weight will be used, but apart from scaling and timing differences, the other size parameters could equally well have been used.

3 Genetic Variation in Asymptotic Size

As mentioned in the introduction, the working hypothesis is that the asymptotic size, rather than growth rate, has a genetic component, while growth can only occur during a limited period in juveniles. The following observations have led to this hypothesis: (1) There is a positive correlation between estimated heritability of fledgling weights and mean fledgling size and there are negative correlations between mean size and phenotypic variance and between heritability and phenotypic variance. A decrease in heritability would result from additional environmental variance with equal genetic variance, but the observed decline in heritability is stronger than expected on this basis. (2) Nestlings that are far smaller than their parents have been observed and they did not show any subsequent compensatory growth for tarsus length and incomplete compensatory growth for weight. In these data there was no resemblance between parents and offspring (van Noordwijk et al. 1987). (3) The second observation implies that the phenotypic value of the parents is generally close to the genotypic value, in contrast to the offspring. This may indeed be so, because the survival after fledging of individuals that are far smaller than expected on the basis of the size of their parents is very low. Therefore, the phenotypic size of the parents is much closer to their genetic value.

This hypothesis has led to experiments involving partial cross-fostering and manipulation of brood sizes. The aim of the cross-fostering is to exclude a resemblance between parents and offspring that is caused by parental care. Brood size manipulation is the easiest way to change the amount of food that individual nestlings receive, unless the parents can fully satisfy the increased demand in larger broods. This can be checked by comparing mean nestling weights between brood sizes. The experiment allows us to compare full sibs that were brought up in different brood sizes and genetically unrelated individuals that were brought up together. It was predicted that heritability estimates for fledgling size should be lower for nestlings raised in enlarged broods than for nestlings raised in reduced broods.

The results from a small pilot experiment pointed in this direction (van Noordwijk 1984). First analyses of a simplified form of this experiment carried out in 1986, show a small insignificant difference in h^2 for cross-fostered offspring for both weight and tarsus length which goes in the predicted direction (Henrich and van Noordwijk, in prep.). For the offspring raised by their biological parents there is an insignificant difference the other way around, but this is difficult to interpret, since there is also clear evidence for non-genetic components in parent-offspring resemblance in these data, especially for weight. Henrich and van Noordwijk (in prep.) found a significant positive regression of nestling weights on the weights of the foster parents in enlarged broods. When winter weights were used instead of the parental weights during the breeding season, this resemblance disappeared completely, but the resemblance between nestlings and their biological parents remained. The experiment will be continued in 1987 and 1988 to obtain the sample size required for detecting a 0.2 difference in heritability as statistically significant.

For the moment I regard this description of genotype environment interaction as an interesting, but insufficiently corroborated hypothesis. This hypothesis will be used, however, in the subsequent simulation studies, because it provides a useful framework.

4 Problems in the Analysis of Nestling Growth

It is easy to collect data on nestling growth. One can visit nests at regular time intervals and measure the nestlings. Apart from the disturbance this might cause to the

normal feeding pattern, there are few problems in the data collection. The interpretation of these data, however, and relating these data to the underlying processes is
much more difficult. This is because variation in intrinsic growth properties, depending on either age or size, and variation in parental feeding rate depending on food
availability, foraging conditions or intrinsic factors are difficult to separate. Moreover,
a substantial part of the food brought to the nest is needed to maintain body temperature. The food available for growth therefore depends on ambient temperature,
female brooding behaviour and brood size (Mertens 1969; Royama 1966). The relation between the amount of food brought to the nest and the food available for growth
is therefore complex. On top of these sources of variation, the distribution of food
items over the nestlings may be uneven, especially when food is scarce. Any analysis
of growth makes it necessary that assumptions are made about all these points, be it
implicitly or explicitly.

4.1 Growth Curves

Richards (1959) formulated a family of growth curves, in which the commonly used
ones are comprised as special cases. White and Brisbin (1980) considered the estimation of these curves in view of the lack of independence of the deviations in subsequent observations. Their approach is given by the following formula for size specific
growth:

$$\frac{\Delta W}{\Delta t} = \frac{2\,(m+1)}{T\,(m-1)}\,(W_\infty^{1-m}\,W_t^m - W_t) + e_t \;.$$

W_t is the weight at time t, W_∞ is the asymptotic weight, m is a shape parameter and
T is a timing parameter that determines the length of the growing period. White and
Brisbin consider situations with ad lib food, but nevertheless observe serial correla-
tions in the deviations e_t.

With fluctuating food availability one may interpret the error term as an environmental deviation (plus an error term). It is reasonable that nestling growth follows
the growth curve when sufficient food is available, but falls short in case of under-
nourishment. This can be modelled by introducing a proportion R of the growth under
optimal conditions that can be realized. One may then consider values for R that are
slightly less than zero for the case of complete starvation. In principle, the R-values
are measurable from food intake rates. Our growth formula now becomes:

$$\frac{\Delta W}{\Delta t} = \frac{2\,(m+1)}{T\,(m-1)}\,(W_\infty^{1-m}\,W_t^m - W_t) * R_t \;.$$

Using this formulation we can reconsider the phenomenon of compensatory growth,
the observation that growth increase per time is greater at the same age after a previous
reduction in growth rate. Some compensatory growth simply results from the shape
of the growth curve. It is conceivable that on top of that individuals react by changing the shape of their growth curve in a way that may be approximated by R-values
greater than one. I would describe the latter case as *true compensatory growth* in
contrast to the *apparent compensatory growth* resulting from the size dependence of

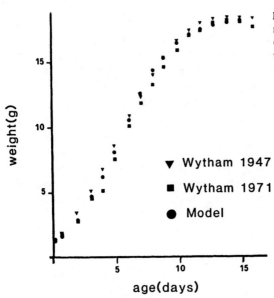

Fig. 1. The growth curves of well-fed nestling great tits from data in Gibb (1950) and Schifferli (1973) together with the basic growth equation used in this study

growth as given by the growth curve. I will not consider the possible occurrence of true compensatory growth.

Let us assume that we can determine the growth curve under ideal conditions and that we can describe the observed growth curve as a series of R-values which indicate the environmental conditions for each nestling. In Fig. 1, two series of daily mean sizes of nestling great tit weights from the literature (Gibb 1950; Schifferli 1973) are given together with the growth curve with the parameters $W_0 = 1.5$ (g), $W_\infty = 18.0$ (g), $T = 13.0$ (days), $m = 1.7$ and $R_t = 1.0$ for all t that was used as a basis for the reported simulations. One can now analyze the effects of growing up in broods in terms of relations between the R-values of nestlings in the same brood or in different broods but in the same period, and thereby presumably undergoing the same gross effects of weather on food availability. The patterns in the R-values can also be analyzed in terms of the resulting relations between the mean weight in a brood and the intra-brood variances.

There are several a priori possibilities:

1. There are strong correlations in R between individuals raised in the same brood. This is to be expected from the sharing of food between the nestlings, but may be counteracted by competition between the nestlings leading to:
2. No correlation in R-values, but with even stronger interactions between the nestlings, also to:
3. Negative correlations in R-values. In these cases it is assumed that the processes leading to these values are always active, if not then:
4. Variable correlations in R-values, depending on e.g. initial size variation between the nestlings. This is likely to occur to some degree.

4.2 Growth Curve Simulations

A first question to be answered qualitatively is whether observed features of growth curves of nestlings can be obtained with strongly correlated environmental factors, but without any interactions between nestlings. For these simulations one needs a series of R-values (per nestling, per day) that are not too far below 1 and are correlated from day to day and between nestlings. This was realized by multiplying the daily growth with a fraction R_t which in turn is the product of a deterministic component RS_t and a random component RR_t. RS_t is given as a time series and

$$RR_t = (1 - S * \sqrt{p \cdot q \cdot r})$$

where p, q and r are random numbers uniformly distributed between 0 and 1, and where S is a scaling factor to change the amount of stochasticity. For S = 1.0 this RR_t has values between 0 and 1, but is skewed with a mode around 0.8. This represents the fact that most days are fairly good. In these simulations p is constant for 5 days and for all nestlings and q changes daily but is the same for all nestlings and r is unique for every nestling and every day. The result from a typical run under poor conditions is given in Fig. 2. In comparison with real data, the variance is far too small for the fairly low mean sizes at ages above 5 days.

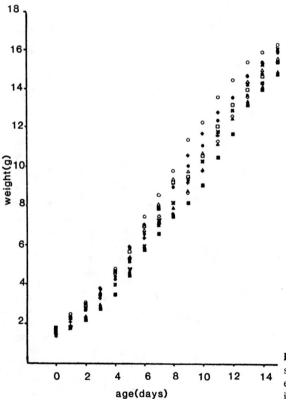

Fig. 2. The growth curves of ten simulated nestlings under fairly poor environmental conditions without interactions between the nestlings

To simulate the effects of competition for food between nestlings, one may assume
the following behaviour shown by nestlings: If you are hungry raise your neck when
you hear a parent coming; the hungrier you are, the higher you should try to raise
your neck. The parent simply feeds the nearest beak. Without going into the details
of nestling behaviour, this description is reasonably close to real behaviour (Löhrl
1968; Bengtsson and Ryden 1983). It leads to a situation where a combination of
size and motivation determines which nestlings will be fed on a particular feeding
visit. This can be modelled by multiplying the factor R_t (i.e. the proportion of maximum growth that is realized on a particular day) with the relative size of the nestling
(i.e. its own size divided by the mean nestling size) with a maximum of 1.0 for the
product. Thus, under poor conditions the whole brood will suffer equally if all nestlings have the same size, whereas in a brood with variable size the large nestlings will
hardly suffer and the smaller ones all the more.

This model is fully capable of showing runt individuals developing. These may
catch up under good feeding conditions but are doomed under poor conditions. A
run with sib competition is shown in Fig. 3.

Good and poor conditions were implemented in two ways, by a scaling factor on
the random component (S = 0.3 for good and S = 1.0 for poor conditions) and additionally by a time series of multipliers (RS_t) for the increases in size (equal to 1.0
for good conditions and 5 days with 0.7 followed by 1.0 for poor conditions). In a

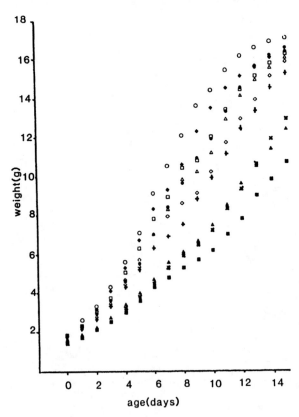

Fig. 3. The growth curves of a simulated brood with ten nestling under
the same conditions as in Fig. 2,
but with the addition of sib competition. The same string of pseudo-
random numbers was used in both
cases, and the symbols correspond.
Note that the increase in variance is
due to the larger nestlings growing
faster as well as the smaller nestlings
growing more slowly. The effect of
nestling competition is especially
strong on individuals that are
genetically large but hatch from a
small egg (e.g. x is the second big-
gest in Fig. 2 and in eighth place
here)

few initial simulations a single asymptotic weight was used for all individuals in a brood. This did not produce satisfactory results.

In the simulations presented in Fig. 2 through 7, the initial sizes of each of ten individuals in a brood were drawn from a uniform distribution between 1.35 and 1.85 g, which is slightly larger than realistic for variation in egg size alone, but may be realistic if hatching spread is also taken into account; and the asymptotic sizes were drawn independently from a uniform distribution between 17.0 and 19.0 g. This spread of asymptotic sizes is roughly what would be expected due to Mendelian segregation within a brood. This spread in asymptotic sizes is, of course, a very simplistic way of representing the genetic variation for size.

Since the number of random draws per run is constant, the use of the same series of pseudo-random numbers allows a one to one matching of the nestlings in both runs. This allows one to see the effects of the systematically varied factors more clearly.

It is difficult to compare many runs in the form of complete growth curves. A convenient way of summarizing the growth curves is in a scatter plot of initial (egg) size and final (15-day) size. These graphs are given in Fig. 4 for both the runs in Figs. 2 and 3 plus two additional ones to have all four combinations of good and poor conditions and with or without sib competition. It is clear that especially under poor environmental conditions in the presence of sib competition a strong correlation between final and initial size arises. This can be shown most clearly if one plots the dif-

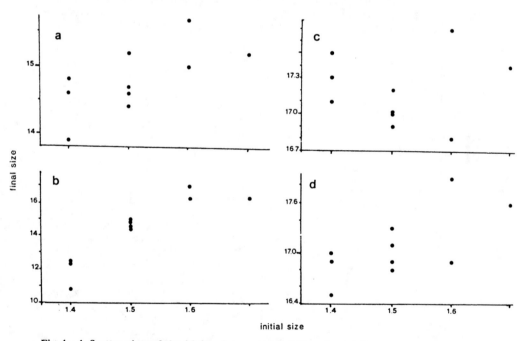

Fig. 4a–d. Scatter plots of the 15-day sizes against hatching sizes of the same individuals in four situations: in **a** and **b** under poor and in **c** and **d** under good environmental conditions. Sib competition is absent in **a** and **c** and present in **b** and **d**. The correlations are a r = 0.67*; b r = 0.89**; c r = 0.00; d r = 0.69*. See Fig. 7 for the correlations obtained in nine other runs (* 0.05 > P > 0.01, ** P < 0.01)

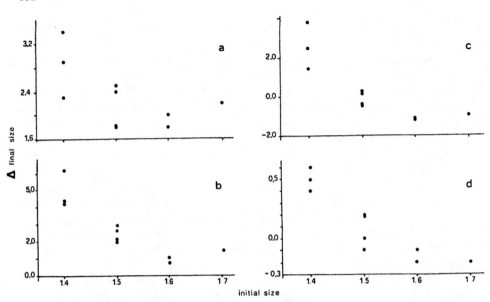

Fig. 5a–d. Scatter plots for the differences in final size from pairs of runs in Fig. 4 plotted against initial size. In a and b a comparison is made for the effect of good versus poor environmental conditions; in a without, in b with sib competition. In c and d a comparison is made for the effect of competition; in c under poor and in d under good environmental conditions. The correlation are: a r = –0.63; b r = –0.84**; c r = –0.85**; d r = –0.85** (* 0.05 > P > 0.01, ** P < 0.01)

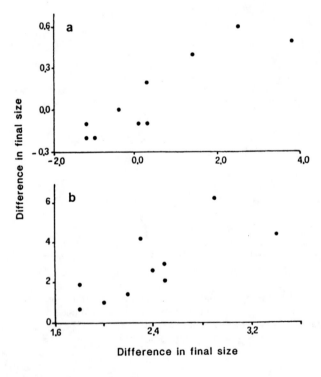

Fig. 6a,b. Scatter plots of differences in final size. In a the differences with and without competition are compared under poor (X, see Fig. 5c) and good (Y, see Fig. 5d) environmental conditions. In b the effect of environmental conditions is compared without (X, see Fig. 5a) and with (Y, see Fig. 5b) sib competition. The correlations are a r = 0.91**; b r = 0.76* (* 0.05 > P > 0.01, ** P < 0.01)

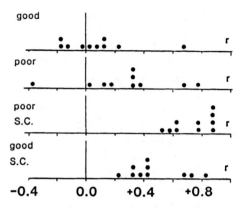

Fig. 7. The correlation coefficients between initial and final size obtained in ten runs including the one depicted in Fig. 4

ference in final size for the same individual in different runs against the initial size, as in Fig. 5. In Fig. 5a the difference between good and poor conditions, in the absence of sib competition is plotted. The range of this difference (from 1.7–3.4 g) is smaller than in the presence of competition (from 0.8–6.2 g). The correlation with initial size is also less clear in the absence of competition. In Fig. 5c and d, it can be seen that both under poor and good environmental conditions, the effect of competition is correlated with initial size, and that competition leads to higher final sizes for individuals with big initial sizes (negative differences).

An impression of the relative effects of competition and environmental conditions may be obtained from Fig. 6 where scatter plots of the vertical axes from Fig. 5 are given. It shows that the differences in final size due to competition (Fig. 6a) under both environments are more strongly correlated than the difference in final size due to environmental conditions (Fig. 6b) in the absence or presence of competition. These figures, using the differences between runs, are useful to gain insight in the relative importance of the processes involved. Apart from the correlation between initial and final size, they cannot be related to field data, however. In most cases, however, the correlation between initial and final sizes is sufficient to find out whether sib competition was important or not, if correlation coefficients can be obtained for several broods. This is illustrated in Fig. 7, where the correlation coefficients belonging to the scatter plots similar to Fig. 4 are given for ten sets of runs. It seems that sib competition always leads to positive correlations that get stronger when conditions get poorer. In the absence of sib competition, poor conditions sometimes cause positive correlations, but with a much wider scatter than in the presence of competition.

This suggests that the correlation between hatching and fledging size of individuals can be used as an index of the strength of sib competition, which will in turn depend on the environmental conditions. It further suggests that experiments in which the variation in initial size is manipulated should provide sensitive tests for the proposed mechanism of sib competition.

5 Further Experiments

This analysis of the role of sibling interaction as an element introducing non-equi-distant reaction norms for body size leads to several suggested experiments. It is ob-vious that increasing the within-brood variance in size at or shortly after hatching is predicted to have little effect when feeding conditions are good. Such conditions can be created experimentally by reducing brood size. The same manipulation should lead to a strong correlation between hatching and fledging size in enlarged broods. This can be done while cross-fostering nestlings, were differences in hatching of up to 2 days may be used to separate any effects from egg size per se from the effects of relative nestling size. Experiments along these lines to investigate the effects of hatch-ing synchrony have recently been performed by Slagsvold (1986) in the pied flycat-cher. He, however, did not consider genetic variation in growth properties, but he did find that the optimal degree of hatching synchrony was dependent on the initial clutch size and weather conditions.

Perhaps an even more elegant experiment can be done by manipulating hatching dates within a clutch. It seems feasible, and we will attempt this experiment in 1987 to remove eggs during laying and then to add them again to the same clutch after the onset of incubation, which should result in their hatching later. Van Balen (1973) found that when eggs are removed before the third egg is laid, this will lead to the production of a larger clutch. Especially for females for which a clutch size from the previous year is known, this could lead to a detectably larger clutch in which the ef-fects of the manipulation should be greater than in the normal sized or reduced clutch.

6 Discussion

The analysis in this chapter can be seen as a case study in integrated population bio-logy. It is not clear which elements are mainly genetic and which elements are main-ly ecological. We started with an analysis of genotype-environment interaction in body size and analyzed the growth of nestlings leading to these body sizes. The idea that competition between nestmates plays an important role in these interactions was then confirmed by simulation models. This model leads in turn to suggested ex-periments that should further elucidate the role of sib competition. In itself, the oc-currence of sib competition in nestling growth is a problem that one would classify as ecological. However, the notion of a genetically determined, individual final size, greatly facilitates the analysis of this competition. The notion of predetermined final sizes is a key element in using the correlation between initial and final size as an indi-cator for sib competition.

Hopefully, the suggested experiments will provide better insight in the ecological process of sib competition, but since sib competition is a key element in growth under poorer conditions, it will at the same time lead to a better understanding of the geno-type-environment interaction in body size. Further, these experiments could tell us about the strategy of hatching synchrony (see Slagsvold 1986) and could test the hypo-thesis that there are separate selection pressures for egg size and within-clutch varia-tion in egg size (van Noordwijk et al. 1981).

The exploration of the consequences of sib competition with this simulation model is by no means complete. The model provides a plausible mechanism that could explain the puzzling observations on the heritability of body size. It is feasible to investigate the consequences of genetic variation in the other parameters of the growth equation. This should lead to further critical experiments. This chapter is therefore a preliminary report on an attempt to bring genetic variation, the physiology of growth and the ecology of sib competition and variable resource availability together in a single framework. At a later stage, it is possible to include more genetics, to have the final size of individuals emerge from genes that are transmitted from the parents. As a first step, however, the notion of individual variation in parameters of the growth curves is sufficient to explore the consequences of genetic variation.

7 The Status of Neo-Darwinism

A true integration of physiological, ecological and genetic aspects is the area of evolutionary biology, where the least advances have been made since the formulation of the neo-Darwinian synthesis. Although mechanisms have received much attention at the molecular genetic level and are all-important in gene physiology, the study of mechanisms has been largely neglected at the level of population genetics. The ecophysiology of natural selection could be built upon existing work on the physiology of reproductive effort (e.g. Drent and Daan 1980; Tinbergen 1986), if the genetics of variation in reproductive effort can be incorporated, if the effects are studied over a lifetime and if the interactions between generations are included. It could alternatively start from analyses of variation in lifetime reproductive success in natural populations, if it is possible to include the ecophysiology and again the genetics.

Case studies in which both the genetics and the physiology have received ample attention are found in the development of resistance against pesticides or pollutants such as heavy metals (see e.g. Bishop and Cook 1981). In these cases a single, strong selection pressure made it relatively easy to study the micro-evolutionary processes. It is not clear, however, to what extent such systems involving extreme environments are fully representative for all micro-evolutionary changes. It is possible that qualitatively different aspects emerge if many different selection pressures are operating intermittently.

Another way of approaching an integration of physiology, ecology and genetics is through the concept of reaction norms. The reaction norms are seen as the transmitted program that "tells" an organism how to react to different environmental situations. This concept is very helpful as a thinking model. It can be used to formulate what the optimal solution to a trade-off between growth and reproduction would look like, given certain aspects of the physiology and given certain selection pressures (Stearns and Koella 1986). It can also be used as a starting point for dealing with continuous environmental variation in a quantitative genetic analysis (van Noordwijk and Gebhardt 1987). It is difficult, however, to measure individual variation in reaction norms unless the study organisms can be easily cloned. It is therefore likely that advances in this direction will come through cyclical parthenogens (e.g. *Daphnia*) and plants where vegetative reproduction allows one to have many copies of a single genotype.

It is my belief that none of these approaches is a priori superior to any of the others. All will have to be taken up and hopefully communication problems will not arise between those who follow these different approaches.

It is urgent that we gain a better understanding of the processes of micro-evolution. We have to reach a level where we can predict how much adaptability one can expect in a particular population. The enormous impact of humans on the environments of almost all species creates strong selection pressures. Up till now, conservationists and those who try to predict environmental impacts are virtually limited to statements about selection pressures that are greater than the potential for adaptation, i.e. extinctions. A better understanding of the interactions between the maintenance of genetic variation, environmental heterogeneity and population structure should make it possible to make more quantitative statements that could also contain information on the accumulation of different impacts.

Acknowledgements. I thank S. Henrich, G. de Jong and S.C. Stearns for helpful discussions and critique. I thank the Swiss Nationalfonds for financial support (grant 3.131-0.85).

References

Balen JH van (1967) The significance of variations in body weight and wing length in the great tit *Parus major*. Ardea 55:1–59

Balen JH van (1973) A comparative study of the breeding ecology of the great tit *Parus major* in different habitats. Ardea 61:1–93

Balen JH van (1980) Population fluctuations in the great tit and feeding conditions in winter. Ardea 68:144–164

Bengtsson H, Ryden O (1983) Parental feeding rate in relation to begging behavior in asynchronously hatched broods of the great tit *Parus major*. An experimental study. Behav Ecol Sociobiol 12:243–251

Bishop JA, Cook LM (1981) Genetic consequences of man-made change. Academic Press, London New York

Drent RH, Daan S (1980) The prudent parent: energetic adjustments in avian breeding. Ardea 68: 225–252

Garnett MC (1981) Body size, its heritability and influence on juvenile survival among great tits *Parus major*. Ibis 123:31–41

Gibb J (1950) The breeding biology of the great and blue tit mice. Ibis 92:507–539

Löhrl H (1968) Das Nesthäkchen als biologisches Problem. J Ornithol 109:383–395

Mertens JAL (1969) The influence of brood size on the energy metabolism and water loss of nestling great tits *Parus major major*. Ibis 111:11–16

Noordwijk AJ van (1984) Quantitative genetics in natural populations of birds illustrated with examples from the great tit, *Parus major*. In: Wöhrman K, Loeschcke V (eds) Population biology and evolution. Springer, Berlin Heidelberg New York, pp 67–79

Noordwijk AJ van (1986) Two stage selection, where the first stage only reduces the environmental variance in body size in the great tit. Proc XIX Congr Int Ornithol, Ottawa 1986 (in press)

Noordwijk AJ van, Gebhardt M (1987) Reflections on the genetics of quantitative traits with continuous environmental variation. In: Loeschcke V (ed) Genetic constraints on adaptive evolution. Springer, Berlin Heidelberg New York, pp 73–90

Noordwijk AJ van, Balen JH van, Scharloo W (1981) Genetic variation in egg dimensions in natural populations of the great tit. Genetica 55:221–232

Noordwijk AJ van, Balen JH van, Scharloo W (1987) Heritability of body size in a natural population of the great tit *(Parus major)* and its relation to age and environmental conditions during growth. Genet Res (in press)

O'Connor R (1977) Differential growth and body composition in altricial passerines. Ibis 119: 147–166

Richards FJ (1959) A flexible growth function for empirical use. J Exp Bot 10:290–300

Royama T (1966) Factors governing feeding rate, food requirement and brood size of nestling great tits *Parus major*. Ibis 108:313–347

Schifferli L (1973) The effect of egg weight on the subsequent growth of nestling great tits *Parus major*. Ibis 115:549–558

Slagsvold T (1986) Asynchronous versus synchronous hatching in birds: experiments with the pied flycatcher. J An Ecol 55:1115–1134

Stearns SC, Koella JC (1986) The evolution of phenotypic plasticity in life-history traits: predictions of reaction norms for age and size at maturity. Evolution 40:893–913

Tinbergen JM (1986) Cost of reproduction in the great tit *(Parus major)*: intraseasonal costs associated with brood size. Ardea 74:111–122

White GC, Brisbin IL, Jr (1980) Estimation and comparison of parameters in stochastic growth models for barn owls. Growth 44:97–111

The Measured Genotype Approach to Ecological Genetics

A. R. TEMPLETON[1] and J. S. JOHNSTON[2]

1 Introduction

Evolution, in its most basic sense, is a change in the frequencies of alleles or gametic complexes over space and time. Because evolution is a genetic process, it is not surprising that advances in the science of genetics, and recombinant DNA techniques in particular, are having a major impact on evolutionary biology. The most immediate impact is in the areas of molecular evolution and phylogenetic reconstruction, but in this chapter we will show that recombinant DNA technology can also be applied to the area of ecological genetics; a subdiscipline of evolutionary biology that at first glance seems far removed from molecular genetics.

Studies in ecological genetics attempt to simultaneously understand both the genetic basis and the ecological significance of phenotypic variation in natural populations. When such an integration of disciplines can be successfully executed, it provides a direct means for studying natural selection and the process of adaptation. Given the central role that adaptation plays in much evolutionary theory, one might predict that ecological genetics would be the central and most vigorous subdiscipline of evolutionary biology. Unfortunately, it has often proven to be very difficult to implement an ecological genetic research program. One of the major obstacles has traditionally been the difficulty of finding an organism on which one could simultaneously perform genetic and ecological studies. One great advantage of recombinant DNA techniques is that they can be applied to most organisms. For example, in the Templeton laboratory, we have performed genetic screens using Southern blotting techniques on both mitochondrial and nuclear DNA in *Daphnia*, many insect species, reptiles, and a large number of mammalian species. Hence, with these techniques, one is free to study genetic variation in an organism with an interesting ecology that is amenable to field studies.

However, this same claim could have been, and was, made for protein electrophoresis during the 1960s. This tool also provided a means for studying genetic variation in natural populations. Yet, with a few exceptions, studies using protein electro-

1   Department of Biology, Washington University, St. Louis, MO 63301, USA
2   Department of Entomology, Texas A&M University, College Station, TX 77843-2475, USA

Population Genetics and Evolution
G. de Jong (ed.)
© Springer-Verlag Berlin Heidelberg 1988

phoresis did not provide much insight into the details of adaptation in natural populations. The basic problem was that the phenotypes of interest to the field biologist (e.g., life history traits, tolerance to environmental stresses, etc.) could usually not be related in any direct fashion to the gene loci amenable to protein electrophoretic analysis. As a result, although one could study the genetics and ecology of an organism, it was difficult to integrate them into a common research program.

Recombinant DNA technology can eliminate this difficulty. Protein electrophoresis was primarily restricted to genes coding for soluble enzymes. Recombinant DNA techniques can be used to study genetic variability for a much broader class of loci. We already have an extensive, albeit incomplete, store of knowledge concerning the genetic control of many basic biochemical, physiological, and developmental processes. Recent advances in molecular genetics are making it easier and easier to study genes that control specific functions.

The ability to assay genetic variation at loci of known phenotypic significance allows a "measured genotype" approach (Boerwinkle et al. 1986) to studying phenotypic variability in natural populations. By studying phenotypic traits of ecological interest, it is sometimes possible to use the accumulated background knowledge referred to above to identify a class of candidate loci that are likely to be involved in the direct control of the phenotypes being investigated. Recombinant DNA techniques can then be used to study genetic variation at these candidate loci, and one can directly test for associations between the genetic variation detectable at the molecular level with the phenotypic effects observed at the biochemical, physiological, and developmental levels. If relevant genetic variation is discovered at this step, the research program can be extended to the population and ecological levels in order to study the adaptive significance of the relevant genetic variation. In this manner, both the proximate and ultimate causes of adaptive traits can be studied in a research program that integrates reductionistic and holistic approaches. Moreover, with such well-defined genetic systems, one can also study directly the role of pleiotropy, developmental constraints, and epistasis in the adaptive process. In this chapter, we will present an example of such an integrated research program dealing with *Drosophila*.

2 The Molecular Biology of Abnormal Abdomen in Drosophila Mercatorum

Natural populations of *Drosophila mercatorum* from the Island of Hawaii are polymorphic for a trait known as abnormal abdomen *(aa)* (Templeton and Rankin 1978). As the name implies, the trait can be recognized from morphological effects. Flies with *aa* tend to retain juvenilized abdominal cuticle as adults, resulting in a disruption of the normal pigmentation, segmentation, and bristle patterns on the adult abdomen. However, the morphological effects are not very penetrant, and under field conditions are rarely expressed. More importantly, *aa* has many effects on life history in both the larval and adult phases (Templeton 1982, 1983). In particular, *aa* prolongs the larval developmental stage by about 3 days, but in the adult it speeds up reproductive maturity to 2 days after eclosion versus about 4 to 5 days in most non-*aa* flies. These early maturing female flies also have increased ovarian output, but greatly decreased longevity under laboratory conditions. These morphological and life history effects are all temperature-dependent, with higher temperatures accentuating the disparity between *aa* and non-*aa* lines.

This suite of temperature-dependent phenotypes resembles similar suites of phenotypic effects in *Drosophila melanogaster* that are associated with the bobbed syndrome. Fortunately, the molecular basis of the bobbed phenotypic syndrome is konwn; it is a deficiency in the number of 18S/28S ribosomal genes (Ritossa et al. 1966). In most *Drosophila* species, the 18S/28S ribosomal genes exist as tandemly duplicated clusters on the X and Y chromosomes. The fundamental units that are duplicated consist of an 18S gene, a transcribed spacer, a 28S gene, and a nontranscribed spacer. About 200 to 250 copies of this basic unit are normally found on the X chromosome. Hence, the phenotypes associated with the *aa* syndrome suggested that the 18S/28S rDNA cluster would be a reasonable candidate set of loci. This hypothesis was strengthened by Mendelian genetic studies that revealed that the *aa* syndrome is controlled by two closely linked X-linked elements that map very near the centromere, the same physical location of the rDNA cluster (Templeton et al. 1985). Moreover, *aa* expression is normally limited to females, but male expression is Y-linked (Templeton et al. 1985). The Y chromosome has very few genes, but as noted earlier it does have an rDNA cluster. Hence, the phenotypic and Mendelian genetic studies strongly suggest that the underlying molecular cause of *aa* lies in the ribosomal DNA.

We therefore began investigating the molecular biology of rDNA in *D. mercatorum* and relating our findings to the presence or absence of *aa*. This work was also facilitated by spontaneous mutations both to and from *aa*. Since this work has already been published (DeSalle et al. 1986; DeSalle and Templeton 1986), we will only present the conclusions here. First, *aa* flies have normal to slightly higher numbers of rDNA repeats than non-*aa* flies. Hence, *aa* is not the molecular equivalent of bobbed, which represents a numerical reduction in the amount of rDNA. However, diploid somatic cells of *aa* flies display increased levels of rDNA. This somatic increase in the amount of rDNA in diploid cells is known as compensatory response and is normally found only in bobbed flies. Hence, although *aa* flies had plenty of rDNA, their diploid somatic cells were behaving as if there were a deficiency of rDNA.

Since the quantity of rDNA was normal, we investigated the qualitative nature of the rDNA present in *aa* flies through restriction endonuclease mapping. We discovered that all *aa* flies have at least a third of their X-linked 28S genes interrupted by a 5-kb insertion. This insert has a transposonlike structure, with long direct repeats on the ends and 14-bp-long inverted terminal repeats on the ends of the direct repeats. However, there is no evidence for this element transposing in the present population, and all copies seem to be confined to the X-linked rDNA. Northern analyses indicate that this insert disrupts normal transcription of the 18S/28S unit in which it is embedded. Hence, this insert creates a functional deficiency of rDNA, although there is no numerical deficiency.

Further studies (DeSalle and Templeton 1986) reveal that the presence of many inserted 28S genes is necessary but not sufficient for the expression of *aa*. The other necessary element for *aa* expression involves the control of underreplication of rDNA in the formation of polytene tissues (DeSalle and Templeton 1986). The process of polytenization in *Drosophila* involves many rounds of endoreplication of nuclear DNA in certain somatic lineages (such as larval salivary glands and fat bodies). Hence, in general, the DNA of the fly is greatly overreplicated in these cells. However, there is heterogeneity within the genome in the amount of overreplication. In general, the rDNA is underreplicated relative to most single copy DNA during the process of polytenization. We discovered an X-linked genetic system that causes the preferential underreplication of inserted 18S/28S units. Hence, even if a fly has a large number of inserted 28S genes, compensatory response circumvents this functional deficiency in diploid, somatic cell lineages, and preferential underreplication of inserted 28S genes circumvents this functional deficiency in polytene, somatic cell lineages. Thus, the expression of *aa* requires two molecular events: (1) a third or more of the 28S genes must bear the insert and (2) there must be no preferential underreplication of inserted 28S genes during the formation of polytene tissue. When these two molecular requirements are satisfied, a functional deficiency of ribosomal DNA can be induced in polytene tissues, thereby explaining the bobbedlike suite of phenotypic effects.

A ribosomal deficiency in critical larval tissues such as the polytene fat body would be expected to slow down protein synthesis in general, and hence explain the overall slowdown in larval development. Moreover, the proteins most effected by a ribosomal deficiency would be those that are normally synthesized very rapidly over a short developmental time period. One such protein is juvenile hormone esterase, a protein made by the larval fat body and that degrades juvenile hormone. This protein is usually made at the end of the larval period. It is also known that the abdominal histoblasts are very sensitive to juvenile hormone titers in the prepupal and early pupal stages, and that phenocopies of abnormal abdomen can be induced by topical application of juvenile hormone at this time (Templeton and Rankin 1978). Studies on *aa* flies reveal a deficiency of juvenile hormone esterase during this critical time period (Templeton and Rankin 1978); thereby providing a straightforward explanation for the morphological effects of *aa*. Polytene tissues are not as important in the adult, so no general slowdown in adult development is expected. However, the alterations in juvenile hormone metabolism induced by *aa* may explain the alterations in adult life history because juvenile hormone stimulates the production of egg-specific proteins and ovarian maturation and output (Wilson et al. 1983). This hypothesis is currently being investigated.

3. The Population Biology of Abnormal Abdomen

We have now defined the molecular basis of *aa* and the effects this syndrome has on individual life history. To understand the ultimate causes of why this life history/ developmental syndrome is polymorphic in natural populations, we must now turn to field studies. Our primary field site is located in the Kohala Mountains on the Island of Hawaii. This site was chosen for several reasons. First, the only repleta group *Drosophila* inhabiting this site are *D. mercatorum* and *D. hydei*, which are readily distinguishable. Hence, problems with sibling species that occur in mainland locations are avoided. Second, unlike on the mainland, *D. mercatorum* has only one host plant in Hawaii, the cactus *Opuntia megacantha*. Thus, only one larval environment needs to be characterized and collecting sites for adults are readily identified and located. Third, the Kohala Mountains have an extremely steep rainfall gradient. This provides the environmental diversity needed to uncover the selective importance of the *aa* syndrome. Fourth, the trade winds blow very strongly in this locality, and these winds greatly limit the opportunity for dispersal between cactus patches (Johnston and Templeton 1982). Hence, if the environmental diversity does create selective differences in *aa*, the limited dispersal insures that local differences in the frequency of *aa* should evolve that reflect these selective differences.

Before studying the ultimate significance of the life history effects of *aa* in natural populations, it is first necessary to measure the age structure in nature and to discover the primary sources of adult mortality. Our previous work (Johnston and Templeton 1982) indicates that one of the primary determinants of mortality in adult *D. mercatorum* is desiccation. We are able to determine the age of wild-caught adults (in days from eclosion) up to about 2 weeks of age by using the techniques described in Johnston and Ellison (1982). During normal weather years, when the top of the mountain is very humid and the bottom very dry, we observe dramatic shifts in age structure as a function of elevation (Johnston and Templeton 1982). In the humid regions, flies are so long-lived, that most are too old to be reliably scored (i.e., greater than 2 weeks of age). As one descends into drier habitats, the age structure shifts towards younger flies such that at the bottom virtually all adult flies are less than 1-week-old. Capture/ recapture studies confirm this pattern. Under dry conditions, the daily mortality of adult flies is estimated to be 19% per day, which is consistent with very few flies living more than a week (Templeton and Johnston 1982). In contrast, no detectable mortality over a 3-day period could be detected under the more humid conditions.

These results on the age structure of natural populations as a function of humidity have several important implications concerning the predicted fitness effects of the *aa* phenotypes. Under low humidity conditions, the decreased innate longevity of *aa* flies is virtually irrelevant because almost all flies are dead from desiccation long before these innate differences should have an impact. However, earlier sexual maturation and increased egg-laying capacity give *aa* flies a fecundity advantage over non-*aa* flies under desiccating conditions (Templeton and Johnston 1982). As the humidity increases and the age structure of the population shifts towards older individuals, the early fecundity advantage of *aa* flies should be counteracted or even reversed by their decreased longevity relative to non-*aa* flies (Templeton and Johnston 1982).

These predictions can be tested by taking advantage of the natural variation in humidity that occurs over space and time. In normal weather years (i.e., years close to the long-term average in rainfall, and with humid conditions at the top of the mountain and dry at the bottom), the above fitness considerations lead to the prediction of a cline in the frequency of *aa*, with *aa* being more common at the bottom of the hill and rarer at the top. A transect of study sites was established on the leeward side of the Kohala mountains, going from the upper elevational limits of the *D. mercatorum* range (site A, 1030 m above sea level) to the lower elevational limit on the transect (site F, 795 m above sea level). In addition, a site at the base of Kohala, in the saddle between Kohala and the volcano Mauna Kea, was also studied (site IV, 670 m above sea level). The humidity at this site is normally very close to that of site F. Table 1 presents the data on the frequency of X chromosomes that allow the expression of *aa* when placed in the genetic background of a laboratory tester stock (Templeton and Johnston 1982). Heterogeneity in the frequencies of *aa* was tested for using the statistical procedures given in Templeton and Johnston (1982).

Table 1. Changes in the frequency of X chromosomes allowing expression of abnormal abdomen in *Drosophila mercatorum* as a function of site location and temporal fluctuations in weather (Straight lines join together those sites with no statistically significant heterogeneity in *aa* frequency)

	Year:	1980	1981	1982	1984	1985
	Weather:	Normal	Drought	Humid	Normal[a]	Normal[a]
Site	Altitude (m)	Frequency of abnormal abdomen X chromosomes				
A	1030	–	–	0.285	0.111	0.222
B	950	0.275	0.486	0.250	0.291	0.265
C&D	920	–	0.380	0.213	0.334	0.429
F	795	–	0.406	–	0.540	0.367
IV	670	0.455	0.438	0.325	0.360[b]	0.276[b]

[a] Site IV was unusually humid due to an alteration in wind patterns.
[b] The site IV frequencies were not significantly different from the upper elevational sites.

As can be seen from Table 1, in 1980 (a normal weather year), there is a statistically significant difference in the frequency of *aa*, with only about 25% of the X chromosomes supporting *aa* near the top of the hill, and almost 50% of the X's being *aa* at the bottom (Templeton and Johnston 1982).

In 1981, the Island of Hawaii suffered a severe drought, and the humidity ecotone disappeared that year. That year, it was dry throughout the entire range, and the age structure cline also disappeared, with almost all individuals being less than a week of age irrespective of their site of capture (Templeton and Johnston 1982). The model now predicts that the cline in *aa* frequency should also disappear, and in particular, that there should be an overall increase of *aa* frequencies at the high elevation sites to the frequencies normally seen at the lower elevations. As can be seen from Table 1, this is precisely what happened. There are now no significant differences in *aa* frequency between sites, and these frequencies are not significantly different from the low elevation frequencies of 1980 but are significantly different from the high elevation frequencies of 1980 (Templeton and Johnston 1982).

Early in 1982, the El Chichon Volcano had an explosive eruption in Mexico. The resulting debris cloud reduced the incidence of solar radiation by 10% that spring in Hawaii. This apparently triggered one of the wettest, most humid springs and early summers in Hawaiian history. Once again, the humidity ecotone did not exist, but this time it was humid both at the top and bottom of the mountain. The *aa*-fitness theory of Templeton and Johnston (1982) predicts no cline in *aa* frequencies, but unlike 1981, the frequencies should now be low at all sites (around 25%). This is exactly what happened (Table 1). As in 1981, there are no significant differences in *aa* frequencies between sites, but unlike 1981, the *aa* frequencies are now significantly different from the 1980 low elevation frequencies (and the 1981 frequencies) and not significantly different from the 1980 high elevation frequencies.

In 1984 and 1985, the weather had reverted to an almost normal pattern. The one exception was the local humidity conditions at site IV. Sites A through F are located well on the side of Kohala, but site IV is located at the base of Kohala, in the saddle between Kohala and the volcano Mauna Kea. Usually, the trade winds blow humid air through the saddle, but shear off before reaching site IV. Hence, normally site IV is quite dry and has a similar humidity to site F. However, starting in 1984 and continuing into 1985, the winds blew further through the saddle, causing site IV to become much more humid than previously. As can be seen from Table 1, the expected cline in *aa* frequency was reestablished in 1984 and 1985, with *aa* X chromosomes being rarer at the top and becoming increasingly common toward the bottom. The exception to this pattern was site IV, which had significantly lower *aa* frequencies than site F but which had homogeneous frequencies to the high altitude site B.

4 Discussion

The abnormal abdomen system tracks the spatial and temporal heterogeneity in the natural environment in exactly the manner predicted by the model given in Templeton and Johnston (1982). Therefore, we conclude that the ultimate causes that maintain this polymorphic life history syndrome relate to its adaptive advantages in the demographic environment imposed by desiccating environments. Note that *aa* does not represent a direct adaptation to desiccation itself; rather, it is favored because the early maturation and fecundtiy of *aa* females confers a fitness advantage under the age structure imposed by desiccation. Moreover, under this age structure, the decreased longevity of the *aa* flies has very little fitness impact. These inferences are possible because the measured genotype approach allowed us to follow the evolutionary fate over space and time of specific genetic variants for which we had generated a priori evolutionary predictions based upon our knowledge of the variant's phenotypic significance.

As mentioned in the introduction, the measured genotype approach to ecological genetics not only allows us to infer the existence of natural selection, but it also provides us with much insight into the details of the adaptive process, details which touch upon some major controversies in modern evolutionary biology. One current controversy concerns the role of fitness epistasis and coadaptation in natural populations (e.g., see Carson and Templeton 1984 vs. Barton and Charlesworth 1984). As mentioned in Section 2, *aa* depends upon two molecular elements: inserts in the 28S genes of the X-linked rDNA complex, and the X-linked control of the failure to preferentially underreplicate 28S genes with inserts during polytenization. Although closely linked, there traits are genetically separable with a recombination frequency of 0.7% (Templeton et al. 1985). Hence, the selected *aa* complex definitely depends upon strong epistatic interactions between loci, and *aa* chromosomes can be regarded as an example of a coadapted supergene. In addition, many autosomal and Y-linked modifiers of *aa* exist that display very strong epistasis with the *aa* supergene (Templeton et al. 1985). It is known that these modifiers are polymorphic in natural populations, although their adaptive significance remains to be studied. Nevertheless, there is no doubt that strong epistasic interactions characterize the genetic basis of the *aa* syndrome, and that at least the epistasis between the X-linked elements plays a critical role in the adaptive response.

Pleiotropy has long received much attention from evolutionary biologists (e.g., Wright 1932; Williams 1957), but its significance is still controversial. For example,

Rose (1982) and Turelli (1985) have recently investigated models of antagonistic pleiotropy (in which some pleiotropic effects contribute in a positive fashion to fitness, but others in a negative fashion). Their models cast serious doubt on some of the predictions of the classical mutation/selection balance models of quantitative genetics and, in particular, provide novel mechanisms responsible for observed levels of heritability in natural populations. The *aa* syndrome provides an excellent example of a polymorphic system characterized by extensive antagonistic pleiotropy. Some pleiotropic effects, when looked at in isolation, appear to be detrimental to fitness (increased larval developmental times, decreased adult longevities), while others appear to be beneficial (increased rate of ovarian maturation, increased fecundity). The predictions of Templeton and Johnston (1982), which have been confirmed by the observations reported here, were generated by examining the shifting trade-offs of these antagonistic effects on fitness as a function of different environmental regimes. If polymorphic systems like *aa* represent a major component of quantitative genetic variation in natural populations, then the evolutionary predictions based upon the mutation/selection balance models must be regarded as suspect.

Finally, *aa* sheds some light on the role of developmental constraints in adaptive evolution. For example, the direct effect of *aa* is with regard to transcriptional control of ribosomal genes in certain polytene tissues. Yet, this cell-specific transcriptional control mechanism can affect many other tissues and alter development and life history in radical ways through the normal developmental responses built into the ontogeny of the organism. Thus, the transcriptional regulation of rDNA in larval fat body can apparently exert a translational control on the protein juvenile hormone esterase, which in turn alters the developmental fate of the abdominal histoblast cells through the normal developmental response to juvenile hormone displayed by these cuticle cells (Templeton and Rankin 1978). Thus, the direct effect of gene action is modulated through the preexisting developmental responses to produce the ultimate phenotype. The *aa* syndrome also illustrates that developmental constraints can cause the evolution of nonadaptive, or even maladaptive, traits as a side effect of adaptive evolution. For example, there is no obvious adaptive benefit to a slow-down in egg-to-adult developmental time, and indeed this trait regarded in isolation should cause fitness to decline. Yet, this trait can be selectively favored when it is placed in the context of the entire developmental syndrome of which it is a part.

In summary, the measured genotype approach to ecological genetics not only provides a method for studying natural selection, but does so in a way that provides valuable insights into the details of the adaptive process.

Acknowledgments. This work was supported by National Institutes of Health grant R01 AG02246.

References

Barton NH, Charlesworth B (1984) Genetic revolutions, founder effects, and speciation. Annu Rev Ecol Syst 15:133–163

Boerwinkel E, Chakraborty R, Sing CF (1986) The use of measured genotype information in the analysis of quantitative phenotypes in man. I. Models and analytical methods. Ann Human Genet 50:181–194

Carson HL, Templeton AR (1984) Genetic revolutions in relation to speciation phenomena: the founding of new populations. Annu Rev Ecol Syst 15:97–131

DeSalle R, Templeton AR (1986) The molecular through ecological genetics of abnormal abdomen. III. Tissue-specific differential replication of ribosomal genes modulates the abnormal abdomen phenotype in *Drosophila mercatorum*. Genetics 112:877–886

DeSalle R, Slightom J, Zimmer E (1986) The molecular through ecological genetics of abnormal abdomen. II. Ribosomal DNA polymorphism is associated with the abnormal abdomen syndrome in *Drosophila mercatorum*. Genetics 112:861–875

Johnston JS, Ellison JR (1982) Exact age determination in laboratory and field-caught *Drosophila*. J Insect Physiol 28:773–779

Johnston JS, Templeton AR (1982) Dispersal and clines in *Opuntia* breeding *Drosophila mercatorum* and *D. hydei* at Kamuela, Hawaii. In: Barker JSF, Starmer WT (eds) Ecological genetics and evolution: The cactus-yeast-*Drosophila* model system. Academic Press, London New York, pp 241–256

Ritossa FM, Atwood KD, Spiegelman S (1966) A molecular explanation of the bobbed mutants of *Drosophila* as partial deficiencies of ribosomal DNA. Genetics 54:818–834

Rose M (1982) Antagonistic pleiotropy, dominance, and genetic variation. Heredity 48:63–78

Templeton AR (1982) The prophecies of parthenogenesis. In: Dingle H, Hegmann JP (eds) Evolution and genetics of life histories. Springer, Berlin Heidelberg New York, pp 75–101

Templeton AR (1983) Natural and experimental parthenogenesis. In: Ashburner M, Carson HL, Thompson JN (eds) The genetics and biology of *Drosophila*, vol 3C. Academic Press, London New York, pp 343–398

Templeton AR, Johnston JS (1982) Life history evolution under pleiotropy and K-selection in a natural population of *Drosophila mercatorum*. In: Barker JSF, Starmer WT (eds) Ecological genetics and evolution: The cactus-yeast-*Drosophila* model system. Academic Press, London New York, pp 225–239

Templeton AR, Rankin MA (1978) Genetic revolutions and control of insect populations. In: Richardson RH (ed) The screwworm problem. Univ Texas Press, Austin, pp 81–111

Templeton AR, Crease TJ, Shah F (1985) The molecular through ecological genetics of abnormal abdomen in *Drosophila mercatorum*. I. Basic genetics. Genetics 111:805–818

Turelli M (1985) Effects of pleiotropy on predictions concerning mutation-selection balance for polygenic traits. Genetics 111:165–195

Williams GC (1957) Pleiotropy, natural selection, and the evolution of senescence. Evolution 11:398–411

Wilson TG, Landers MH, Happ GM (1983) Precocene I and II inhibition of vitellogenic oocyte development in *Drosophila melanogaster*. J Insect Physiol 29:249–254

Wright S (1932) The roles of mutation, inbreeding, crossbreeding and selection in evolution. Proc 6th Int Congr Genet 1:356–366

Is Population Genetics in Its Present Scope Sufficient for a Theory of Evolution?

Adaptation

What Is the Progress Towards Understanding the Selection Webs Influencing Melanic Polymorphisms in Insects?

P. M. BRAKEFIELD[1]

1 Introduction

Industrial melanism refers to a correlation between high frequencies of melanic forms of an insect and regions of industrialization. This phenomenon in the peppered moth *Biston betularia* (L.) is the classic textbook example of the evolution of an adaptive trait in response to a changing environments involving the spread of adapted phenotypes by natural selection. The accounts include information about the central hypothesis of a change in the relative crypsis of non-melanic and melanic phenotypes due to blackening of the resting background of the moths by industrial air pollution. The entomologist, J.W. Tutt, writing at the end of the last century, presented particularly graphic descriptions of the essential features of this hypothesis (e.g. Tutt 1896). We can now ask: how much further have we actually progressed in our understanding since then and what has the theory of population genetics contributed to this understanding?

2 Industrial Melanism: The Background

Many factors have been shown or postulated to influence industrial melanism in *B. betularia* or other moths (Table 1). Kettlewell (1955, 1956) demonstrated that bird predators in an industrial environment found fewer black peppered moths on tree trunks than the pale non-melanics which were, in contrast, at an advantage in terms of crypsis and visual predation in a rural wood. Further experimental work over the last 30 years has concentrated on (1) monitoring the association between high frequencies of melanics and industrial regions; (2) investigating the survivorship, and sometimes the dynamics, of cohorts of melanic and non-melanic moths in different environments and (3) performing crosses to examine the genetics of the polymorphism and segregation ratios. This approach has led to extensive survey data, some estimates of ecological parameters and information about the co-allelic series, controlling the phenotypes. Yet several workers have emphasized the inadequacies in our understand-

1 Department of Evolutionary Biology, RU Leiden, Schelpenkade 142, 2313 ZT Leiden, The Netherlands

Population Genetics and Evolution
G. de Jong (ed.)
© Springer-Verlag Berlin Heidelberg 1988

Table 1. A summary of factors which have been considered to influence examples of industrial melanism in insects in which the adults are cryptic

Category	Factor
Relative crypsis	Resting behaviour: choice of site
	Nature of resting site (incl. colour and texture)
	Light intensity and wetness of resting surface
	Nature of phenotypic differences
	Phenotype combination in mating pairs
Predation	Hunting tactics and behaviour of predator
	Frequency-dependent effect (sympatric moth spp.)
Biology	Migration rate
	Population density
	Mating behaviour
	Randomness of mating
	Adult activity, including thermal effects
	Timing of adult eclosion, phenology
	Genotype viability and pre-adult development

ing of the mechanisms of natural selection involved in patterns of spatial differentiation in phenotype frequencies and in maintenance of the polymorphism (see Bishop et al. 1978a; Mikkola 1984; Howlett and Majerus 1987; Liebert and Brakefield 1987).

Such inadequacies are well illustrated by our inability when applying simple mathematical models developed from population genetics theory and based on field estimates of migration and differential predation (with a weak frequency-dependent component) to obtain reasonable fits between expected and observed patterns of spatial differentiation (Cook and Mani 1980; Mani 1980, 1982). For example, the model of the steep cline in the frequencies of *carbonaria* melanics and *typica* non-melanics from urban Liverpool to rural North Wales, studied in the field by Bishop (1972), shows a point of inflexion which is markedly displaced towards Liverpool. The matching is greatly improved when non-visual differences in fitness between the phenotypes are incorporated into the models using an iterative type of procedure (Mani 1980). Some support for such differences in *B. betularia*, although not in the precise form suggested by the models, comes from analysis of deviations from expected segregation ratios in reared material (Creed et al. 1980). Heterozygous advantage has also been postulated by some workers, for example to account for the region of high melanic frequencies in rural East Anglia (Lees and Creed 1975). An association of high melanic frequencies and normal or rich epiphytic vegetation occurs in the north of the Netherlands (Fig. 1). I discuss below ways in which non visual selection could be investigated. It will be necessary to develop an assay for genotypes before we can establish whether heterozygous advantage occurs in natural populations, a point emphasized by Ford (1975). However, Cook and Mani's studies showed that spatial variation could be satisfactorily modelled without introducing heterozygous advantage.

The inadequate description of spatial changes in phenotype frequencies given by a balance between visual selection and gene flow could arise either because of the exclusion of some other relevant factor or because of errors in our estimates of the model's parameters. This chapter describes some more recent work which shows

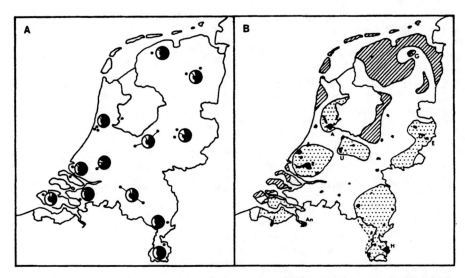

Fig. 1. Maps of the Netherlands: **A** pie diagrams representing frequency of *carbonaria* (*black*), *insularia* (*halved*) and *typica* (*white*) forms of *Biston betularia* in samples from the sites indicated by *small dots* (data of BJ Lempke for 1969 from Kettlewell 1973). **B** Epiphytic vegetation in the 1960s with "deserts" *dotted*; transitional zone in *white* with poor to locally subnormal vegetation; normal, rich or luxuriant areas *hatched*; larger towns and main industrial areas in *black* (After Barkman 1969; see also De Wit 1983)

that our estimates of ecological parameters in *B. betularia* do need refining. This situation has arisen partly because field studies have concentrated on the ecological genetics of adult populations without due attention to the ecology and behaviour of individuals and to the complete life cycle.

In Britain the frequency of *carbonaria* melanics is declining in many industrial regions in response to reduced air pollution since the implementation of clean air legislation in 1956 (Clarke, Mani and Wynne 1985; Cook, Mani and Varley 1986). This phenomenon has been termed "evolution in reverse" by Clarke et al. and provides particularly compelling evidence for the general basis of industrial melanism in air pollution. However, if we do not act soon we may yet again miss a golden opportunity to quantify the mechanisms of natural selection underlying a classic example of evolutionary change.

The occurrence of industrial melanism has also been documented in many other insects (see Kettlewell 1973; Lees 1981). A particularly well-researched example is the melanic polymorphism in the two-spot ladybird beetle *Adalia bipunctata* (L.). As in *B. betularia*, melanism in this species is controlled by alleles at a single gene locus with melanics dominant to non-melanics. The polymorphism has been ascribed by Creed (1966, 1971a) to industrial melanism with the melanic forms being favoured in some unidentified, direct way by particulate air pollution. Some evidence has been obtained for the alternative theory of thermal melanism which predicts that the black forms are favoured in conditions of low sunshine because of more efficient absorption of solar radiation (Lusis 1961). I shall discuss our understanding of natural selection acting on this polymorphism and compare the responses of *Adalia* ladybirds (Creed 1971b; Brakefield and Lees 1987) and *B. betularia* to declining air pollution.

3 Pollution, Epiphytes and Selection in *Biston* moths

The bark of deciduous trees in rural woods is generally pale and variegated due to the absence of dark, particulate pollutants and the presence of luxuriant epiphytes. The *typica* form of *B. betularia* at rest on tree trunks in such environments is the most cryptic phenotype, closely matching the background at least in the eyes of human observers (Lees and Creed 1975; Steward 1985). In constrast, the completely blackened resting surfaces in extreme industrial environments with no epiphytes and high levels of the black particulate pollutants of smoke and soot can only favour the *carbonaria* melanics. The results of various types of predation experiments (see Bishop et al. 1978b) are qualitatively in keeping with this switch in relative crypsis between the extreme environments. More problems arise in environments influenced by moderate levels of air pollution and when the intermediate melanic phenotypes of the *insularia* complex are considered.

Mikkola (1979, 1984) gives an important critique of the predation experiments performed on *B. betularia* using multiple mark-release-recapture techniques or the exposure of dead-frozen moths on trees to predators, to estimate the relative survivorship of different phenotypes. In particular, he argues that they are all based wholly or largely on the assumption that the moths rest on tree trunks in nature. Mikkola's observations of male moths resting on partly paint-sprayed, horizontal and vertical branch sections in an outdoor cage suggested that the normal resting sites of the species were actually beneath narrow branches in the tree canopy! Furthermore, some of the predation experiments have involved substantially higher densities of moths than those characteristic of *B. betularia* in natural populations (Bishop et al. 1978b); a factor which could readily influence the extent of frequency-dependent predation. Howlett and Majerus's (1987) observations of some moths found at rest in the wild provide some support for branch resting (see also Kettlewell 1958). Their pilot experiment involving dead moths placed on trunks or on the joints between trunks and branches indicated that this difference in resting site influenced the proportion of moths disappearing, in their case over the subsequent 72 h.

Liebert and Brakefield (1987) extended the scope of Mikkola's cage experiments by developing a technique for observing the behaviour of *living* females in the *wild*. Moths were held without mating for three nights after eclosion before release. Observations showed that this procedure did not substantially influence resting, mating or egg-laying behaviour. Rather, it overcame an initial dispersal flight on the night after eclosion and probably reduced the tendency to climb after settling. Over 250 moths were released on trunks, branches and branchlets in a rural wood in Somerset and in a semi-natural wood within the city of Cardiff influenced by moderate levels of air pollution. No moths were released within 50 m of a known survivor. The behaviour of 131 pairings and of surviving egg-laying moths was observed.

Liebert and Brakefield's (1987) results support Mikkola's (1984) conclusion that *B. betularia* predominantly rests on branches and shows an appropriately specialized

Fig. 2A–D. Photographs of *Biston betularia* at rest. **A** *typica* and **B** *carbonaria* moths pairing below oak branchlets in Cardiff darkened by the fungus *Torula herbarum* (with camera angle near vertical); **C** *typica* pairing on a trunk with lichens in Somerset; **D** *typica* female resting against a thallus of foliose lichens below which she laid eggs over a 5-day period (photos **C** and **D** by TG Liebert)

resting attitude. This is illustrated by the photographs in Fig. 2. Many moths will rest underneath, or on the side of branchlets in the tree canopy. Once females have settled and paired after dark they only move quite short distances. They remain *in copula* for nearly 24 h. The field observations and additional experiments performed in an outdoor cage showed that the presence of foliose lichens has a profound influence on the positioning of pairings and egg-laying females, and therefore on crypsis. Many moths on branches tended to be associated with a boundary between rather clear-cut, longitudinal "zones" of epiphyte cover or bark colouration. In rural areas with luxuriant foliose lichens, especially on upper branch surfaces, single or paired *typica* moths may "mimic" the lichen thalli. This effect is probably enhanced by the overlapping of wings exhibited by many pairings. Moths on trunks usually rest with their wings parallel to the vertical axis and against a lichen thallus or some irregularity such as an old branch scar. Females of all phenotypes show a strong preference to oviposit beneath a thallus of a foliose lichen (Fig. 2). They rest against or close to the thallus during the day. In the absence of foliose lichens, eggs are laid in cracks in the bark and females tend to search more widely for oviposition sites. Overall the observations indicated a more varied choice of resting position than proposed by Mikkola (1984). Some moths will rest on non-horizontal branchlets and others on main branches or trunks (see Howlett and Majerus 1987).

In the absence of air pollution, the nature of epiphytic floras and therefore, of the colour and texture of bark, is dependent on such factors as species of tree, age and acidity of bark, light intensity and effects of the growth form of trees on drainage of rainwater and nutrients. This results in considerable heterogeneity in resting backgrounds within trees as well as between trees, although descriptive data are required. The upper canopy branches with high light intensities are frequently associated with especially luxuriant growth of foliose and fruticose lichens. Variability, which may be even more marked, also occurs within trees in regions influenced by moderate or declining levels of air pollution. The most polluted environments produce bark which is uniformly black because of particulates and denuded of epiphytes because of the susceptibility of these organisms to gaseous pollutants.

4 Estimating Visual Selection and Migration

When the observations which indicate that *B. betularia* moths rest in a variety of sites within trees, especially on narrower branches in the canopy, are considered, any heterogeneity in the resting backgrounds within trees must be critical in determining the distributions of relative crypsis for phenotypes. We need to quantify the frequency distribution of resting sites used by the species, or of the phenotypes if these differ in their choice of site in the wild (see Mikkola 1984; Liebert and Brakefield 1987). The next step will then be to estimate visual selection on the polymorphism over the differing types of resting site.

The technique of releasing 3-day-old, unmated females can provide the basis for a method of estimating the relative survivorship of the different phenotypes as *living* moths, either *in copula* or when egg-laying, at *natural* resting sites. Relative crypsis can also be measured by analyzing background matching in photographic records

(Endler 1984). Experiments would involve releases of cohorts of different phenotypes in standardized sites on trees in environments with differing loads of air pollution. The analysis of small data sets obtained by Liebert and Brakefield (1987) for moths released under less standardized conditions suggested that overall mortality was higher in the urban, than the rural environment. There was no evidence of differential predation within habitats. As Kettlewell (1973) has suggested, predation of mating pairs in some resting environments may be non-random with respect to phenotype composition. In particular, the survivorship of a certain phenotype *in copula* may differ for assortatively and disassortatively mating individuals. Liebert and Brakefield's analysis suggests that at Cardiff melanic x melanic pairings have a higher mortality than other combinations.

Capture-recapture experiments performed by Clarke and Sheppard (1966) and Bishop (1972) demonstrated that male moths may fly considerable distances. Bishop's data have been used to estimate the parameter of migration rate incorporated in modelling studies. However, the restriction of this parameter to movement of adult males may lead to serious underestimates of migration. Kettlewell (1973), Bishop et al. (1978b) and others have noted the possibility of passive wind dispersal of larvae. Liebert and Brakefield (1987) examined the behaviour of larvae hatching from the eggs massed under foliose lichens or in cracks. The very small larvae immediately suspend themselves on silk threads and are dispersed by any air current. Many eggs will be oviposited in the upper canopy and hatching larvae in June to August may be transported upwards on thermals. It is necessary to quantify the frequency distribution of dispersal by larvae, for example by using wind tunnel experiments. Any substantial long-distance dispersal may help to account for the occurrence of high melanic frequency in rural East Anglia (Kettlewell 1973) and northern Holland (Fig. 1). These areas are flat and not well wooded. In the case of East Anglia, the prevailing winds in July are from the south and west and are predominantly light (Shellard 1976). Thus, they may carry larvae from industrial regions with high melanic frequencies in the Midlands and the London area.

5 Sunshine, Mimicry and Selection in *Adalia* Ladybirds

The lower part of Fig. 3 indicates the wide variety of factors postulated or shown to influence melanism in *A. bipunctata* (refs. in Muggleton 1978; Brakefield 1985). An elegant analysis has been made of the genetics of female mating choice for melanic males found in a stock from northern England (see Majerus et al. 1986). Here, I shall concentrate on selection acting on the colour and pattern per se of the dorsal cuticle of the beetle.

Lusis (1961) suggested that the predominantly black forms of *A. bipunctata* were at an advantage over the red form with two black spots in regions of low sunshine because of a more efficient absorption of solar radiation. The basis of this theory of thermal melanism in differing thermal properties of the cuticles has been confirmed experimentally (Muggleton et al. 1975; Brakefield and Willmer 1985). In addition, there are a number of lines of evidence supporting an influence of thermal melanism in natural populations: (1) negative correlations in Britain and the Netherlands be-

SELECTIVE PREDATION ?

Mimicry

RED/BLACK 'ring'   Polymorphic   BLACK/RED 'ring'

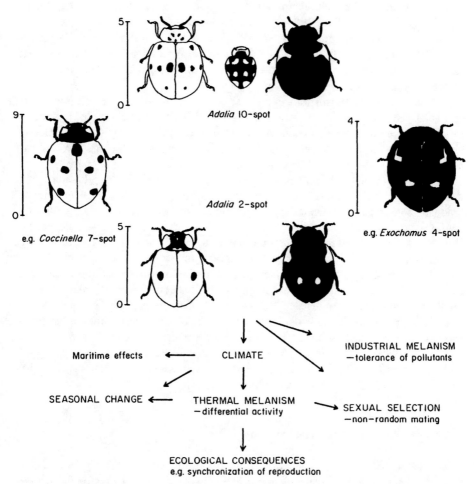

Fig. 3. The most abundant melanic and non-melanic forms of *Adalia bipunctata* and *A. decem-punctata* (intermediate form in reduced size) together with examples of monomorphic ladybird members of Müllerian mimicry rings in Western Europe (red areas of beetles are depicted in *white* and size scales are in mm). An indication of selective factors which have been considered to influence the polymorphism in *A. bipunctata* is given (Brakefield 1985)

tween hours of sunshine during the peak period of adult reproductive activity and the frequency of melanics (Muggleton et al. 1975, Brakefield 1984a); (2) an earlier average date of eclosion of melanics from pupae within populations in the Netherlands (Brakefield 1984b); (3) an earlier average date of mating, dispersal to oviposition habitats and death in the parental populations of these pupae (Brakefield 1984b); (4) a general mating advantage to melanics and a corresponding increase in melanic frequency between the post-hibernation populations of adults and their offspring in the Netherlands (Brakefield 1984c). However, an experimental manipulation is still required to confirm the proposed chain of causality between sunshine, thermal budgets and reproduction. The effects on timing of reproduction may lead to spatial variation in the synchronization of the egg laying of the phenotypes and the annual increase in numbers of their aphid prey.

Neither thermal melanism nor any other selective factor indicated in the lower part of Fig. 3 can account for the colour pattern per se of the phenotypes. If thermal melanism were the overriding selective influence favouring melanism then completely black forms would be expected. The idea that mimetic relationships are involved has recently been reviewed and expanded upon by Brakefield (1985). Many ladybird beetles are protected from bird predators by a well-developed chemical defence and are more or less monomorphic, tending to conform to one of a small number of "modal" colour patterns which are conspicuous and contrasting. This distribution conforms to a series of rather loose Müllerian mimicry rings. Some polymorphic species, including *Adalia*, exhibit forms which are apparently components of different putative mimicry rings (Fig. 3). Brakefield (1985) suggests that such species may be less unpalatable to birds and have responded to prolonged disruptive selection acting in a frequency-dependent way to promote, on a global scale, polymorphic mimicry intermediate in nature between classical Batesian and Müllerian mimicry. The likely occurrence of very tightly linked loci, forming a "supergene", controlling visible polymorphisms in ladybirds is consistent with such an explanation (Komai 1956; Houston and Hales 1980). An alternative hypothesis is that some predators avoid ladybirds and others feed on them, thereby promoting polymorphism through a balance between aposematic and apostatic selection (see Thompson 1984).

6 Declines in Melanic Frequency in *Biston*

Cook et al. (1986) describe a general decline in the frequency of the *black carbonaria* form of *B. betularia* in Britain with a contraction in the range of the region of high frequency in northern England. The best documented change for a single locality is at West Kirby on the Wirral peninsula south of Liverpool where the frequency has declined steeply from about 90% in the 1960s to below 60% in 1985 (Clarke et al. 1985).

The establishment of smokeless zones in Britain after implementation of clean air legislation in 1956 resulted in a rapid fall in smoke emissions which were halved nationally within 10 years (see Lees 1981). This led to a lightening of tree surfaces in many areas such as the Wirral towards the margins of industrial centres. There is some indication that a slight fall in the frequency of *carbonaria* occurred at West Kirby following the introduction of a smokeless zone in the area during 1962 and 1963 (Clarke and Sheppard 1966) suggesting that any initial lightening of resting surfaces may have played an early, if perhaps relatively minor, role in the overall change (cf. in Manchester, Cook et al. 1970). The frequency of *carbonaria* remained more or less stable in the later 1960s before declining more dramatically in the 1970s up to the present day (Clarke et al. 1985).

The decline in national emissions of sulphur dioxide was much more gradual than for smoke (Lees 1981) although this may not have been so evident on the Wirral (see Clarke et al. 1985). Epiphytes vary in their susceptibility to gaseous components of air pollution, especially sulphur dioxide (e.g. Hawksworth and Rose 1970; Seaward and Hitch 1982; De Wit 1983). Several crustose lichens can tolerate higher levels of surface acidity than the most tolerant foliose-species. The falls in air pollution are associated with a form of successional change in epiphytic floras. Furthermore, Liebert and Brakefield (1987) consider that epiphytes including foliose lichens will re-establish themselves on the upper surface on branches in the upper canopy before the trunks. They suggest that one reason for this is that in marginal industrial regions this area of a tree includes much new growth unaffected by past high levels of pollutants and are unlikely to have become dominated by more aggressive epiphytes which may inhibit establishment of foliose lichens (M.R.D. Seaward in Clarke et al. 1985). The tolerance limit to sulphur dioxide characteristic of the more abundant foliose lichens, such as *Hypogymnia physodes*, lies at about the level reached at West Kirby in 1970 (Clarke et al. 1985). Sulphur dioxide continued to fall up to the late 1970s. Indeed, the closest mathematical fit to the dramatic decline in *carbonaria* is given when the fitness of *typica* is linearly correlated with observed sulphur dioxide concentrations. Clarke et al. consider that foliose lichens are still virtually absent at West Kirby and suggested the explanation that *typica* is favoured by some non-visual effect. Liebert and Brakefield (1987) argue that if foliose lichens began to re-establish themselves in the upper tree canopy (largely out of sight of ground observers) their presence could account for the decline in *carbonaria* through changes in relative crypsis and visual selection. They also argue that *carbonaria* may suffer a quite abrupt switch in relative fitness early in the re-establishment of foliose lichens because of the preference of each phenotype for foliose lichens as oviposition sites and the habit of females to rest in close proximity to them (Fig. 2). We need to know much more about the interactions between pollution, epiphytes and resting backgrounds of *B. betularia* before we can be confident about our understanding of the processes of natural selection underlying such declines. We also require much more precise data on the availability of matching backgrounds for *insularia* phenotypes in areas of more moderate air pollution in the south of Britain. These areas are likely to be characterized by a high heterogeneity in resting backgrounds available to the moth.

7 Declines in Melanic Frequency in *Adalia*

Creed (1971b) detected a decline in the frequency of melanic *A. bipunctata* at five sites in or near Birmingham in the 1960s. Brakefield and Lees (1987) found that the decline continued in an approximately linear form until about 1978 when it levelled off. Although there were differences in melanic frequency between sites, there was no significant heterogeneity in the rate of declines. The common regression slope was compared with declines in air pollutants measured in Birmingham during the annual periods of peak adult reproductive activity. The pattern of change in melanic frequency showed a close correspondence to levels of smoke ($r = +0.91$) but a weaker association with sulphur dioxide ($r = +0.72$) which, in particular, continued to decline after 1978. The relationship is consistent with the hypothesis that as smoke concentration falls, solar radiation reaching the ground increases and melanic beetles lose their thermal advantage. It could also be accounted for if smoke favours melanics in some direct way which is as yet unidentified. The related species, *A. decempunctata*, which has a phenotypically similar polymorphism (Fig. 3) shows no evidence of parallel changes in melanic frequency in Birmingham. Brakefield (1984a, 1985) found that this species exhibits no geographical variation in the Netherlands, whereas *A. bipunctata* shows steep clines. He also suggested that the strictly arboreal habit of this species, in contrast to *A. bipunctata*, predicts that thermal melanism is unlikely to have any significant influence on adult reproduction. One might expect any direct influence of smoke to influence both polymorphisms in a qualitatively similar manner.

The observed declines in melanic frequency in *A. bipunctata* and *B. betularia* are associated with similar, and more or less constant, disadvantages to melanics in comparison to the earlier periods (Cook et al. 1986, Brakefield and Lees 1987). The disadvantage is estimated at 10 and 12% in these species respectively. The declines began about 10 years later in *B. betularia* probably because of the indirect effect of gaseous pollutants acting through epiphytic growth in this species and the later fall of these components than of particulates.

8 Non-Visual Differences in Fitness

The yellow to red pigments of ladybirds are carotenoids. Britton et al. (1977) identified a wide variety of these hydrocarbons in *Coccinella septempunctata*. Insect melanins are the end products of two branches of the tyrosine oxidation pathway, the first step of which is synthesis of dopa from the amino acid catalyzed by a tyrosinase enzyme (see Needham 1978). Dopa melanins are polymers of indolic quinones synthesized in one branch via dopa. Dopamine melanins are synthesized via the other branch following conversion of dopa to dopamine by the enzyme dopa decarboxylase. The latter type of pigment predominates in melanic morphs of the moth *Manduca sexta* (Hiruma et al. 1985) and the beetle *Tribolium castaneum* (Kramer et al. 1984). The detailed understanding of the regulation of melanization in these systems provides an excellent potential basis for work on *Biston* and *Adalia* designed to identify differences between phenotypes and genotypes in pigments or their precursors and understand their regulation. This could lead to identification of modes of gene action and development of an assay for genotypes to test for heterozygous advantage. Preliminary work by P.M. Brakefield and L. Dinan has developed techniques for quantifying tyrosine and dopa in individual pupae of *B. betularia* and carotenoids in single elytra of coccinellids using high pressure liquid chromatography. Melanic *A. bipunctata* have substantially lower concentrations of carotenoids in their elytra than non-melanics (P.M. Brakefield and L. Dinan, unpub. data). The red pigments are laid down primarily under areas of the cuticle which are not melanized.

9 Conclusions

The tracing of an example of industrial melanism in mechanistic terms back to its genetic basis by examining the gene to phenotype relationship using biochemical and, perhaps, molecular genetics techniques is necessary to exclude linkage disequilibrium and investigate the full range of effects of the gene locus. Table 2 illustrates the diversity of factors which could be associated with non-visual effects of genes controlling melanism. We need to be able to identify all genotypes so that their relative fitnesses can be measured. Work to date has concentrated on crucial features of the adaptations at the ecological and population level. I have indicated that more rigorous estimates of ecological parameters are required in the light of insights about the behaviour of individual insects before we can confidently apply the theory of population genetics in any complete way. Present-day declines in melanic frequencies provide us with excellent field systems to combine with laboratory studies of gene action and non-visual effects in an integrated approach to understanding the selection webs influencing these melanic polymorphisms.

Table 2. Summary of some of the non-visual effects which have been, or are likely to be associated with genes controlling melanic polymorphisms in insects

Effect	Comments	Species	Reference
Viability	Lower for homozygous melanics	*Phigalia pilosaria*	Lees (1981)
Body size and fecundity	Smaller for a black, recessive larval mutant	*Manduca sexta*	Safranek and Riddiford (1975)
Development rate	Melanic adults emerge later. Cost of synthesis of nitrogenous pigments (see Graham et al. 1980) may be involved	*Odontoptera bidentata*	Cook and Jacobs (1983)
Behavioural	*Ebony* polymorphism involves dopamine (a neurotransmitter) and β-alanine biology; resting site choice in moths	*Drosophila melanogaster;* Various moths	Kyriacou et al. (1978); e.g. Steward (1985)
Activity	Differential timing of adult activity	*Phigalia titea*	Sargent (1983)
Toughness of cuticle	Pupae of melanics are stronger	*Xylophasia monoglipha*	M. Rothschild (pers. commun.)
Susceptibility to attack	By fungae, micro-organisms or parasites. Involves host phenols, quinones or melanization (e.g. in encapsulation)	Various insects	Refs. in Charnley (1984), Götz and Boman (1986)

References

Barkman JJ (1969) The influence of air pollution on bryophytes and lichens. Air pollution, PUDOC, Wageningen, pp 197–209

Bishop JA (1972) An experimental study of the cline of industrial melanism in *Biston betularia* (L.) (Lepidoptera) between urban Liverpool and rural North Wales. J Anim Ecol 41:209–243

Bishop JA, Cook LM, Muggleton J, Seaward MRD (1975) Moths, lichens and air pollution along a transect from Manchester to North Wales. J Appl Ecol 12:83–98

Bishop JA, Cook LM, Muggleton J (1978a) The response of two species of moths to industrialisation in northwest England I. Polymorphism for melanism. Philos Trans R Soc London Ser B 281:489–515

Bishop JA, Cook LM, Muggleton J (1978b) The response of two species of moths to industrialisation in northwest England II. Relative fitness of morphs and population size. Philos Trans R Soc London Ser B 281:517–540

Brakefield PM (1984a) Ecological studies on the polymorphic ladybird *Adalia bipunctata* in The Netherlands. I. Population biology and geographical variation in melanism. J Anim Ecol 53: 761–774

Brakefield PM (1984b) Ecological studies on the polymorphic ladybird *Adalia bipunctata* in The Netherlands. II. Population dynamics, differential timing of reproduction and thermal melanism. J Anim Ecol 53:775–790

Brakefield PM (1984c) Selection along clines in the ladybird *Adalia bipunctata* in The Netherlands: a general mating advantage to melanics and its consequences. Heredity 53:37–49

Brakefield PM (1985) Polymorphic Müllerian mimicry and interactions with thermal melanism in ladybirds and a soldier beetle: a hypothesis. Biol J Linn Soc 26:243–267

Brakefield PM, Lees DR (1987) Melanism in *Adalia* ladybirds and declining air pollution in Birmingham. Heredity 59:273–277

Brakefield PM, Willmer PG (1985) The basis of thermal melanism in the ladybird *Adalia bipunctata*: differences in reflectance and thermal properties between the morphs. Heredity 54:9–14

Britton G, Goodwin TW, Harriman GE, Lockley WJS (1977) Carotenoids of the ladybird beetle, *Coccinella septempunctata*. Insect Biochem 7:337–345

Charnley AK (1984) Physiological aspects of destructive pathogenesis in insects by fungi: a speculative review. In: Anderson JM, Rayner ADM, Walton DWH (eds) Invertebrate microbial interactions. Cambridge Univ Press, pp 229–270

Clarke CA, Sheppard PM (1966) A local survey of the distribution of industrial melanic forms in the moth *Biston betularia* and estimates of selective values of-these in an industrial environment. Proc R Soc London Ser B 165:424–439

Clarke CA, Mani GS, Wynne G (1985) Evolution in reverse: clean air and the peppered moth. Biol J Linn Soc 26:189–199

Cook LM, Jacobs ThMGM (1983) Frequency and selection in the industrial melanic moth *Odontoptera bidentata*. Heredity 51:487–494

Cook LM, Mani GS (1980) A migration-selection model for the morph frequency variation in the peppered moth over England and Wales. Biol J Linn Soc 13:179–198

Cook LM, Askew RR, Bishop JA (1970) Increasing frequency of the typical form of the peppered moth in Manchester. Nature (London) 227:1155

Cook LM, Mani GS, Varley ME (1986) Postindustrial melanism in the peppered moth. Science 231:611–613

Creed ER (1966) Geographic variation in the two-spot ladybird in England and Wales. Heredity 21:57–72

Creed ER (1971a) Melanism in the two-spot ladybird *Adalia bipunctata* in Great Britain. In: Creed ER (ed) Ecological genetics and evolution. Blackwell, Oxford, pp 134–151

Creed ER (1971b) Industrial melanism in the two-spot ladybird and smoke abatement. Evolution 25:290–293

Creed ER, Lees DR, Bulmer MG (1980) Pre-adult viability differences of melanic *Biston betularia* (L.) (Lepidoptera). Biol J Linn Soc 13:251–262

Endler JA (1984) Progressive background matching in moths and a quantitative measure of crypsis. Biol J Linn Soc 22:187–231

Ford EB (1975) Ecological genetics, 4th edn. Chapman & Hall, London

Götz P, Boman HG (1986) Insect immunity. In: Kerkut GA, Gilbert LI (eds) Comprehensive insect physiology, biochemistry and pharmacology, vol 8. Pergamon, Oxford, pp 453–484

Graham SM, Watt WB, Gall LF (1980) Metabolic resource allocation vs. mating attractiveness: adaptive pressures on the 'alba' polymorphism of *Colias* butterflies. Proc Natl Acad Sci USA 77:3615–3619

Hawksworth DL, Rose F (1970) Qualitative scale for estimating sulphur dioxide air pollution in England and Wales using epiphytic lichens. Nature (London) 227:145–148

Hiruma K, Riddiford LM, Hopkins TL, Morgan TD (1985) Roles of dopa decarboxylase and phenoloxidase in the melanization of the tobacco hornworm and their control by 20-hydroxyecdysone. J Comp Physiol B 155:659–669

Houston KJ, Hales DF (1980) Allelic frequencies and inheritance of colour pattern in *Coelophora inaequalis* (F.) (Coleoptera: Coccinellidae). Aust J Zool 28:669–677

Howlett RJ, Majerus MEN (1987) The understanding of industrial melanism in the peppered moth *(Biston betularia)* (Lepidoptera: Geometridae). Biol J Linn Soc 30:31–44

Kettlewell HBD (1955) Selection experiments on industrial melanism in the Lepidoptera. Heredity 9:323–342

Kettlewell HBD (1956) Further selection experiments on industrial melanism in the Lepidoptera. Heredity 10:287–301

Kettlewell HBD (1958) The importance of the microenvironment to evolutionary trends in the Lepidoptera. Entomologist 91:214–224

Kettlewell HBD (1973) The evolution of melanism. Clarendon, Oxford

Komai T (1956) Genetics of ladybeetles. Adv Genet 8:155–188

Kramer KJ, Morgan TD, Hopkins TL, Roseland CR, Aso Y, Beeman RW, Lookhart GL (1984) Catecholamines and β-alanine in the red flour beetle, *Tribolium casteneum*. Roles in cuticle sclerotization and melanization. Insect Biochem 14:293–298

Kyriacou CP, Burnet B, Connolly KJ (1978) The behavioural basis of overdominance in competitive mating success at the *ebony* locus of *D. melanogaster*. Anim Behav 26:1195–1207

Lees DR (1981) Industrial melanism: genetic adaptation of animals to air pollution. In: Bishop JA, Cook LM (eds) Genetic consequences of man-made change. Academic Press, London New York, pp 129–176

Lees DR, Creed ER (1975) Industrial melanism in *Biston betularia*: the role of selective predation. J Anim Ecol 44:67–83

Liebert TG, Brakefield PM (1987) Behavioural studies on the peppered moth *Biston betularia* and a discussion of the role of pollution and epiphytes in industrial melanism. Biol J Linn Soc 31:129–150

Lusis JJ (1961) On the biological meaning of colour polymorphism of lady beetle *Adalia bipunctata* L. Latv Ent 4:3–29

Majerus MEN, O'Donald P, Kearns PWE, Ireland H (1986) Genetics and evolution of female choice. Nature (London) 321:164–167

Mani GS (1980) A theoretical study of morph ratio clines with special reference to melanism in moths. Proc R Soc London Ser B 210:299–316

Mani GS (1982) A theoretical analysis of the morph frequency variation in the peppered moth over England and Wales. Biol J Linn Soc 17:259–267

Mikkola K (1979) Resting site selection of *Oligia* and *Biston* moths (Lepidoptera:Noctuidae and Geometridae). Ann Entom Fenn 45:81–87

Mikkola K (1984) On the selective forces acting in the industrial melanism of *Biston* and *Oligia* moths (Lepidoptera: Geometridae and Noctuidae). Biol J Linn Soc 21:409–421

Muggleton J (1978) Selection against the melanic morphs of *Adalia bipunctata* (two-spot ladybird): a review and some new data. Heredity 40:269–280

Muggleton J, Lonsdale D, Benham BR (1975) Melanism in *Adalia bipunctata* L (Col Coccinellidae) and its relationship to atmospheric pollution. J Appl Ecol 12:451–464

Needham AE (1978) Insect biochromes. In: Rockstein M (ed) Biochemistry of insects. Academic Press, London New York, pp 253–305

Safranek L, Riddiford LM (1975) The biology of the black larval mutant of the tobacco hornworm, *Manduca sexta*. J Insect Physiol 21:1931–1938

Sargent TD (1983) Melanism in *Phigalia titea* (Lepidoptera: Geometridae): A 14-year record from central Massachusetts. J NY Entomol Soc 91:75–82

Seaward MRD, Hitch CJB (eds) Atlas of the lichens of the British Isles, vol 1. Inst Terrest Ecol, Cambridge

Shellard HC (1976) Wind. In: Chandler TJ, Gregory S (eds) The climate of the British Isles. Longman, London New York, pp 39–73

Steward RC (1985) Evolution of resting behaviour in polymorphic 'industrial melanic' moth species. Biol J Linn Soc 24:285–293

Thompson V (1984) Polymorphism under apostatic and aposematic selection. Heredity 53: 677–686

Tutt JW (1896) British moths. Routledge

Wit T de (1983) Lichens as indicators for air quality. Environ Monit Assess 3:273–282

Acknowledgements: I thank Dr. L.M. Cook for helpful comments during the course of developing this paper and Mr. T.G. Liebert for many stimulating discussions about industrial melanism.

Ethanol Adaptation and Alcohol Dehydrogenase Polymorphism in *Drosophila*: From Phenotypic Functions to Genetic Structures

J. R. DAVID [1]

1 Introduction

Besides its central position for any biological study, evolutionary biology is also known for its many debates, controversies and dilemmas. Possibly the most conspicuous controversy of the last 2 decades is the neutralist-selectionist controversy concerning the significance of biochemical polymorphism (Kimura 1983). Also, the widespread use of the word adaptation, because of its several but imprecise meanings, has been also the subject of ancient (Williams 1966) or recent criticisms (Gould and Vrba 1982; Krimbas 1984).

Trying to show that allozyme polymorphism is maintained by selective forces, numerous biologists have correlated genetic structures to environmental, natural or experimental, variables (Ayala 1975; Nevo 1978). Doing so, they mainly asked the question: is the genetic polymorphism beneficial to fitness? It is, however, possible to put the problem in a reverse way and ask: does better fitness favor, or lead to, genetic polymorphism? In the latter case, we recognize that studying any ecological adaptation one must not start from a genetic structure but from physiological or ethological traits. For reaching a complete understanding of an enzymatic polymorphism, the chosen model has to fulfill several favorable conditions:

1. The system must be polymorphic;
2. The protein has to be related to one or several functions;
3. Differences in enzyme activity must exist between alleles and genotypes;
4. The physiological functions must have some adaptive significance in natural populations;
5. It is also favorable, but not compulsory, that genetic differences exist between geographic populations.

The alcohol dehydrogenase (ADH) system in *Drosophila melanogaster* has long been recognized as a favorable case (Clarke 1975) and is presumably the most studied among animals (see reviews by David 1977; Van Delden 1982). However, active investigations have progressively uncovered the great complexity of the model. We will

1 Laboratoire de Biologie et Génétique Evolutives, CNRS, F-91190-Gif-sur-Yvette, France

Population Genetics and Evolution
G. de Jong (ed.)
© Springer-Verlag Berlin Heidelberg 1988

see that if environmental alcohol is a strong selective factor in nature, and ADH is a key enzyme for this adaptation, the role of the polymorphism of this enzyme still appears controversial, calling for new and deeper investigations.

2 Environmental Alcohol as a Selective Factor

The evolutionary success of angiosperms and the concomitant appearance of sweet fruits has produced a large amount of sugar-rich resources. When decaying, these resources are mainly attacked by yeasts which excrete alcohol in their environment. Alcohol is primarily a toxic product which repels most possible consumers, among which insects are the most frequent.

Drosophila species are saprophagous and their basic ecological niche consists of decaying plant materials and fungi. Numerous species, belonging to various evolutionary radiations in the family, have adapted themselves to decaying sweet resources and are known as fruit breeders.

If the presence of ethanol in fermenting fruits is a significant selective pressure, we expect to find a higher tolerance in fruit breeders than in non-fruit breeders. Experimental studies have confirmed this expectation (David and Van Herrewege 1983). On the average, the tolerance (expressed by the lethal concentration killing 50% of flies: LC 50) was found higher in the former case (3.50±0.25) than in the latter (1.76±0.18). Especially, some non-fruit breeders are extremely sensitive to alcohol, with an LC 50 of about 1%. However, an overlap exists between the two groups and some non-fruit breeders, such as *D. busckii*, *D. funebris* and several species of the *D. repleta* group are quite tolerant (unpublished results). The amount of alcohol found in fermenting fruit remains generally below 4% (Van Delden 1982; McKechnie and Morgan 1982) but, in the present global environment artificial, man-made fermentations may contain more than 10% alcohol. *D. melanogaster* is well known for its capacity to proliferate in wine cellars during vintage time, and it exhibits a large amount of geographic variability with respect to alcohol tolerance. For example, French populations, which are often found in wine cellars, have an LC 50 of 17%, while Afrotropical populations, which mainly breed on fruits, and are ancestral for the species, have a tolerance of 7% (David and Bocquet 1975). Two other species are presently known to exhibit a high alcohol tolerance and to bread in man-made fermentation: they are *D. virilis*, found in breweries (David and Kitagawa 1982) and *D. lebanonensis*, the most tolerant species of all (28%), which is found in Spanish wine cellars (David et al. 1979).

3 Larval Versus Adult Adaptation

The previous observations strongly suggest that alcohol, as a selective force, acts during the larval development while tolerance is generally measured, for convenience, on adults. Ecological studies showed that very sensitive adults may be attracted and can feed on fermenting fruits without harmful effects (David and Van Herrewege 1983). On the other hand, significant differences between larvae and adults could be expected, especially since molecular studies have shown that the transcription of the *Adh* gene is different in larvae and adults (Benyajati et al. 1983). Comparative studies were thus undertaken to compare larval and adult tolerance in geographic races of *D. melanogaster* and a strong parallelism has been found in both stages (Van Herrewege and David 1985). Also, we have evidence that many non-fruit breeders are very sensitive as larvae, and are killed by fermentation when put on a sweet laboratory medium seeded with baker's yeast. It seems therefore that larval and adult tolerances are strongly correlated, although some exceptions may be found.

Drosophila larvae and pupae are hosts to numerous parasitoid wasp species (Carton et al. 1986) which may encounter resources with a high level of alcohol. Only wasp females are attracted to fermenting resources and they may remain in contact with alcohol for several hours. In this case, the parasitoid larva is protected from alcohol by the detoxificating capacity of the host and only the adult female seems directly exposed to selection. Experimental analyses of various species confirmed these expectations (Boulétreau and David 1981): only the species breeding in *D. melanogaster* exhibited a strong tolerance (up to 10%) to alcohol. More surprisingly, in this latter case, only the females were tolerant, while the males remained sensitive with an LC 50 of about 2.5%. The study of *Drosophila* parasites confirms the selective effect of ethanol but surprisingly, it has been impossible, up to now, to show the occurrence of an ADH enzyme in these hymenoptera.

4 Physiological Role of ADH

From a large number of experimental studies, a consensus exists for believing that in the fruitfly, like in humans, ethanol metabolism occurs in the following pathway:

ethanol → acetaldehyde → acetate → acetyl-coA → energy .

The first step is mediated, exclusively, by ADH, as shown by the comparison of wild-type and ADH-negative flies (David et al. 1976). Acetaldehyde, a highly toxic product (LC 50 about 0.5%) is not found in the environment and, in the live fly, is immediately detoxified into acetate. This second step is still controversial. It was first believed that the very active aldehyde oxidase (AO) found in all *Drosophila* species could mediate this transformation. This hypothesis was clearly ruled out by the analysis of AO negative mutants which are able to metabolize ethanol and acetaldehyde (David et al. 1978, 1984). From in vitro studies, it was also suggested that ADH had a dual function and was responsible for both steps (Heinstra et al. 1983). This seems unlikely to occur in vivo since the lack of ADH prevents any metabolic utilization of ethanol but not of acetaldehyde. The occurrence of a specific, not yet clearly identified acetaldehyde dehydrogenase seems the most likely hypothesis (Lietaert et al. 1985).

In *D. melanogaster*, the presence of an active ADH is responsible for two different phenotypic traits: alcohol tolerance and alcohol utilization.

Tolerance is measured by exposing the flies to high, toxic concentrations of alcohol. The biological activity of ADH can be measured by subtracting the LC 50 of an ADH negative strain from that of wild-type flies with similar genetic backgrounds (David et al. 1976). Methanol is not detoxified, and this makes sense if we consider that this alcohol is not a substrate for ADH. Results for longer chain alcohols are more surprising since there is no correlation between enzymatic and biological activity. The main physiological effect concerns ethanol (LC 50 increases from 2.5 up to 17%), while this alcohol is a poor enzyme substrate (Day et al. 1974; Hovik et al. 1984). Longer chain alcohols are better substrates but they are poorly detoxified. Finally, the best substrates are secondary alcohols, like isopropanol and isobutanol, which are almost not detoxified (Van Herrewege et al. 1980) and which are strong inhibitors of the enzyme activity (Papel et al. 1979; Van Herrewege et al. 1980). This last point is reasonable if we consider that secondary alcohols are converted into ketones which are metabolic dead ends. From an evolutionary point of view, we could suggest that the specificity of *Drosophila* ADH for secondary alcohols reflects some ancestral, but still unknown function, which also exists in non-fruit breeders (David and Van Herrewege 1983). The role of ADH in ethanol metabolism, which clearly has an adaptive value in nature, may be considered as a fairly recent evolutionary event, i.e. an exaptation in the sense of Gould and Vrba (1983): the physiological response is not provided by a better catalytic activity, but by an increase in the amount of the protein.

The metabolic capacity to use alcohols is a phenotypic trait which can be completely dissociated from tolerance by treating adult flies with a very low, non-toxic concentration. In *D. melanogaster*, for example, a linear response, i.e. an increase of life span with ethanol concentration, is observed between 0 and 2% (Van Herrewege and David 1974). The effects of adding a low amount of alcohol is spectacular: with 0.4%, life duration was extended by 2 to 5 days (David and Van Herrewege 1983), while no difference was observed in ADH negative flies. Indeed, this metabolic test is the best way to demonstrate that species very sensitive to ethanol are, however, able to use very low amounts of ethanol. By analogy with *D. melanogaster*, it is assumed that this utilization is based on the presence of an active ADH, although its very low amount cannot be detected by the usual spectrophotometric analysis. Further investigations on such sensitive species would be welcome in order to confirm the presence of ADH and to provide a further argument on an ancestral, still unknown, function for that enzyme.

5 Selection for Better Tolerance or Metabolic Utilization

It was hypothesized that *D. melanogaster* has adapted itself to artificially fermented resources in the fairly recent past, and that the very high tolerance of French populations is a novel evolutionary event (David and Bocquet 1975). According to this hypothesis, European populations, with respect to alcohol tolerance, could be considered as African populations submitted to directional selection and having reached a selective plateau, so that any further increase would be impossible. Several natural populations with various levels of tolerance were submitted to artificial selection and, in all cases, a strong response, typical of any quantitative trait was observed: rapid increase at the beginning during the first ten generations; then an attenuation and a plateau after about 30 generations (David and Bocquet 1977; David et al. 1977). It was shown that the French strain harbored a large amount of additive genetic variability since, in 30 generations, its tolerance increased from 17 to 27%, so that some balancing selection should exist in nature, as for other quantitative traits.

The most ethanol-tolerant species, *D. lebanonensis* (LC 50: 28%) is also remarkably efficient in its use of very low amounts of alcohol for increasing life duration (David et al. 1979). It was thus suspected that adaptation occurred primarily for better utilization and that the very high tolerance, which does not seem necessary in nature, could be a by-product of a selection for better utilization. This hypothesis was checked in *D. melanogaster*. First, the three lines selected for higher tolerance were studied for metabolic capacity and the conclusions were divergent (Van Herrewege and David 1980). In one case ethanol utilization was clearly better in the tolerant-selected line than in controls, but in two other cases it was not. A reciprocal investigation was undertaken by selecting an African and a French population for better ethanol utilization (Van Herrewege and David 1984). In about ten generations, the mean survival in the presence of 2% ethanol was clearly improved, from 7 to 12 days. Measurement of tolerance of the selected strains again produced divergent results: an increase of tolerance in the African strain but no variation in the French one.

From these experiments, we may conclude that ethanol tolerance and ethanol utilization behave like any other quantitative trait. Both are based on an active ADH but the two phenotypes are, at least in part, genetically independent.

6 Geographic Variability of ADH Polymorphism in Natural Populations

Natural populations of *D. melanogaster* harbor two widespread alleles, F and S, and it has been known for a long time that frequencies vary quite regularly with latitude. The fact that similar patterns, i.e. an increase of the F allele with latitude, are observed on different continents such as North America (Johnson and Schaffer 1973), Australasia (Oakeshott et al. 1982) and between Europe and Africa (David et al. 1986), is a powerful argument for assuming some adaptive significance of the cline, especially if we consider that numerous climatic variables, among which temperature and rainfall, correlate with latitude.

Since ethanol tolerance is also highly correlated to latitude, it is not surprising to find that tolerance and the Adh^F frequency are also highly correlated: r = 0.89 for European and African populations (David et al. 1986). This observation raises a major problem: is there a causal relationship between the two traits?

7 Genetic Basis of Geographic Variability in Ethanol Tolerance

It has been known for a long time (Day et al. 1974 and many other investigators) that the F allele exhibits, on average, a higher enzyme activity than S. Also numerous investigators have treated laboratory populations with ethanol and observed, with some exceptions, an increase in F frequency (see Van Delden 1982, for a review). From these observations the following adaptive scenario could be suggested in the case of natural populations. Utilizing more alcoholic resources, French populations were submitted to a selective pressure which resulted in a higher tolerance itself mediated by a higher frequency of the F allele.

To check this hypothesis, several investigations were undertaken. Firstly, homozygous FF and SS lines were extracted from a French and an Afrotropical population:

no significant difference was observed between genotypes from the same geographic area (David et al., in prep.). More precisely, French SS flies were tolerant to alcohol, while FF African flies were still very sensitive. Secondly, SS flies from the same geographic area were submitted to selection for increased ethanol tolerance (David et al., in prep.). A clear, progressive response was observed in both lines, and the increase in tolerance was even faster than starting from a polymorphic population. Thirdly, crosses were undertaken between a very tolerant French line and a very sensitive African one. Reciprocal F1 were identical and intermediate between parents, but the F2 were more sensitive than the F1 (David et al., in prep.). Such a result suggests some epistatic interactions and thus a complex genetic determinism. Fourthly, chromosome transfers were made using balancer stocks between a sensitive and a tolerant line but the results failed to yield any clear-cut conclusion: in particular, there was no apparent linkage between ethanol tolerance and the second chromosome which carries the *Adh* gene (David et al., in prep.). All these observations are concordant in showing that the great genetic difference observed between Afrotropical and European flies is not primarily based on the great difference observed in *Adh* allelic frequencies.

8 An Attempt to Falsify the Hypothesis of a Balanced Polymorphism at the Adh Locus

There is still some possibility that the latitudinal clines of *Adh*, on different continents, could be due to stochastic events and thus reflect the history of colonizations. We may suppose, for example, that when Europe was colonized from African populations, a bottleneck occurred and led to an increase of *Adh-F* frequency. The parallel cline observed in North America would then reflect several independent colonizations followed by a partial mixing of natural populations: tropical America would have been colonized by African populations and temperate America by European populations. Under this neutralist hypothesis, the genetic stability of geographic populations, e.g. in France, would be maintained as a Hardy-Weinberg equilibrium. In such a case, an artificial manipulation of the frequencies in a natural population would lead to a new, more or less permanent, equilibrium. An experiment was done in a fairly remote and isolated locality of southern France, close to Saint-Girons (Capy et al. 1987). Homozygous SS lines were extracted from the native population, then pooled into a mixed population to restore genetic variability except at the *Adh* locus, and a large number of artificially grown flies was released in the locality in late June, at the beginning of the breeding season. During the release, native flies, identified by FF genotype, were totally outnumbered. After the release, SS and FS flies were collected over a period of about 3 months, but the overall frequency of S alleles dropped to reach the initial level of 5% at the beginning of November. From a temperature survey it is likely that the released flies persisted for three generations before their disappearance. Incidentally, the region of the experiment was quite distant from vineyards and artificial fermentations.

These results, which confirm a previous, similar experiment made in Tropical Africa (David 1983) do not confirm the neutralist hypothesis and show that the released flies and their offspring were counterselected. We cannot, however, conclude that a balanced polymorphism in the native population existed with an equilibrium frequency of 5% of S. The elimination of SS flies may be due to inbreeding, to a modification of their behavior consecutive to laboratory culture, to long-distance dispersal or to some ecological accident. An alcohol selection seems to be ruled out since no vineyards exist in or near the place of release.

9 Concluding Remarks and Prospects

Present knowledge concerning the effect of environmental ethanol are summarized in Fig. 1. Facing a high alcohol concentration, flies have to increase their tolerance which can be achieved by several processes: decreasing the toxic penetration via either ingestion or by respiration; increasing the metabolic detoxification or excretion; increasing the tolerance of the nervous system to hemolymph concentration of ethanol. This last point has been insufficiently investigated, although some recent results (Cohan and Graf 1985) could involve such a mechanism. If the environmental con-

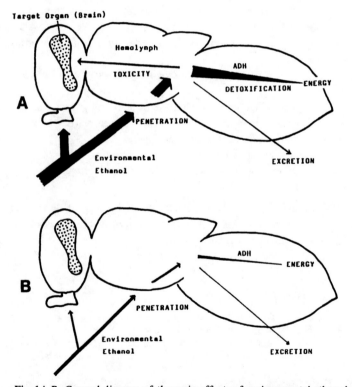

Fig. 1A,B. General diagram of the main effects of environmental ethanol upon the fruitfly. **A** Influence of a high concentration: tolerance may result from a decreased penetration, from greater detoxification through ADH and greater excretion, all these processes decreasing the hemolymph concentration. A modification of the nervous system is also possible, leading to higher tolerance of the target organs. **B** Influence of a low concentration: the main observed effect is energy production mediated by ADH and further metabolic steps

centration is low, ethanol appears mainly as a resource, provided an active ADH is present. In this case, natural selection could favor an increased penetration and a decreased excretion. Any of these processes can be genetically modified and this overall complexity explains why alcohol tolerance and alcohol utilization can be selected as quantitative traits.

It therefore appears that the problem of *Adh* polymorphism in *D. melanogaster*, which at the beginning seemed quite simple, has progressively turned into a very complex story. From the results of numerous investigations, several conclusions can be drawn:

1. Alcohol is a significant environmental pressure;
2. Alcohol adaptation involves several, interacting genetic systems and thus appears as a quantitative trait;
3. *Adh* polymorphism is presumably selected;
4. Selective pressures on this locus may involve alcohol but also other unknown forces.

Besides these conclusions, numerous unanswered questions remain, including:

1. What is the amount of ethanol in the larval resources of various species?
2. Is there, on the average, more ethanol in the larval resources used by temperate populations of *D. melanogaster*?
3. What is the role of isopropanol, which appears quite abundantly in some fermentations?
4. What is the intra- and interspecific relationship between ADH activity and physiology?
5. Are there regular physiological differences between *Adh* alleles and genotypes?
6. Are there functions for ADH other than metabolization of environmental alcohols?
7. Is the tolerance of the nervous system genetically variable?
8. What is the significance of behavioral variations towards environmental alcohol?
9. Are there other selective forces (e.g. acetic acid production) related to alcoholic fermentation?

It is hoped that most of these questions will receive a satisfying answer during the next decade, thus improving the status of alcohol adaptation in drosophilids as an important model in the current evolutionary paradigm.

References

Ayala FJ (1975) Genetic differentiation during the speciation process. Evol Biol 8:1–78

Ayala F (ed) (1976) Molecular evolution. Sinauer, Sunderland, Mass

Benyajati C, Spoerel N, Haymere H, Ashburner M (1983) The messenger RNA for alcohol dehydrogenase in *Drosophila melanogaster* differs in its 5' end in different developmental stages. Cell 33:125–133

Boulétreau M, David JR (1981) Sexually dimorphic response to host habitat toxicity in *Drosophila* parasitic wasp. Evolution 35:395–399

Capy P, David JR, Delay B, Louis J (1987) Essai de modification de la structure génétique d'une population naturelle de *Drosophila melanogaster* dans le sud de la France. C R Coll Biol Popul (Lyon) (pp 300–305)

Carton Y, Boulétreau M, Lenteren J van, Alphen J van (1986) The *Drosophila* parasitic wasps. In: Ashburner M, Carson H, Thompson JN (eds) The genetics and biology of *Drosophila*, vol 3. Academic Press, London New York, pp 348–394

Clarke B (1975) The contribution of ecological genetics to evolutionary theory: detecting the direct effect of natural selection on particular polymorphic loci. Genetics 79:101–113

Cohan FM, Graf JD (1985) Latitudinal cline in *Drosophila melanogaster* for knockdown resistance to ethanol fumes and for rate of response to selection for further resistance. Evolution 39:278–293

David JR (1977) Signification d'un polymorphisme enzymatique: la déshydrogénase alcoolique chez *Drosophila melanogaster*. Ann Biol 16:451–472

David JR (1983) An attempt to modify allelic frequencies at the *Adh* locus of a *Drosophila melanogaster* population in a tropical environment. Genet Sél Evol 15:495–500

David JR, Bocquet C (1975) Similarities and differences in latitudinal adaptation of two *Drosophila* sibling species. Nature (London) 257:588–590

David JR, Bocquet C (1977) Genetic tolerance to ethanol in *Drosophila melanogaster*: increase by selection and analysis of correlated responses. Genetica 47:43–48

David JR, Herrewege J van (1983) Adaptation to alcoholic fermentation in *Drosophila* species: relationships between alcohol tolerance and larval habitat. Comp Biochem Physiol 74A:283–288

David JR, Kitagawa O (1982) Possible similarities in ethanol tolerance and latitudinal variations between *Drosophila virilis* and *D. melanogaster*. Jpn J Genet 57:89–95

David JR, Bocquet C, Arens MF, Fouillet P (1976) Biological role of alcohol dehydrogenase in the tolerance of *Drosophila melanogaster* to aliphatic alcohols: utilization of an ADH-null mutant. Biochem Genet 14:989–997

David JR, Bocquet C, Fouillet P, Arens MF (1977) Tolérance génétique à l'alcool chez *Drosophila*: comparaison des effets de la sélection chez *D. melanogaster* et *D. simulans*. C R Acad Sci Paris 285:405–408

David JR, Bocquet C, Van Herrewege J, Fouillet P, Arens MF (1978) Alcohol metabolism in *Drosophila melanogaster*: the most active aldehyde oxidase produced by the *Aldox* locus is useless. Biochem Genet 16:203–211

David JR, Herrewege J van, Monclus M, Prevosti A (1979) High ethanol tolerance in two distantly related *Drosophila* species: a probable case of recent convergent adaptation. Comp Biochem Physiol 63:53–56

David JR, Daly K, Herrewege J van (1984) Acetaldehyde utilization and toxicity in *Drosophila* adults lacking alcohol dehydrogenase or aldehyde oxidase. Biochem Genet 22:1015–1029

David JR, Mercot H, Capy P, McEvey S, Herrewege J van (1986) Alcohol tolerance and *Adh* gene frequencies in European and African populations of *Drosophila melanogaster*. Genet Sel Evol 18:405–416

Day TH, Hillier PC, Clarke B (1974) The relative quantities and catalytic activities of enzymes produced by alleles at the alcohol dehydrogenase locus in *Drosophila melanogaster*. Biochem Genet 11:155–165

Delden W van (1982) The alcohol dehydrogenase polymorphism in *Drosophila melanogaster*: selection at an enzyme locus. Evol Biol 15:187–222

Gould SJ, Vrba E (1982) Exaptation – a missing term in the science of form. Paleobiology 8:4–15

Heinstra PWH, Eisses KT, Schoonen WGEJ, Aben W, Winter AJ de, Horst DJ van der, Marrewijk WJA van, Beenakkers AMT, Scharloo W, Thorig GEW (1983) A dual function of alcohol dehydrogenase in *Drosophila*. Genetica 60:129–137

Herrewege J van, David JR (1974) Utilisation de l'alcool éthylique dans le métabolisme énergétique d'un insecte: influence sur la durée de survie des adultes de *Drosophila melanogaster*. C R Acad Sci Paris 279:335–338

Herrewege J van, David JR (1980) Alcohol tolerance and alcohol utilization in *Drosophila*: partial independence of two adaptive traits. Heredity 44:229–235

Herrewege J van, David JR (1984) Extension of life duration by dietary ethanol in *Drosophila melanogaster*: response to selection in two strains of different origins. Genetica 63:61–70

Herrewege J van, David JR (1985) Ethanol tolerance in *Drosophila melanogaster*: parallel variations in larvae and adults from natural populations. D I S 61:180–181

Herrewege J van, David JR, Grantham R (1980) Dietary utilization of aliphatic alcohols by *Drosophila*. Experientia 36:846–847

Hovik R, Winberg JO, McKinley-McKee J (1984) *Drosophila melanogaster* alcohol dehydrogenase. Substrate stereospecificity of the Adh^F alleloenzyme. Insect Biochem 14:345–351

Johnson FM, Schaffer HE (1973) Isozyme variability in species of the genus *Drosophila*. VII: Genotype-environment relationships in populations of *Drosophila melanogaster* from eastern United States. Biochem Genet 10:149–163

Kimura M (1983) The neutral theory of molecular evolution. Cambridge Univ Press

Krimbas C (1984) On adaptation, neo-Darwinian tautology and population fitness. Evol Biol 17: 1–57

Lietaert MC, Libion-Mannaert M, Wattiaux de Coninck S, Elens A (1985) *Drosophila melanogaster* aldehyde dehydrogenase. Experientia 41:57–58

McKechnie SW, Morgan P (1982) Alcohol dehydrogenase polymorphism of *Drosophila melanogaster*: aspects of alcohol and temperature variation in the larval environment. Aust J Biol Sci 35:85–93

Nevo E (1978) Genetic variation in natural populations: patterns and theory. Theor Popul Biol 13:121–177

Oakeshott JG, Gibson JB, Anderson PR, Knibb NR, Anderson DG, Chambers GK (1982) Alcohol dehydrogenase and glycerol-3-phosphate dehydrogenase clines in *Drosophila melanogaster* on different continents. Evolution 36:86–96

Papel I, Henderson M, Herrewege J van, David JR, Sofer W (1979) *Drosophila* alcohol dehydrogenase activity in vitro and in vivo: effects of acetone feeding. Biochem Genet 17:553–563

Williams GC (1966) Adaptation and natural selection. A critique of some current evolutionary thought. Princeton Univ Press, Princeton

Multigenic Selection in *Plantago* and *Drosophila*, Two Different Approaches

W. van Delden[1]

1 Introduction

The transformation of Darwin's original theory of evolution into what is currently known as the neo-Darwinian theory has been accomplished by the insertion and integration of the findings of population genetics and ecology. Especially population genetics theory has contributed substantially to the understanding of the mechanisms of evolution and has provided the tools for a quantitative approach of micro-evolutionary processes. Population genetics models supply the tools for quantitative predictions of the effects of natural selection, genetic drift, gene flow, mutation and nonrandom mating on the genetic composition of populations. We now know, e.g. that selection is not necessarily only purifying as Darwin thought, but that particular selection mechanisms, like overdominance and some cases of frequency-dependent selection will lead to stable allele frequency equilibria and thus contribute to the maintenance of genetic variation. Population genetics theory has even generated quite revolutionary new ideas with respect to the mechanisms of evolution, as in the neutral theory of molecular evolution. In this theory, so brilliantly summarized in Kimura's (1983) recent book, the main driving forces in molecular evolution are thought to be mutation and genetic drift, while selection only takes a subordinate position.

It thus appears that population genetics provides the framework for a reliable and relatively easy way to analyze evolutionary processes in nature. The tasks to fulfil seem obvious and imply the estimation of relevant parameters in field populations and the subsequent introduction of the figures at the proper places in the appropriate formula. In practice, however, severe difficulties and complications arise, which, unfortunately, lead to less optimistic prospects. The problems involved are generally not due to inaccurate theory but have other origins. Among those problems a prominent one is the difficulty in obtaining reliable estimates of the necessary parameters like mutation rates, effective population sizes, gene flow rates and selection coefficients. Especially the last item is of paramount importance for the study of natural selection, the cornerstone of the neo-Darwinian theory. However, estimation of fit-

---

1 Department of Genetics, University of Groningen, Kerklaan 30, NL-9751 NN Haren, The Netherlands

Population Genetics and Evolution
G. de Jong (ed.)
© Springer-Verlag Berlin Heidelberg 1988

ness values in natural populations with enough precision appears extremely difficult and even under well-controlled laboratory conditions small fitness differences are hard to detect (Lewontin 1974).

A further problem comes from the increasing complexity associated with multilocus situations. Theoretical, analytic approaches of selection in multilocus cases soon become very intricate with increasing numbers of loci. Simulation studies in this field, as e.g. by Franklin and Lewontin (1970), provide results which are not easily understood intuitively. Experiments dealing with polymorphisms at several loci are an arduous task, the more so by the trivial fact of the drastic increment of the numbers of genotypes associated with an increase in the number of loci. In cases where genotypic fitnesses have to be estimated this may invoke nearly insolvable practical problems. Still the interactions of loci are common facts as they arise from the organization of the genetic material, both in a structural and a functional sense. The former association of genes comes from their linear organization on chromosomes, creating physical bonds. This may lead to disturbing phenomena as the case of hitchhiking where, due to linkage disequilibrium between two genes, of which one is under selection, it may appear that the other, neutral gene is also under selection, because its allele frequencies change in time. In the case of functional relations between genes, pleiotropy and epistatic interactions will occur. The events at a particular locus are then dependent on those at other loci.

Complexes of interacting loci are often hard to analyze, still they are often significant from an evolutionary viewpoint (Hedrick et al. 1978). A special case is the loci involved in the determination of quantitative characters which are often important fitness traits. The analysis of these genes each with small individual effects, which are generally strongly influenced by the environment, occurs with the methods of quantitative genetics. In S. Wright's shifting balance theory of evolution the multilocus approach is emphasized as interacting loci with pleiotropic effects constituting the core of the theory (Wright 1977).

In the next sections I will treat two quite different approaches to multigene studies, originating from recent work in progress in our laboratory. One comes from the study of the population genetics of plant species belonging to the genus *Plantago*; in this case considerable parts of the total phenotype are involved. The other study was originally focussed on the alcohol dehydrogenase *(Adh)* polymorphism in *Drosophila melanogaster*, but extended to greater parts of the genome.

2 Population Genetics of Plantago Species

2.1 Interspecific Differences in Genetic Structure

The five *Plantago* species occurring in the Netherlands were studied in the framework of a multidisciplinary project involving demographic, ecophysiological and population genetics aspects. A basic question in the project is why some plant species are able to preserve themselves in particular habitats, where other species are unable to do so. The aim is thus to analyze the specific adaptations to particular habitats. For an understanding of the population genetics of the species under study, knowledge of

Table 1. Breeding systems in the *Plantago* species occurring in the Netherlands

	P. major	*P. lanceolata*	*P. coronopus*	*P. media*	*P. maritima*
Selfcompatibility	+	−	+	−	−
Gynodioecy	−	+	+	−	+
Diploidy	+	+	+	−[a]	+

[a] Tetraploid.

the breeding system is essential. Table 1 shows that considerable differences occur among the five plantain species. Two species (*P. major* and *P. coronopus*) are self-compatible and thus self-fertilization is possible, while *P. lanceolata*, *P. media* and *P. maritima* are unable to self-fertilize, due to a gametophytic self-incompatibility system. *P. major* and *P. media* are true hermaphrodites, as the other species have both hermaphrodite and male-sterile plants due to a complex genetic system, involving both nuclear and cytoplasmic genes (Van Damme and Van Delden 1982; Van Damme 1983). *P. media* is tetraploid in the Netherlands, but in other parts of Europe also diploids occur. As the breeding system has a profound impact on, among other things, the level of inbreeding and the rate of gene flow, the amount, the organization and the geographic distribution of genetic variation will predictably vary among the *Plantago* species. This is illustrated by the work of H. Van Dijk, who extensively surveyed the allozyme variation in *P. major*, *P. lanceolata* and *P. coronopus* (Van Dijk and Van Delden 1981; Van Dijk 1985a,b). A summary of his results is given in Table 2 together with some preliminary data for *P. media* (P. van Dijk, unpublished) and *P. maritima*. *P. major* is characterized by low $\bar{P}$ values ($\bar{P}$ is the mean fraction of polymorphic loci over populations) and low $\bar{H}_e$ values (H_e is the mean gene diversity per locus: $H_e = 1 - \Sigma p_i^2$, where $p_1, p_2 ... p_i$ are the allele frequencies) compared with *P. lanceolata* and *P. coronopus*. The low variability in *P. major* is not surprising as the species is self-compatible. *P. coronopus* is also capable of self-fertilization, however, the greater genetic variability found in this species may be associated with the occurrence of gynodioecy, which implies at least cross-fertilization in females (male-steriles). The F-statistics show profound differences between the three species: F_{IS}, describing the relation between the observed fraction of heterozygotes and the fraction expected

Table 2. Levels of allozyme variation and F-statistics for the *Plantago* species in the Netherlands

	$\bar{P}$	$\bar{H}_e$	F_{IS}[b]	F_{ST}[b]
P. major	0.15	0.047	0.79	0.14
P. lanceolata	0.33	0.127	0.12	0.04
P. coronopus	0.31	0.088	0.07	0.11
P. media[a]	0.50	0.320	−	−
P. maritima	0.30	0.100	−	−

[a] Tetraploid.
[b] − Not determined.

from Hardy-Weinberg ratios, is high in *P. major*. In this species nearly completely homozygosity for allozyme variants is found. *P. major* further shows high F_{ST} values, indicating considerable population differentiation. The genetic distances between populations are much greater in *P. major* than in *P. lanceolata*, while *P. coronopus* is intermediate in this respect (Van Delden 1985).

2.2 Ecotypic Differentiation in Plantago Major

In *P. major* two subspecies are distinguished: *P. major major* and *P. major pleiosperma* (Pilger 1937; Mølgaard 1976). Van Dijk and Van Delden (1981) found that the two subspecies can be discriminated by subspecies-specific alleles at two allozyme loci: a glutamate-oxaloacetate-transaminase *(Got-1)* locus and a phosphoglucomutase *(Pgm-1)* locus. The subspecies differ in quite a number of morphological and life-history characters. They also differ in the habitat type in which they occur. Ssp *major* is found predominantly on roadsides and lawns, while ssp *pleiosperma* inhabits unstable sites like riversides and arable land. Conspicuous differences between the two subspecies are e.g. the number of seeds per capsule and the total number of seeds per spike which are higher in ssp *pleiosperma*. Individual seed weight, however, is higher in ssp *major*, while the relative growth rate is higher in ssp *pleiosperma*. Flowering time starts earlier in ssp *pleiosperma* than in ssp *major*. These differences between the subspecies are apparently associated with the habitats in which they occur: the characteristic traits in ssp *pleiosperma* provide rapid growth and reproduction which is advantageous in unstable habitats, while the specific traits found in ssp *major* are of selective advantage in the stable, though stressful, environments that this subspecies has to cope with.

Both within ssp *major* and ssp *pleiosperma* different ecotypes can be distinguished. In ssp *major* the most conspicuous differences are among genotypes which occur on lawns, and roadsides, while in ssp *pleiosperma* riverside and ruderal types can be distinguished. The ecotypes differ for quite a number of morphological and life-history characters which interact in such a way that specific phenotypes come into being, each of which is adapted to a particular habitat. Apparently, several morphological and physiological characters are involved, based on many genes. The total integrated phenotype, adapted to particular stress factors, is thus accomplished by the adjustment of several traits, bringing about the specific plant dimensions, characteristic for a particular habitat. The plant form in the lawn ecotype of ssp *major* e.g. is prostrate, which provides escape from mowing and grazing (Warwick and Briggs 1980), while the roadside ecotype is highly trampling-resistant due to its particular phenotypic constitution (Blom 1977). Van Dijk (1985a) has performed reciprocal transplantation experiments of seedlings of different ecotypes of both ssp *major* and ssp *pleiosperma* in four habitats where *P. major* occurs: in a path, a lawn, a riverside and a dune valley. Substantial differences in survival and reproduction were found. Ecotypes survived best or even exclusively in their own sites; seed production occurred only in a riverside and the corresponding ecotype of ssp *pleiosperma* produced the highest number of seeds. This illustrates the specific adaptations of each ecotype to its definite site.

2.3 Localization of Ecologically Important Traits in Plantago Major

Van Dijk (1984) has performed an analysis of the localization of genes controlling ecologically important quantitative traits. For this purpose he has used allozyme variants as markers. In preliminary crosses the allozyme markers had been classified into a number of linkage groups. In subsequent crosses genotypes, different both for allozyme alleles at a number of loci and for several morphological and life-history characters, were taken as parents. The F_1 was self-fertilized and the resulting F_2 progeny was assayed both for allozyme genotype and for the quantitative traits. The traits studied were: leaf length, leaf blade length, leaf blade width, petiole length, inflorescence length, inflorescence position, spike length, scape length (and the ratio of a number of these traits), seed number per capsule, number of inflorescences and flowering time. In this way associations, if any, between allozyme loci and quantitative characters could be established. Figure 1 gives, as an example, the outcome of such an F_2 analysis for a cross between an ssp *major* lawn ecotype (G_1) and an ssp *pleiosperma* riverbank ecotype (Z_2). The parents were homozygous for different alleles at seven allozyme loci. They were further different for a number of quantitative traits of which only leaf length is shown in Fig. 1. The ssp *major* ecotype has short, round leaves, while the ssp *pleiosperma* ecotype has longer, oval leaves as shown by the two columns at the left. The mean F_1 and F_2 values for leaf length are also given. In the F_2 generation segregation for the various allozyme variants occurs and, as shown here for *Pgm-1*, the mean values for leaf length for each separate genotype can be scored.

In the case of *Pgm-1*, genotypes differ significantly in leave length, but for other allozyme loci no differences are found among the genotypes. The most probable explanation for this phenomenon is that the genes controlling leaf length are located in the vicinity of *Pgm-1*. Van Dijk (1984) has performed a number of comparable crosses, involving several traits. The resulting overall picture of the associations between allozyme loci and quantitative traits, derived from his analysis, is given in Table 3.

It turns out that most of the quantitative traits involved in ecotype-specific adaptations are localized in linkage group 1, while a few are associated in linkage group 2 and 4. The genes controlling these characters are thus clustered and linked with particular allozyme loci, especially with *Pgm-1*. The characters which contribute to the composition of a specific ecotype are apparently combined in a gene complex.

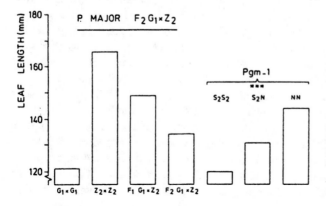

Fig. 1. Association of leaf length and *Pgm-1* genotype in the F_2 of a cross between an ssp *major* lawn ecotype (G_1) and an ssp *pleiosperma* riverbank ecotype (Z_2). Means for parents, F_1 and F_2 are given for comparison. The differences between the *Pgm-1* genotypes are significant $(P < 0.001)$

Table 3. The association of quantitative traits with linkage groups in *P. major* (+ indicates a significant correlation; – indicates no correlation)

Characters	Linkage group		
	1 (Pgm-1)	2 (Got-1)	4 (6 Pgd-2/Got-2/Est-4)
Leaf length	+	–	+
Blade length/width ratio	+	+	–
Petiole length/blade length ratio	+	–	–
Inflorescence position	–	–	–
Spike/scape ratio	+	–	+
Seed number/capsule	+	–	–
Flowering time	+	–	–

2.4 Localization of Ecologically Important Traits in Plantago Lanceolata

It is interesting to contrast the findings for *P. major* with the situation met with in *P. lanceolata*. Ecotypic differentiation in this species seems to be absent at first sight in view of the small interpopulation differences. Wolff and Van Delden (1987), however, have shown by means of quantitative genetic analyses that consistent genetic differences exist among populations. The most pronounced ecotypic differences occur between hayfield and pasture populations. The former are, compared to the latter, characterized by, among other things, a small leave angle, a low number of leaves and spikes, long scapes, early flowering dates, heavy seeds and long cotyledons. The differences between the two main ecotypes in the Netherlands in this species are in majority adaptations related with the height and the density of the vegetations in which they occur. In hayfields, e.g. light transmission is impaired and strong competition for light will occur and consequently plants with long, erect leaves will be better competitors. In pastures, on the other hand, light is not a limiting factor but plants have to cope with regular defoliating through grazing. In this case prostrate plants with short leaves and spikes have higher chances for survival and reproduction.

Wolff (1987) has analyzed the chromosomal localization of some of the quantitative characters involved in the ecotypic differentiation in *P. lanceolata* comparable to the way as described before for *P. major*. The results are quite contrary in both species. In *P. lanceolata* closely related characters are localized in different linkage groups, while in *P. major* they appear to be clustered in a few complexes. It can be argued that integrated gene complexes, as found in *P. major*, procure high fitness due to the well-balanced sets of alleles. Clustering of genes will diminish the destabilizing effects of recombination. The reason for such profound differences between the species is unclear. It is tempting to relate the clustered location of correlated characters in *P. major* and the scattered locations in *P. lanceolata* to the differences in breeding systems between both species. However, in *P. major* the self-fertilization rate is very high and individuals are homozygous to a high extent. Destruction of favourable gene complexes through recombination in local populations is thus not likely. It may be that ecotypic differentiation in *P. major* is more pronounced than in *P. lanceolata* (which is evident from the morphology) which will make much higher demands upon the genetic architecture of the correlated traits. Further research (Wolff, unpublished) is directed to this problem.

3 The Association Between Allozyme Polymorphisms and Inversions in Drosophila Melanogaster

3.1 The Alcohol Dehydrogenase Polymorphism

In the previous case of *Plantago* species it was the phenotype that formed the starting point in a multilocus study, with the final aim to penetrate the genetic organization of ecologically important traits. Such a procedure may, at the end, lead to an analysis of the effects of the individual genes and their interactions. The opposite approach was taken in a study originally aimed at an individual allozyme locus, the alcohol dehydrogenase *(Adh)* locus, in *Drosophila melanogaster*.

The *Adh* polymorphism has been extensively studied (reviewed in: Van Delden 1982; Zera et al. 1984). Two common alleles, *Fast (F)* and *Slow (S)* occur in nature. The three corresponding genotypes differ in their in vitro biochemical properties. Adh^{FF} homozygotes exceed Adh^{SS} homozygotes in ADH activity, while heterozygotes possess intermediate enzyme activity. The *Adh* genotypes further differ in in vitro enzyme stability at temperatures over 40 °C. ADH from Adh^{SS} homozygotes is more stable than that of Adh^{FF} homozygotes and heterozygotes. The *Adh* polymorphism is one of the well-documented cases of allozyme polymorphisms where in vitro kinetic differences among genotypes proved to be associated with fitness differences under relevant stress conditions (Van Delden 1982). In this respect it has been shown that great differences in survival occur among phenotypes on ethanol supplemented food. Survival is positively correlated with in vitro ADH activity. In polymorphic populations under ethanol stress this leads to a rapid increase in frequency of the Adh^F allele. On regular food, on the other hand, some kind of balancing selection seems to operate (Bijlsma-Meeles and Van Delden 1974; Van Delden et al. 1978). Differential survival of *Adh* genotypes at high temperature could possibly be inferred from the previously mentioned genotypic differences in in vitro temperature stability. Survival tests at high temperature, however, have provided equivocal results (Johnson and Powell 1974; Van Delden and Kamping 1980; Kohane and Parsons 1986).

3.2 Genetic Changes in a Greenhouse Population

In a long-term experiment, extending over 13 years (1972-1985), we have tracked a seminatural *D. melanogaster* population in a closed compartment of a tropical greenhouse for allele frequency changes at a number of allozyme loci (Van Delden 1984; Van Delden and Kamping 1987). Substantial allele frequency changes at the *Adh* locus were observed, though the direction of change was reversed after 1976 (Fig. 2). It was further found that at the *αGpdh* locus also considerable allele frequency changes

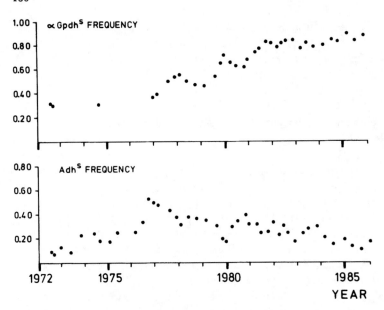

Fig. 2. Allele frequency changes at the *Adh* and *αGpdh* loci in the greenhouse population. Assays were generally based on samples of 150 flies

occurred. The *Adh*[S] and *αGpdh*[S] frequencies were negatively correlated (r = 0.82; *P* < 0.001) during 1976–1985 and linkage disequilibrium was found between both genes from 1976 up to and including 1983. During that period the presence of *In(2L)t*, a long inversion on the left arm of the second chromosome, together with at least one short inversion in that chromosome region, were noticed in the greenhouse population. As *αGpdh* is included in *In(2L)t* and *Adh* lies just outside, while *In(2L)t* is invariably associated with the *Adh*[S]*αGpdh*[F] configuration, the occurrence of linkage disequilibrium between the two loci in the population can be explained. An intriguing question is what is the reason for the genetic changes in the population? In view of the large population sizes in the greenhouse, genetic drift can be excluded. As selection can be considered as a potential force in bringing about allele frequency changes, it may operate either on each of both allozyme loci, separately, on multiple *Adh-αGpdh* genotypes or on *In(2L)t*. The greenhouse population will frequently experience high temperature stress during hot summers and an association of temperature and genetic composition is thus not unlikely. This hypothesis was tested in 1981 by starting laboratory populations at 20°, 25° and 29.5 °C (populations were also set on medium supplemented with 18% ethanol). These laboratory populations were founded with flies from the greenhouse. Figure 3 shows that depending on temperature, quite different responses occur as far as allele frequency changes are concerned, proving that temperature is a potent selective force. It is striking that in the 29.5 °C population, when *Adh*[S] vs *αGpdh*[S] is plotted, a linear regression line is obtained that is highly similar to the comparable regression line in the greenhouse population during 1976 to 1985 (y = −0.46 x + 0.66 and y = −0.51 x + 0.63 respectively). The directions of the allele frequency changes, however are opposite!

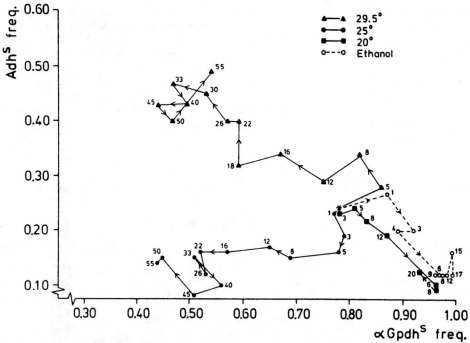

Fig. 3. Allele frequency changes at the *Adh* and *αGpdh* loci in laboratory populations derived from the greenhouse populations in 1981 and thereafter kept at different temperatures and on ethanol supplemented food. The *numbers* indicate the generations in which a sample was taken for allele frequency determination (sample size: 150 flies)

A further finding is that at 29.5 °C linkage disequilibrium is preserved, while at the lower temperatures linkage equilibrium is restored after 20 generations. Determination of the frequencies of *In(2L)t* and the short inversion after 35 generations at 29.5 °C (Table 4) shows a significant increase in the frequencies of both inversion types compared to the situation in the greenhouse in 1981, when the laboratory populations were founded. An explanation for these findings is selective advantage of *In(2L)t* at high temperature.

3.3 The Role of In(2L)t

Egg-to-adult survival was measured at 20°, 25°, 29.5° and 33 °C. Six chromosomal genotypes were tested: homozygotes for *In(2L)t* (LL); homozygotes for a small inversion of the left arm of the second chromosome (RR); homozygotes for standard second chromosomes (NN) and the three chromosomal heterozygotes LR, LN and RN. All chromosomes carried the $Adh^S \alpha Gpdh^F$ alleles. At 20° and 25 °C no differences in survival among the genotypes occurred, at 29.5 °C only the extremes LN and NN differed significantly (survival percentages 82.3 and 64.1 respectively). Table 5 gives survival at 33 °C (in angles). Overall egg-to-adult survival is given together with egg-to-pupa and pupa-to-adult survival. The most conspicuous differences are found for

Table 4. The occurrence of inversions in the left arm of the second chromosome in samples of flies taken from the greenhouse population in 1981 and in samples taken from the 29.5 °C population in generation 35 (n is the number of chromosomes tested)

Allozyme genotype (Adh-αGpdh)	n	Greenhouse population (1981) Fraction of chromosomes with:		
		Standard sequence	Short inversion	In(2L)t
FF	15	0.057	0.021	0
FS	129	0.521	0.151	0
SF	20	0.026	0.021	0.057
SS	28	0.125	0.021	0
Totals	192	0.729	0.214	0.057
Allozyme genotype (Adh-αGpdh)	n	29.5 °C population (generation 35) Fraction of chromosomes with:		
		Standard sequence	Short inversion	In(2L)t
FF	34	0.110	0.060	0
FS	84	0.225	0.195	0
SF	59	0.025	0.025	0.245
SS	23	0.070	0.045	0
Totals	200	0.430	0.325	0.245

Table 5. Mean overall egg-to-adult, egg-to-pupa and pupa-to-adult survival of different chromosome genotypes (symbols explained in text) at 33 °C[a]

Egg-adult survival:	LR	LN	LL	RR	RN	NN
	56.5	55.7	45.9	39.3	38.7	32.1
Egg-pupa survival:	LN	LR	RR	LL	RN	NN
	63.4	60.4	52.7	52.3	51.3	50.5
Pupa-adult survival:	LR	LN	LL	RN	RR	NN
	73.7	68.5	65.4	53.3	52.7	43.6

[a] All chromosomes carried $Adh^S \alpha Gpdh^F$. The genotypes are listed in decreasing order of survival. Genotypes not significantly different at the 5% level (Tukey's test for multiple comparisons) are underlined by a common line. Survival is expressed in angles.

pupa-adult survival and the genotypes with $In(2L)t$ survive more often than those lacking the inversion. The results confirm the hypothesis that inversion genotypes perform better at high temperature.

Egg-to-adult survival is of course only a single fitness component and for a full understanding of the genetic changes in the greenhouse populations more comprehensive fitness estimates are needed. However, the magnitude of the survival differences indicate that a rapid increase in inversion frequency at high temperature is to be expected.

This explains the situation in the greenhouse. Initially present at low frequencies at the introduction into the greenhouse, the inversions must have increased in frequency during hot summers, when temperature easily exceeds 40 °C on warm summer days. Weather registrations show that from 1972 up to and including 1976 the summers became warmer, culminating in the exceptional warm summer of 1976. This is witnessed by the rapid increase of the frequency of Adh^S, associated with $In(2L)t$ ($\alpha Gpdh$ frequencies were only rarely assayed till 1977). The following summers were relatively cold and the decline in Adh^S and $\alpha Gpdh^F$ frequencies indicates a gradual decrease in inversion frequency, probably due to a selective disadvantage of the inversion-carrying genotypes. The laboratory experiments at different temperatures confirm this view.

The data are in conformity with the geographic distribution of $In(2L)t$ (and other inversion also), which shows a decline in frequency with latitude and thus with temperature. This clinal distribution has been observed on the three continents North America, Asia and Australia which have been studied so far (Mettler et al. 1977; Inoue and Watanabe 1979; Knibb et al. 1981), and the same relation seems to exist for Africa and Europe (D. Sperlich, personal communication). For the Adh^S and $\alpha Gpdh^F$ alleles comparable negative correlations with latitude have been observed (reviewed in Van Delden and Kamping 1987), which may originate from the unique association of these alleles with $In(2L)t$.

This leads to the compelling question whether the allele frequency changes observed in our experiments are merely a reflection of the alleles being carried along passively with $In(2L)t$. The results obtained from the laboratory populations seem to contradict this suggestion (Fig. 3). The $\alpha Gpdh^F$ frequency change is probably temperature-dependent, in view of the opposite behaviour found at 20° and 25 °C. At 20 °C the $\alpha Gpdh^F$ allele is apparently disadvantageous to its bearer, but at 25° and 29.5 °C it is advantageous. A further argument comes from Table 4 when the frequencies of Adh^S and $\alpha Gpdh^F$ in standard chromosomes, without inversions, are calculated. These frequencies for $\alpha Gpdh^F$ are 0.11 and 0.31 for the greenhouse and the 29.5 °C population respectively, strongly pointing to a selective advantage at 29.5 °C in the absence of inversions. The Adh^S frequencies in standard chromosomes, on the other hand, are 0.21 and 0.22 respectively, indicating the absence of high temperature effects for this gene. The changes in Adh frequencies, apart from the ones caused by the association with $In(2L)t$ are probably independent of temperature. In the absence of inversions the Adh^S frequency will decline till 0.20 irrespective of temperature and stay around that value.

The presence of polymorphisms both for chromosomal rearrangements and allozyme variants thus leads to complex situations as far as the effects of selection are concerned. We have tried to unravel this knot and to separate the chromosomal from the allozyme effects. This was accomplished by the construction of populations which are either polymorphic for $In(2L)t$ and the standard chromosome or monomorphic for the standard second chromosome arrangement. Within each of both categories three different allozyme constitutions were constructed: (1) populations polymorphic for Adh only, (2) polymorphic for $\alpha Gpdh$ only and (3) polymorphic for both Adh and $\alpha Gpdh$. Also, an additional population type was created: polymorphic for $In(2L)t$ and the standard second chromosome and monomorphic for both Adh and $\alpha Gpdh$.

Monomorphic populations possessed the S allele for Adh and the F allele for $\alpha Gpdh$. The initial frequencies in the polymorphic populations were 0.50. The resulting seven population types were each set under seven different environmental regimes: four different temperatures; two high larval density levels (one with a generation interval of 14 days, the other with a 21-day interval) and medium supplemented with ethanol (12%). $In(2L)t$ and allozyme frequencies were determined in the polymorphic populations in the course of time. The comparison of the changes in the various population types enables an analysis of the selection effects due to inversions and allozymes (and their interactions) in each environment. Table 6A gives $In(2L)t$ frequencies in generations 6 and 12 for the population type with no variation for Adh and $\alpha Gpdh$. This allows an estimation of selection for $In(2L)t$ on a standard $Adh^S \alpha Gpdh^F$ background. Table 6A confirms the conclusions obtained from the greenhouse population and the populations derived from it, that at lower temperatures $In(2L)t$ is at a disadvantage. It is remarkable that the two populations with high crowding levels which differ in the time of transfer (14 or 21 days) are so distinct in $In(2L)t$ frequency. Further research into the reason for this has proved that individuals homozygous for the inversion have a much longer development time than homozygotes for the standard chromosome sequence (Van Delden and Kamping, unpublished). This explains the rapid decline in $In(2L)t$ frequency under the 14-day scheme as here genotypes with rapid development are favoured.

Table 6B gives, as an example, the results of an assay of the frequencies of $In(2L)t$, Adh^S and $\alpha Gpdh^S$ in generation six at 20 °C. The high similarity between the two replicate populations proves that the different frequencies observed for the population types cannot be ascribed to sampling effects. Without going into details Table 6B

Table 6. A Frequency of $In(2L)t$ in seven environments. HD_{14} and HD_{21} refer to populations with generation intervals of 14 and 21 days respectively (sample size: 150 flies)
B $In(2L)t$ and allozyme frequencies in generation 6 in populations kept at 20 °C (+ is polymorphic; – is monomorphic; I and II are replicate populations)

A

| Generation | Environment | | | | | | |
	20 °C	25 °C	29 °C	32 °C	HD_{14}	HD_{21}	Ethanol
6	0.28	0.36	0.56	0.51	0.19	0.42	0.19
12	0.18	0.29	0.52	0.47	0.16	0.41	0.11

B

| Population type | | | I | | | II | | |
$In(2L)t$	Adh	$\alpha Gpdh$	$In(2L)t$	Adh^S	$\alpha Gpdh^S$	$In(2L)t$	Adh^S	$\alpha Gpdh^S$
+	–	–	0.28	–	–	0.28	–	–
+	+	+	0.44	0.44	0.57	0.46	0.46	0.48
+	+	–	0.29	0.29	–	0.34	0.34	–
+	–	+	0.43	–	0.57	0.36	–	0.64
–	+	+	–	0.68	0.32	–	0.57	0.41
–	+	–	–	0.38	–	–	0.32	–
–	–	+	–	–	0.39	–	–	0.42

shows that allele frequency changes at one locus are not independent from the genetic constitution at the other locus, while also the chromosomal constitution interferes. An association of the *Adh* and *αGpdh* polymorphisms has been described by Cavener and Clegg (1978, 1981). They proposed a fitness interaction between the two loci related with the NADH/NAD$^+$ ratio. The preliminary data obtained for the each of the seven environmental conditions point to an association of inversion and allozyme frequencies in some cases, but complete independence in others (Van Delden and Kamping, unpublished).

4 Concluding Remarks

Comparison of the *Plantago* and the *Drosophila* cases proves that two quite different approaches may provide highly useful information with respect to the organization and significance of genetic variation. In the *Plantago* study allozymes merely served as markers and the emphasis is on ecologically important quantitative traits which are of paramount importance for the local adaptation of the plants. The *Drosophila* case shows that single gene polymorphisms are under selection, while interactions with other genes and with chromosomal variants occur. A basic question here is how these variants contribute to important fitness traits.

Though the study of multilocus situations is arduous, it will be rewarding in the end as it contributes to the necessary understanding of phenotypic variation, which is the point of impact for natural selection. In this field population genetics can certainly provide the necessary background for further analyses. As the phenotype as a whole is under selection it is of paramount importance to study the interaction of genes in relation to the environment. Basically, population genetics in its present form is sufficient for the understanding of evolutionary processes. Both the *Plantago* and the *Drosophila* cases, however, demonstrate the need for an approach in which population genetics is integrated with physiology and developmental biology.

References

Bijlsma-Meeles E, Delden W van (1974) Intra- and interpopulation selection concerning the alcohol dehydrogenase locus in *Drosophila melanogaster*. Nature (London) 255:148–149

Blom CWPM (1977) Effects of trampling and soil compaction on the occurrence of some *Plantago* species in coastal sand dunes. II. Trampling and seedling establishment. Oecol Plant 12:363–381

Cavener DR, Clegg MT (1978) Dynamics of correlated genetic systems. IV Multilocus effects of ethanol stress environments. Genetics 90:629–644

Cavener DR, Clegg MT (1981) Multigenic response to ethanol in *Drosophila melanogaster*. Evolution 35:1–10

Damme JMM van (1983) Gynodioecy in *Plantago lanceolata L.* II Inheritance of three male sterility types. Heredity 50:253–273

Damme JMM van, Delden W van (1982) Gynodioecy in *Plantago lanceolata L.* I Polymorphism for plasmon type. Heredity 49:305–320

Delden W van (1982) The alcohol dehydrogenase polymorphism in *Drosophila melanogaster*. Selection at an enzyme locus. Evol Biol 15:187–222

Delden W van (1984) The alcohol dehydrogenase polymorphism in *Drosophila melanogaster*, facts and problems. In: Wöhrmann K, Loeschcke V (eds) Population biology and evolution. Springer, Berlin Heidelberg New York, pp 127–142

Delden W van (1985) The significance of genetic variation in plants as illustrated by *Plantago* population. In: Haeck J, Woldendorr JW (eds) Structure and functioning of plant populations 2. Phenotypic and genotypic variation in plant populations. Elsevier/North Holland Biomedical Press, Amsterdam, pp 219–240

Delden W van, Kamping A (1980) The alcohol dehydrogenase polymorphism in populations of *Drosophila melanogaster* IV. Survival at high temperature. Genetica 51:179–185

Delden W van, Kamping A (1987) The association between the polymorphisms at the *Adh*, *αGpdh* loci and the *In(2L)t* inversion in *Drosophila melanogaster* in relation with temperature. Evolution (in press)

Delden W van, Boerema AC, Kamping A (1978) The alcohol dehydrogenase polymorphism in *Drosophila melanogaster* I. Selection in different environments. Genetics 90:161–191

Dijk H van (1984) Genetic variability in *Plantago* species in relation to their ecology 2. Quantitative characters and allozyme loci in *P. major*. Theor Appl Genet 68:43–52

Dijk H van (1985a) Genetic variability in *Plantago* species in relation to their ecology. Thesis, Univ Groningen

Dijk H van (1985b) Allozyme genetics, self-incompatibility and male sterility in *Plantago lanceolata*. Heredity 54:53–63

Dijk H van, Delden W van (1981) Genetic variability in *Plantago* species in relation to their ecology I. Genetic analysis of the allozyme variation in *P. major* subspecies. Theor Appl Genet 60:285–290

Franklin I, Lewontin RC (1970) Is the gene the unit of selection? Genetics 65:707–734

Hedrick P, Jain S, Holden L (1978) Multilocus systems in evolution. Evol Biol 11:101–184

Inoue Y, Watanabe TK (1979) Inversion polymorphisms in Japanese natural populations of *Drosophila melanogaster*. Jpn J Genet 54:69–82

Johnson FM, Powell A (1974) The alcohol dehydrogenases of *Drosophila melanogaster*: frequency changes associated with heat and cold shock. Proc Natl Acad Sci USA 71:1783–1784

Kimura M (1983) The neutral theory of molecular evolution. Cambridge Univ Press

Knibb WR, Oakeshott JG, Gibson JB (1981) Chromosome inversion polymorphisms in *Drosophila melanogaster*. I. Latitudinal clines and associations between inversions in Australasian populations. Genetics 98:833–847

Kohane MJ, Parson PA (1986) Environment-dependent fitness differences in *Drosophila melanogaster*: temperature, domestication and the alcohol dehydrogenase locus. Heredity 57:289–304

Lewontin RC (1974) The genetic basis of evolutionary change. Columbia Univ Press, New York

Mettler LE, Voelker RA, Mukai T (1977) Inversion clines in natural populations of *Drosophila melanogaster*. Genetics 87:169–176

Mølgaard P (1976) *Plantago major* ssp *major* and ssp *pleiosperma*. Morphology, biology and ecology in Denmark. Bot Tidsskr 71:31–56

Pilger R (1937) Plantaginaceae. In: Engler A (ed) Das Pflanzenreich 4 (Heft 102) Leipzig

Warwick SI, Briggs D (1980) The genecology of lawn weeds. 5. The adaptive significance of different growth habit in lawn and roadside populations of *Plantago major* L. New Phytol 85:289–300

Wolff K (1987) Genetic analysis of ecological relevant morphological variability in *Plantago lanceolata* L. II Localization of quantitative trait loci. Theor Appl Genet 73:903–914

Wolff K, Delden W van (1987) Genetic analysis of ecological relevant morphological variability in *Plantago lanceolata* L. I Population characteristics. Heredity 58:183–192

Wright S (1977) Evolution and the genetics of populations, vol. 3. Univ Chicago Press

Zera AJ, Koehn RK, Hall JG (1984) Allozymes and biochemical adaptation. In: Kerkut GA, Gilbert LI (eds) Comprehensive insect physiology, biochemistry and pharmacology, vol 10. Pergamon, New York, pp 633–674

The Functional Significance of Regulatory Gene Variation: The α-Amylase Gene-Enzyme System of *Drosophila melanogaster*

A. J. KLARENBERG[1]

1 Genetic Variation in Gene-Enzyme Systems

One of the main topics in population and evolutionary genetics has been the question whether electrophoretic variants, encoded by a single structural gene and involved in the same biochemical reaction, are selectively neutral (Nei and Koehn 1983). Only for a very few enzyme loci have selective differences between enzyme variants been established; α-amylase in the fruitfly, *Drosophila melanogaster*, is among the best characterized (Scharloo 1984). The significance of regulatory variation has only been recently recognized (MacIntyre 1982; Wilson 1985).

2 Regulatory Gene Variation

With respect to the spatial organization of gene-enzyme systems, a gene is composed of structural parts and its tightly linked regulatory elements. The structural parts of a gene code for a specific enzyme with a specific function in a biochemical pathway in metabolism. Regulatory elements are always tightly linked to the structural parts of a gene and in most instances *cis*-acting. Distant regulatory genes, located on the same chromosome or on other chromosomes, nearly always show *trans*-action on the structural parts of the gene and its closely linked *cis*-acting regulatory elements. In population genetics studies, regulatory genes are defined by their function (MacIntyre 1982) as revealed in the phenotype. The most important characteristics of regulation in gene-enzyme systems are: (1) quantitative regulation, (2) developmental regulation, (3) tissue-specific regulation, (4) environmental regulation.

---

1 Vakgroep Populatie- en Evolutiebiologie, Rijksuniversiteit Utrecht, Padualaan 8,
  NL-3584 CH Utrecht, The Netherlands

Population Genetics and Evolution
G. de Jong (ed.)
© Springer-Verlag Berlin Heidelberg 1988

3 The α-Amylase Gene-Enzyme System of Drosophila Melanogaster

Selection on the α-amylase gene-enzyme system in *Drosophila melanogaster* operates only when starch is the limiting factor for survival in the food medium; selective differences are caused by differences in starch digestion, which can be related to large differences in total α-amylase activity associated with different structural α-amylase variants (de Jong and Scharloo 1976; Hoorn and Scharloo 1980). A main question is whether selection acts on structural differences of the enzyme generating differences in total activity or effects on kinetic parameters or stability, or that regulatory aspects of gene action are involved.

The *trans*-acting regulatory gene *map*-PMG affects the tissue-specific expression of α-amylase in the posterior midgut (PMG) of the adult (Abraham and Doane 1978). Figure 1a shows the α-amylase midgut activity pattern *(map)* variants of adults. Map-PMG was localized at 1.2 centi-Morgans (*map*-PMG, 2–79.0; Klarenberg et al. 1986b) distal to the two structural loci for α-amylase (*Amy*, 2–77.8; Bahn 1967). The regulation of α-amylase midgut patterns in the adult anterior midgut (AMG) is under control by a regulatory gene which is in the same region of the second chromosome as *map*-PMG (Abraham and Doane 1978). α-Amylase midgut patterns in third instar larvae are regulated by *cis*-acting elements which map less than 0.1 cM from the *Amy* structural region (Klarenberg et al. 1986b). The two very closely linked *Amy* genes (e.g. *Amy*-4,6) represent a gene duplication (Boer and Hickey 1986; Levy et al. 1985).

4 Selection on Regulatory Variants of α-Amylase

Analysis of strains homozygous for second chromosomes extracted from three cage populations of different geographic origin revealed a consistent non-random association between *Amy* and midgut activity pattern *(map)* variants of α-amylase in adults and third-instar larvae (Klarenberg and Scharloo 1986). *Amy*-1 and *Amy*-4.6 are the major variants in these populations. From the frequencies of *map* variants (Fig. 1b) it is evident that the *Amy*-1 variant was more frequently found in combination with α-amylase activity in the adult AMG only. In contrast, *Amy*-4.6 was recorded with

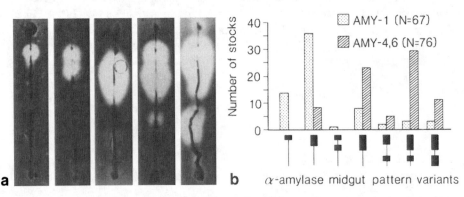

Fig. 1. a Adult α-amylase midgut patterns of *D. melanogaster*. **b** Frequencies of α-amylase midgut pattern variants

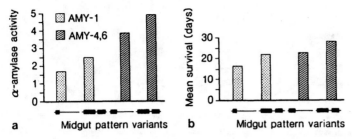

Fig. 2. a α-Amylase activities of different *Amy* and *map* variants. **b** Mean survival times at 50% mortality on starch of the *Amy* and *map* variants

higher frequencies of *map* variants with α-amylase activity both in AMG and PMG. D values for *Amy* and *map*-PMG show a moderate intensity of linkage disequilibrium, all positive and all in the same range. The observation of significant linkage disequilibria between *Amy* and adult *map*-PMG variants, in the same direction and of the same magnitude, occurring in populations of *D. melanogaster* of different geographic origin, meets an important criterion for a selective cause for linkage disequilibrium (Lewontin 1974). Linkage between *Amy* and the *map*-PMG gene has certainly contributed to linkage disequilibrium, when selection is acting on one or both genes. Because most combinations between the structural variants and the adult *map* variants are present in all three populations, one would not expect that genetic drift would produce linkage disequilibrium in the same direction and of similar strength.

Selection experiments performed with adults suggest indeed that some of the combinations of *Amy* and adult *map* variants tested show differences in viability under conditions when selection is operating on the α-amylase system (Klarenberg et al. 1986a). Our data show that within *Amy*-1 and *Amy*-4.6 isogenic strains there is a correlation between total α-amylase activity and *map* variation in adults (Fig. 2).

5 Species Differences, Other Gene-Enzyme Systems

Midgut pattern variation of several other carbohydrases was found to be independent of the α-amylase patterns (Klarenberg 1986). Figure 3 shows differences in α-amylase and α-mannosidase midgut patterns in larvae and adults of different *Drosophila* species respectively.

6 Conclusion

We have shown that regulatory variants for α-amylase are subject to selection. The genetic variation of regulatory genes and elements, the independent regulation of parts of the *Amy* locus in different tissues and in different life stages and the production of new regulatory patterns by recombination suggests a large potential for rapid evolution of regulatory systems in *D. melanogaster*. Other digestive gene-enzyme systems also show variation in regulatory midgut patterns.

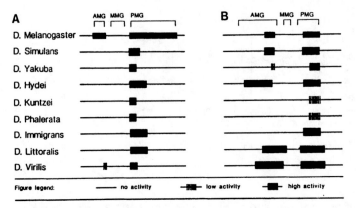

Fig. 3. A α-Amylase midgut patterns of third-instar larvae in different *Drosophila* species. **B** Adult α-mannosidase midgut patterns in different *Drosophila* species

References

Abraham I, Doane WW (1978) Genetic localization of tissue-specific expression of amylase structural genes in *Drosophila melanogaster*. Proc Natl Acad Sci USA 75:4446–4450

Bahn E (1967) Crossing over in the chromosomal region determining amylase isozymes in *Drosophila melanogaster*. Hereditas 58:1–12

Boer PH, Hickey DA (1986) The alpha-amylase gene in *Drosophila melanogaster*: nucleotide sequence, gene structure and expression motifs. Nucl Acids Res 21:8399–8411

De Jong G, Scharloo W (1976) Environmental determination of selective significance or neutrality of amylase variants in *Drosophila melanogaster*. Genetics 84:77–94

Hoorn AJW, Scharloo W (1980) Functional significance of amylase polymorphism in *Drosophila melanogaster*. III. Ontogeny of amylase and some α-glucosidases. Biochem Genet 18:51–63

Klarenberg AJ (1986) Genetic variation in regulation of α-amylase expression in *Drosophila melanogaster*. Thesis, Rijksuniv Utrecht

Klarenberg AJ, Scharloo W (1986) Nonrandom association between structural *Amy* and regulatory *map* variants in *Drosophila melanogaster*. Genetics 114:875–884

Klarenberg AJ, Sikkema K, Scharloo W (1986a) Functional significance of regulatory *map* and structural *Amy* variants in *Drosophila melanogaster*. Heredity (in press)

Klarenberg AJ, Visser AJS, Willemse MFM, Scharloo W (1986b) Genetic localization and action of regulatory genes and elements for tissue-specific expression of α-amylase in *Drosophila melanogaster*. Genetics 114:1131–1145

Levy JN, Gemmill RM, Doane WW (1985) Molecular cloning of the α-amylase genes from *Drosophila melanogaster*. II. Clone organization and verification. Genetics 110:313–324

Lewontin RC (1974) The genetic basis of evolutionary change. Columbia Univ Press, New York London

MacIntyre RJ (1982) Regulatory genes and adaptation: past, present and future. Evol Biol 15: 247–285

Nei M, Koehn RK (1983) Evolution of genes and proteins. Sinauer, Sunderland, Mass

Scharloo W (1984) Genetics of adaptive reactions. In: Wöhrmann K, Loeschcke V (eds) Population biology and evolution. Springer, Berlin Heidelberg New York, pp 5–15

Wilson AC (1985) The molecular basis of evolution. Sci Am 253:148–157

Population Structure

Clonal Niche Organization in Triploid Parthenogenetic *Trichoniscus pusillus:* A Comparison of Two Kinds of Microevolutionary Events

H. NOER[1]

1 Introduction

The phenomenon of parthenogenesis confronts current neo-Darwinistic thought with two major paradoxes. The so-called cost of sex implies that through production of males a sexually reproducing species may reduce its natural rate of increase by as much as 50% in comparison with an asexually reproducing but otherwise similar clone, while the so-called cost of meiosis implies that for each generation individual genotypes are broken and reshuffled with more or less randomly chosen genetic material from other individuals belonging to the same species. If it is taken as axiomatic that natural selection is a phenomenon operating solely at the level of the individual, these two paradoxes become very hard to resolve when at the same time it has to be explained why the vast majority of species in nature reproduce sexually (Williams 1975; Maynard Smith 1978; Bell 1982, 1985).

In order to explain the preponderance of sexual reproduction it is commonly argued that parthenogenesis is an evolutionary "dead end". As put by White (1973), „The rigidity of all genetic systems in which there is no recombination of genes present in different individuals probably dooms them, in all instances, to a relatively brief evolutionary career". During the last decade, however, one of the foundations of this belief has been challenged by the consistent finding that virtually all parthenogenetic morphospecies possess abundant genetic variation (Christensen et al. 1976; Suomalainen et al. 1976; Schultz 1977; Vrijenhoek 1978; Christensen 1979; Mitter et al. 1979; Hebert and Crease 1980; Jaenike et al. 1980; Ochman et al. 1980; Turner et al. 1983; Stoddart 1983). Though the general notion of parthenogenesis being an evolutionary dead end does not appear to be seriously threatened by the existence of this variation, it is at least clear that the microevolutionary capabilities of parthenogenetic species have to be reassessed. The purpose of the present chapter is to discuss some particular aspects of this, exemplified by a study of the triploid parthenogenetic form of the woodlouse *Trichoniscus pusillus*.

1 Institute of Population Biology, Universitetsparken 15, DK-2100 Copenhagen 0, Denmark

Population Genetics and Evolution
G. de Jong (ed.)
© Springer-Verlag Berlin Heidelberg 1988

2 Clonal Diversity

The existence of genetic variation in parthenogens is of course not in itself evidence that such organisms are capable of microevolution. Therefore, studies of both the origin and ecological consequences of the genetic variation are needed in order to investigate the evolutionary meaning of clonal diversity.

2.1 Origins of Clonal Diversity

Genetic variation in parthenogenetic organisms is thought to have three possible sources: (1) repeated origin of parthenogenetic clones from one or two ancestral sexual species (in the following termed polyphyletic origin), (2) meiotic recombination and (3) mutation (which together with 2 is termed monophyletic origin). For ameiotic species, which constitute a large fraction of the known parthenogenetic species (White 1973), only the first and last of these pathways are possible. Clonal diversity by way of polyphylesis has been demonstrated in a number of cases (e.g. Vrijenhoek 1979; Harshman and Futuyma 1985). Clonal diversity of monophyletic origin (i.e. microevolution in a strict sense) is presently more questionable, and the main themes of the following are to be (1) Does adaptively significant monophyletic origin of clones occur? and (2) What are the ecological implications of clonal variation arising in this way? Clearly, the evaluation of the meaning of genetic variation in parthenogenetic species depends critically on the answers to these questions.

2.2 Ecological Implications: The Hypotheses

The ecological implications of clonal diversity has attracted some theoretical consideration. However, apart from the assumption that there are such implications points of view have differed widely. Some have maintained that clones have increased niche widths, either through polyploidization and the associated increase of heterozygosity (Vandel 1931; Suomalainen et al. 1976) or through evolution of "general purpose" genotypes (e.g. Lynch 1984). Roughgarden (1972, 1979) predicted that this increase of the within-phenotype component of niche width should only be expected under certain ecological conditions. In contrast, a number of other authors have assumed just the opposite effects of clonal variation: That individual clones have narrow niches in comparison with the ancestral sexual species, due to the genotypic variability in the latter (e.g. Vrijenhoek 1984). These opposing views appear to reflect differences in the assumed sources of the clonal variation. Monophyletic origin is underlying the expectations of increased niche widths of clones, while polyphyletic origin is explicit in Vrijenhoek's (1984) "Frozen niche" variation model. Relatively narrow niches of clones are also implicitly assumed in Parker's (1979) hypothesis that the strength of selection between clones depends on their origin. Selection is, according to this hypothesis, expected to be stronger between clones of polyphyletic origin due to greater overall differences in the genome, while clones of monophyletic origin are assumed to have nearly identical fitness, at least in the early phase of a microevolutionary course. In general, however, these two sets of expectations, based on the two kinds of origin of clonal diversity, do not allow tests of the two ways of origin by field investigation (see discussion).

3 The Trichoniscus pusillus Case

In an attempt to provide data addressing these problems, the triploid parthenogentic form of the woodlouse *Trichoniscus pusillus* is presently being studied by our group (Fig. 1; Table 1). This small isopod is an extremely abundant animal, found virtually

Table 1. Descriptive statistics for the six *Trichoniscus pusillus* clones discussed in the text

	Clone					
	I	II	III	IV	V	VI
Mean[a]	0.536	0.136	0.188	0.093	0.014	0.033
Variance[b]	0.068	0.030	0.041	0.016	0.001	0.011
Variation[c] (%)	40.72	17.96	24.55	9.75	0.60	6.59

[a] Simple means of relative frequencies of clones.
[b] Variance of relative frequencies of clones.
[c] Percent of total variation (sum of variances) in relative frequencies contributed by each clone.

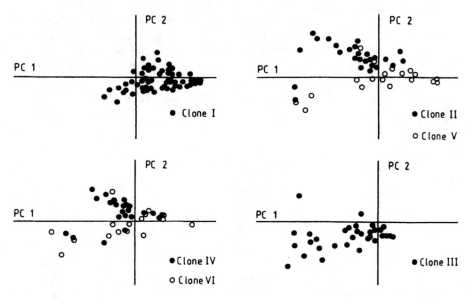

Fig. 1. Patterns in clonal composition of triploid parthenogenetic *Trichoniscus pusillus* over the 100 investigated stations. Stations are plotted according to their scores on first and second principal components (*PC*) extracted from the covariance matrix of relative frequencies of clones. For each clone, only stations with relative frequency above a certain value are shown (*clone I:* 0.5; *clones II, III* and *IV:* 0.1; *clones V* and *VI:* 0.0). Notice that clonal groups *II* and *IV* versus *I, III, V* and *VI* show relatively small spatial overlaps

everywhere in woods provided humidity conditions are not too extreme. In optimal habitats, densities of 2,000 individuals m^{-2} have been reported (Sutton 1968). It has one generation per year, though it appears that it is not strictly univoltine (Sutton 1968; Standen 1973; Christensen et al. 1987).

3.1 Clonal Diversity

Christensen (1979) found four clones of the form marked by the Pgm locus. Based on much larger material collected since then (amounting to about 30,000 individuals), eight to ten different electrophoretic profiles at this locus are presently known, of which six are included in this chapter. Detailed descriptions of these new clones and their distributions will appear elsewhere. Pgm marks all the known clonal diversity in *T. pusillus*.

3.2 Clonal Distributions

The material presented here includes 100 stations from North Zealand, some other Danish islands, Jutland (especially the southern parts) and seven stations from northern Germany. A detailed analysis of the material from northern Zealand is given in Christensen and Noer (in press). Patterns in clonal distributions are investigated by means of principal components analyses of relative frequencies of clones. Results for the presented stations are given in Tables 1, 2, and 3 and in Fig. 1. Clone I is dominant in most samples (Table 1), and the major part of the variation in clonal frequencies is made up by Clones I, II and III (Table 1). This is reflected in the PCA (Tables 2 and 3). One remarkable feature of these patterns is that Clones II and III show a relatively small amount of spatial overlap (Fig. 1).

Though variation in frequencies of Clones IV, V and VI is not accounted for to any significant extent by the first two principal components (Table 3), they all show

Table 2. Loadings of the first two principal components (PC) extracted from the data set summarized in Table 1

	PC 1	PC 2
I[a]	−0.2568	0.0315
II	0.0768	−0.1343
III	0.1394	0.1393
IV	0.0281	−0.0388
V	0.0001	−0.0001
VI	0.0139	0.0051
Variation[b] (%)	55.1	23.9

[a] Clone number.
[b] Percent of total variation in relative clone frequencies extracted by each principal component.

markedly non-random distributions in relation to Clones I, II and III (Fig. 1). Clone IV is nearly always found together with Clone II (cf. Christensen and Noer 1986), while Clones V and VI, though not known very well yet, show a tendency for positive association, and a marked tendency to be found at stations dominated by Clones I and/ or III. Thus, in spite of large overlaps, the six clones show tendencies for pairwise associations: many stations are dominated by Clones I and III, others by Clones II and IV, and Clones V and VI tend to be found together. Though not shown in Fig. 1, extreme values of the first two principal components have mainly been observed at stations in southern Jutland, where Clones I, II and III come close to fixation in a number of cases.

3.3 Factors Affecting Clonal Distributions

There is good evidence that the clonal variation in triploid *T. pusillus* is adaptive. This evidence can be summarized as:

1. Clones show well-marked and constant habitat associations (Christensen and Noer 1986);
2. Eleven stations investigated over 5 years showed simultaneous fluctuations in clone frequencies; Clone I and to some extent Clone IV increasing in 2 years after drought conditions (Christensen and Noer, in press); and
3. Clones I–IV have small but significant brood size differences (Christensen et al., in press).

Strictly speaking, these differences are only known for Clones I–IV. In the following, I assume that Clones V and VI are not neutral in relation to any other clone either.

3.3.1 Moisture Conditions. In Fig. 2, stations scores on PC 1 and PC 2 are shown with indications of elevational level. Stations at higher levels are dominated by Clone I and to some extent by Clone III, while low-lying stations are dominated by either Clones II and IV or by Clone III (in a very few cases by Clone IV). Moisture conditions are thought to be a major variable underlying these patterns. For instance, there is an increasing probability of summer drought from left to right across the figure. In order

Table 3. Communalities of relative clone frequencies from principal components (PC) given in Table 2

	Clone					
	I	II	III	IV	V	VI
PC 1[a]	98.14	19.96	47.52	4.84	0.06	1.75
PC 2	1.45	61.02	47.46	9.22	0.03	0.24
Total	99.59	80.98	94.98	14.06	0.09	1.99

[a] For each clone, the entries give the percentages of the variance in relative frequency extracted by the first and second principal component (PC) respectively.

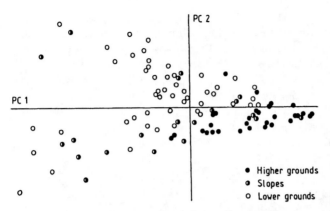

Fig. 2. The investigated stations plotted according to their scores on the first two principal components *(PC)* (i.e. same plot as in Fig. 1). For each station, height in relation to surroundings is indicated. The category *"slopes"* is intermediary. Low-lying stations (high clonal diversity) are overrepresented

to test this hypothesis, a large number of preference experiments have been carried out, measuring the sequence in which individuals leave a slowly drying filter paper. Details of these experiments are given in Christensen et al. (in prep.), while pooled results are shown in Fig. 3. Differences between clones are highly significant and subdivide the four investigated clones into two pairs: Clones I and IV versus Clones II and III, in accordance with the observed temporal fluctuations. In brief, these results suggest that the better ability of Clones I and IV to cope with drought conditions is caused by a more prudent behaviour. Clones V and VI are presently only poorly known with respect to this trait, but preliminary results suggest that Clone V has the pattern of Clones II and III, while Clone VI certainly behaves like Clones I and IV.

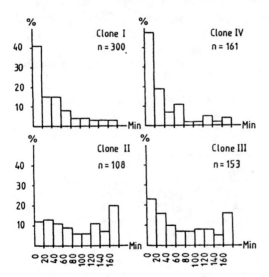

Fig. 3. Results of preference experiments measuring the temporal sequence in which *T. pusillus* individuals leave a piece of slowly drying filter paper. Each experiment included 25 individuals (time = 0 at departure of first individual)

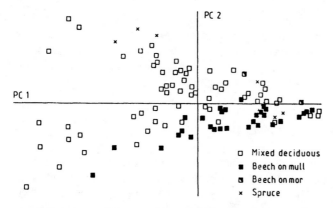

Fig. 4. The investigated stations plotted according to their scores on the first two principal components (*PC*) (i.e. same plot as in Fig. 1). For each station, main habitat type is indicated. The habitat type *"mixed deciduous"* is always found on mull soil and overrepresented in the samples

3.3.2 Habitat. Clonal distributions are clearly correlated with habitat type (Fig. 4), but because habitats are closely associated with elevational level it is difficult to separate the effects of these two sets of variates. However, a number of observations suggest that habitat plays an important role in itself. For instance, the distribution of clonal groups II and IV versus III, V and VI is related to habitat type. On acid soils (mor soils), only the first of these two groups has been observed, while on beech mull only the second has been found (in particular Clone III). Differential habitat preferences between clones have been demonstrated in laboratory preference experiments, but the contribution of preferences to clonal co-existence is probably of small magnitude (Christensen et al., in prep.).

3.3.3 Influence of Biotic Interactions. There are good reasons to assume that stations having high frequencies of Clone I are in marginal habitats, dominated by the influence of abiotic factors (mainly drought stress). In contrast, stations dominated by either Clones II and IV or by Clone III are presumably favourable habitats where biotic interactions play the major role (see Christensen and Noer 1986). This interpretation is supported by a tendency for *T. pusillus* density to vary across the PC diagrams. Absolute densities have not been measured, but samples from stations in the right half of the diagrams often contain fewer individuals. Also, numbers of individuals and diversity of other soil arthropods decrease from left to right across the diagrams. Stations where Clone II or III comes near fixation are from rather isolated woodlots, where virtually all tree species have easily decomposible leaves, suggesting that resource depletion may occur. Accordingly, the observed patterns of clonal distributions are best explained by assuming that competition influences clonal composition at optimal habitats. Under circumstances where nearby marginal habitats allow a more or less continuous recruitment, in particular of Clone I, all clones may co-exist (North Zealand stations), but when conditions prevent immigration and increase the likelihood of resource depletion situations may approach competitive exclusion. This scenario implies that Clones II and III are competitively superior, while Clone IV and in particular Clone I are more "fugitive", i.e. are able to survive under marginal conditions. It is, of course, rather hypothetical and certainly in need of further empirical support, but at present it represents by far the simplest possible hypothesis that is consistent with all our observations. Distributions of Clone I versus Clone IV and II versus III cannot be explained as yet.

3.4 Origins of Clonal Diversity in T. pusillus

Twelve loci have been investigated electrophoretically for Clones I–IV. Eight of these are identical for all four clones. Est, Idh and Mdh show two types, Clones I and IV always belonging to the one and Clones II and III to the other (Christensen 1983; Christensen and Noer 1986). Within these two groups, the only known markers are at the Pgm locus and can be accounted for by hypothetical point mutations. Accordingly, Christensen and Noer (1986) have suggested that the two groups are of monophyletic origin. This hypothesis is supported by the behavioural patterns (Fig. 3). At least five point mutations are necessary for a hypothesis of monophyletic origin of the two groups, and in conjunction with the differences in behaviour this may be taken to suggest that the two groups are of polyphyletic origin. Alternatively, a hypothesis of monophylesis implies that separation between the two groups is an older event. In both cases, the results suggest that differences between the two groups involve a larger part of the genome than differences within groups.

4 Discussion

Detailed discussions of the interpretations of the results for *T. pusillus* are given elsewhere (Christensen and Noer 1986; Christensen et al. 1987). In the following, these results will therefore be addressed in more general terms.

4.1 Mono- and Polyphyletic Origin of Clonal Diversity

Though there seems to be a general agreement that clonal diversity in natural populations is of both mono- and polyphyletic origin, there is some differences in opinions with respect to their relative importance (see Sects. 2.1 and 2.2). In general, the best documental cases involve polyphylesis (e.g. Vrijenhoek 1979; Harhsman and Futuyma 1985), while a number of cases where the variation is assumed to be of monophyletic origin are less certain (Suomalainen et al. 1976; Ochman et al. 1980; Jaenike et al. 1980; Christensen and Noer, in press). However, in the comparison of these two lines of evidence it should be taken into consideration that generally it will be problematic to establish a case of monophyletic origin with the same amount of certainty as a corresponding case of polyphylesis. In the latter, it will suffice to find one critical argument, while in the former it is necessary to show that no such differences exist. In the *Trichoniscus pusillus* case, this difficulty is evident in the interpretation of the results given in Sect. 3.3.

Though the studies quoted for assumed monophyletic origin are not very convincing when taken separately, I think that when taken together they show that monophylesis cannot be ignored as a source of variation. This conclusion is certainly supported by the studies of Lynch (1985), who demonstrated evolution in important life history characteristics in obligately parthenogenetic strains of *Daphnia pulex* at a rate of 2% of the phenotypic value per generation.

Given that both sources contribute to the clonal diversity in natural populations, Harshman and Futuyma (1985) have suggested that the clonal diversity is higher in cases where clones originate polyphyletically. This could be the case, but it is presently not very certain whether actual data support such a conclusion or not. For instance, Suomalainen et al. (1976) found more than 75 clones of possible monophyletic origin in tetraploid *Otiorrhyncus scaber*, though it is not known how much of this variation is adaptive. Moreover, a number of studies in which high clonal diversity of polyphyletic origin has been observed (Vrijenhoek 1979, 1984; Christensen 1980; Turner et al. 1983; Harshman and Futuyma 1985) involve gynogenetic (pseudogamic) species. Given that these have to co-exist with their sexual parent species (or related ones, see Booij and Guldemond 1984) as sperm parasites, they may differ from "independent" asexual species in several aspects. Thus, a pooling of results for these two groups is presently not very justified. At least, all cases of supposed monophylesis appear to involve "independent" parthenogenetic species.

4.2 The Ecological Implications of Clonal Diversity: The Evidence

In a number of cases clones appear to be of relatively small ecological amplitudes (Vrijenhoek 1979, 1984; Mitter et al. 1979; Christensen 1980). These cases, however, are as already stressed of gynogenetic species and may differ from results for other parthenogenetic species, where at least some clones appear to be of quite a large ecological amplitude (Suomalainen et al. 1976; Jaenike et al. 1980). For *Trichoniscus pusillus*, Clone I appears to be able to maintain populations over the complete ecological range of the asexual species, though this ability could well depend on its larger selectivity with respect to microenvironment. In contrast, the associations between Clones II and IV respectively V and VI suggest that some other clones are rather "narrow" in an ecological sense.

These ecological differences between clones of *Trichoniscus pusillus* appear to be critical in relation to the predictions of clonal niche widths quoted in the introduction. Evolution towards greater ecological amplitudes or "general purpose" genotypes would − granted that the clones within the two specified groups are of monophyletic origin − demand that Clone I evolved from Clone IV and probably also that Clone III evolved from Clone II. It should, however, be emphasized that in the opposite case (Clone IV from Clone I and II from III), which is equally likely, quite another interpretation would be implied: A microevolutionary trend towards increased fitness in optimal habitats and a corresponding decrease of niche width.

The results for *Trichoniscus pusillus* confirm Parker's (1979) hypothesis in the sense that the strongest selection indeed seems to take place between two groups of suspected polyphyletic origin (temporal fluctuations in clonal frequencies, see Sect. 3.3), but on the other hand, it is clones from different groups that tend to co-exist. This can be taken to indicate that at least with respect to the habitat part of the niche monophyletic microevolution leads to ecological shifts of a magnitude comparable to polyphyletic microevolution (Christensen and Noer 1986), but it is also possible that the distributional patterns might be explained by an elaboration of Parker's hypothesis. One explanation that presently cannot be excluded is that clones of polyphyletic origin are sufficiently different to co-exist, while the stronger similarity between clones of monophyletic origin leads to eventual competitive exclusion through small, but decisive, fitness differences. Even in the latter case, however, the results for *Trichoniscus pusillus* suggest that microevolution of clonal organisms can lead to ecologically significant changes in natural populations.

Acknowledgements. B. Christensen and B. Friis Theisen are thanked for their cooperation, many critical and helpful discussions, for placing material at my disposal and for criticizing an earlier draft of the manuscript.

References

Bell G (1982) The masterpiece of nature. Croom Helm, London

Bell G (1985) Two theories of sex and variation. Experienta 41:1235–1245

Booij CJH, Guldemond JA (1984) Distributional and ecological differentiation between asexual gynogenetic planthoppers and related sexual species of the genus *Muellerianella*. Evolution 38:163–175

Christensen B (1979) Differential distribution of genetic variants in triploid parthenogenetic *Trichoniscus pusillus* (Isopoda, Crustacea) in a heterogeneous environment. Hereditas 91:179–182

Christensen B (1980) Constant differential distribution of genetic variants in polyploid parthenogenetic forms of *Lumbricillus lineatus* (Enchytraeidae, Oligochaeta). Hereditas 92:193–198

Christensen B (1983) Genetic variation in coexisting sexual diploid and parthenogenetic triploid *Trichoniscus pusillus* (Isopoda, Crustacea). Hereditas 98:201–207

Christensen B, Noer H (1986) Spatial and temporal components of genetic variation in triploid parthenogenetic *Trichoniscus pusillus* (Isopoda, Crustacea). Hereditas 105:277–285

Christensen B, Berg U, Jelnes J (1976) A comparative study on enzyme polymorphism in sympatric diploid and triploid forms of *Lumbricillus lineatus* (Enchytraeidae, Oligochaeta). Hereditas 84:41–48

Christensen B, Noer H, Theisen BF (1987) Differential reproduction of coexisting clones of triploid parthenogenetic *Trichoniscus pusillus* (Isopoda, Crustacea). Hereditas 106:89–95

Harshman LG, Futuyma DJ (1985) The origin and distribution of clonal diversity in *Alsophila pometaria* (Lepidoptera: Geometridae). Evolution 39:315–324

Hebert PDN, Crease TJ (1980) Clonal coexistence in *Daphnia pulex* (Leydig): another planctonic paradox. Science 207:1363–1365

Jaenike J, Selander RK (1985) On the coexistence of ecologically similar clones of parthenogenetic earthworms. Oikos 44:512–514

Jaenike J, Parker ED Jr, Selander RK (1980) Clonal niche structure in the parthenogenetic earthworm *Octolasion tyrtaeum*. Am Nat 116:196–205

Lynch M (1983) Ecological genetics of *Daphnia pulex*. Evolution 37:358–374

Lynch M (1984) The limits to life history evolution in *Daphnia*. Evolution 38:465–482

Lynch M (1985) Spontaneous mutations for life-history characters in an obligate parthenogen. Evolution 39:804–818

Maynard Smith J (1978) The evolution of sex. Cambridge Univ Press

Mitter C, Futuyma DJ, Schneider JC, Hare JD (1979) Genetic variation and host plant relations in a parthenogenetic moth. Evolution 33:777–790

Ochman H, Stille B, Niklasson M, Selander RK, Templeton AR (1980) Evolution of clonal diversity in the parthenogenetic fly *Lonchoptera dubia*. Evolution 34:539–547

Parker ED, Jr (1979) Ecological implications of clonal diversity in parthenogenetic morphospecies. Am Zool 19:753–762

Roughgarden J (1972) Evolution of niche width. Am Nat 106:683–718

Roughgarden J (1979) Theory of population genetics and evolutionary ecology: an introduction. Macmillan, New York

Schultz RJ (1977) Evolution and ecology of unisexual fishes. In: Hecht MK, Steere WC, Wallace B (ed) Evolutionary biology, vol 10. Plenum, New York, pp 277–331

Standen V (1973) The life cycle and annual production of *Trichoniscus pusillus pusillus* (Crustacea:Isopoda) in a Cheshire wood. Pedobiologia 13:273–291

Stoddart JA (1983) The accumulation of genetic variation in a parthenogenetic snail. Evolution 37:546–554

Suomalainen E, Saura A, Lokki J (1976) Evolution of parthenogenetic insects. Evolutionary biology, vol 9. Plenum, New York, pp 209–257

Sutton SL (1968) The population dynamics of *Trichoniscus pusillus* and *Philoscia muscorum* (Crustacea, Oniscoidea) in limestone grassland. J Anim Ecol 37:425–444

Turner BJ, Balsano JS, Monaco PJ, Rasch EM (1983) Clonal diversity and evolutionary dynamics in a diploid breeding complex of unisexual fishes (Poecilia). Evolution 37:798–809

Vandel A (1931) La parthenogenese. Doin Freres, Paris

Vrijenhoek RC (1978) Coexistence of clones in a heterogeneous environment. Science 199:549–552

Vrijenhoek RC (1979) Factors affecting clonal diversity and coexistence. Am Zool 19:787–797

Vrijenhoek RC (1984) Ecological differentiation among clones: the frozen niche variation model. In: Wöhrmann K, Loeschcke V (eds) Population biology and evolution. Springer, Berlin Heidelberg New York, pp 217–231

White MJD (1973) Animal cytology and evolution, 3rd edn. Cambridge Univ Press, London

Williams GC (1975) Sex and evolution. Princeton Univ Press

Microgeographic Variation of Genetic Polymorphism in *Argyresthia mendica* (Lep.: Argyresthiidae)

A. Seitz[1]

1 Introduction

Field studies on the genetic structure of populations show a considerable amount of heterogeneity in space and time. In many cases, these heterogeneities can be related to structures in the environment, such as properties of soil, availability of special food resources, topographic conditions or climate. In other cases the genetic structure can be explained by properties of the plant and animal species under study, e.g. ability and speed of migration and colonization (Karlin and Nevo 1976; Endler 1977; Nevo 1978; Nevo and Yang 1979; Nevo et al. 1981; Seitz and Komma 1984; Wöhrmann 1984).

Interactions between physical environment and an animal species can be studied, if the biotic environment has little or no variability. Such a case can be investigated in highly specialized phytophagous insects which rely only on a limited range of food resources or even only one host plant.

In the years 1977 to 1983, a research project on the ecology of hedges (Zwölfer et al. 1984) was carried out at the University of Bayreuth. It showed the special properties of hedges as typical elements of a landscape which is influenced by man: hedges are highly structured in space with a patchy or netlike arrangement. Animals living on hedges are therefore suitable objects to study populations with a small population size, distributed over a number of subpopulations with varying gene flow.

2 Material and Methods

One species, rather common on hedges is the microlepidopteran *Argyresthia mendica*. The ecology of this species was investigated by G. Heusinger (unpublished), and he provided me with a large amount of individuals. *A. mendica* is univoltine. It overwinters as a larva inside the egg and emerges in spring during early flowering season of *Prunus* species. The larvae feed preferably on flower buds of slews *(Prunus spinosa)*. The last instar larvae pupate in the soil directly under the host plants. The moths

1 Institut für Zoologie, Johannes Gutenberg-Universität, Saarstraße 21, D-6500 Mainz, FRG

Population Genetics and Evolution
G. de Jong (ed.)
© Springer-Verlag Berlin Heidelberg 1988

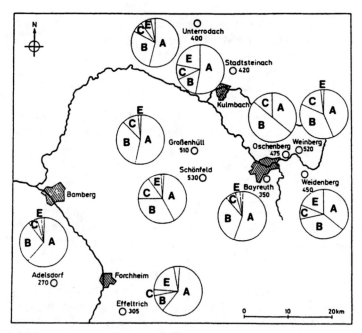

Fig. 1. Study area in Upper Franconia (northern Bavaria). The sampling stations are given with altitude readings in metres. The pie diagrams show frequencies of representative alleles of aconitase

develop during the next 2 weeks and show little flight activity (Rezac 1963, 1964; Heusinger, pers. commun.).

The study area was located in the northeast of Bavaria (FRG) and covers localities with different altitudes ranging from the valley of the Main River up to the borders of the Fichtelgebirge (Fig. 1). This vertically structured area infers a seasonal separation of the flowering period of the host plants and, consequently, of the reproductive cycle of the animals living at different localities. The interaction of the patchy distribution of the host plants and their short availability during the flowering season makes this system ideal for the study of the effects of spatially and seasonally isolated populations.

A. mendica was collected by sampling flowering branches of *Prunus spinosa* and these were kept in the laboratory until the last instar larvae left the plant for pupation. The pupae were reared within soil to adults and these were stored in liquid nitrogen for later gel electrophoresis. Horizontal starch gel electrophoresis was used to study the polymorphism of 12 enzyme loci. The staining procedures were that of Shaw and Prasad (1970) with minor modifications. Twelve populations were sampled but only ten of them resulted in sample sizes for the enzyme aconitase that were large enough for reliable results.

Gene frequencies were computed for all sampling stations and all loci which were studied. These were analyzed for homogeneity by means of a X^2-test and for correlation with altitude by means of a Pearsonian correlation. Genetic distances between populations were estimated according to Nei (1972) and these were analyzed with respect to topographic distances and differences in elevation of the populations. Partial regression analysis was used to separate the effect of topography and altitude.

3 Results and Discussion

All loci under study were polymorphic. The loci coding for glucose-3-phosphate dehydrogenase, phosphoglucose isomerase, malate enzyme, aconitase and two loci coding for isocitrate dehydrogenase showed at least one allele besides the most frequent with frequencies of more than 5%. With the exception of aconitase *(acon)*, in all these cases the frequency of the most frequent allele was higher than 90%. Therefore only *acon* was studied in more detail.

Table 1 shows the allele frequencies of *acon* within the ten sampling stations. It reveals a remarkable variation of allele frequencies between the sampling stations. A X^2-test for homogeneity corroborated the hypothesis of unequal genetic constitution of the studied populations $(P \ll 0.001)$.

The differences, however, are not the result of random genetic drift of isolated populations. The correlation analysis showed a significant correlation between altitude and frequency of the three most frequent alleles of *acon*. The correlations are negative for the allele A $(r = 0.75, P < 0.01)$ and positive for the alleles B $(r = 0.55, P < 0.05)$ and C $(r = 0.91, P \ll 0.001)$. The regression of the allele C on altitude is obviously non-linear (Fig. 2).

These dependencies were interpreted to reflect the genetic adaptation of local subpopulations to the microclimate of the different altitudes. Temperature differences, e.g., are high enough to cause differences of more than 2 weeks in the flowering period of the host plants and the emergence of the larvae of *A. mendica* within the study area. In this way gene flow between populations living at different altitudes is reduced. This enables a good adaptation of the populations to their habitat even under low selection pressure.

Table 1. Allele frequencies of aconitase

Allele R.F.[a]	A 1.00	B 0.88	C 0.81	E 1.07	Other	N[b]
Sampling station						
Adelsdorf	0.61	0.28	0.06	0.03	0.02	100
Bayreuth	0.55	0.33	0.07	0.03	0.02	156
Effeltrich	0.60	0.11	0.04	0.21	0.04	70
Großenhüll	0.53	0.34	0.11	0.02	0.00	122
Oschenberg	0.36	0.50	0.14	0.00	0.00	28
Schönfeld	0.41	0.33	0.15	0.09	0.02	66
Stadtsteinach	0.52	0.15	0.10	0.22	0.01	100
Unterrodach	0.54	0.29	0.07	0.07	0.03	390
Weidenberg	0.36	0.35	0.09	0.20	0.00	56
Weinberg	0.43	0.38	0.17	0.02	0.00	40

[a] R.F. = Relative electrophoretic mobilities in relation to allele A.
[b] N = Number of genes analyzed.

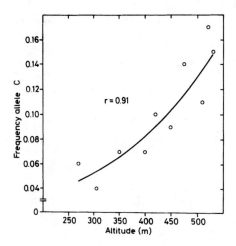

Fig. 2. Relation between frequency of allele C of aconitase and altitude of the station where the populations were sampled. The graph shows the curve of best fit which was estimated by a regression on the logarithms of gene frequencies

Table 2. Genetic distances between populations of *Argyresthia mendica* [a]

Sampling station	(1)	(2)	(3)	(4)	(5)	(6)	(7)	(8)	(9)	(10)	(11)
(1) Adelsdorf											
(2) Bayreuth	8										
(3) Effeltrich	11	8									
(4) Großenhüll	9	2	9								
(5) Haßlach	11	3	1	2							
(6) Oschenberg	18	8	19	6	1						
(7) Schönfeld	22	14	18	13	1	13					
(8) Schwingen	6	2	1	1	1	1	12				
(9) Stadtsteinach	10	7	2	7	2	16	13	1			
(10) Unterrodach	7	2	7	1	3	9	15	1	5		
(11) Weidenberg	17	9	15	7	8	11	20	3	9	6	
(12) Weinberg	15	5	15	3	5	6	16	2	11	4	4

[a] The values are in 10^{-3} Nei units.

Genetic distances between local populations are small. They range from 0.001 to 0.022 and are in accordance with values given by Ayala (1975) for genetic distances between populations within species (Table 2). This has been expected from the analysis of gene frequencies. To test the effect of "isolation by distance", which is to be expected for a series of partially isolated, coupled populations (Wright 1943), a correlation analysis has been performed. This shows a weak, but significant correlation between topographic distance and genetic distance ($r = 0.27, P < 0.05$).

Since the locus *acon* has the highest degree of polymorphism and is the main source of genetic variation within the populations, it has the strongest influence on the genetic distances. Therefore, it was expected that its dependency on altitude causes a correlation between genetic distances and the distances of altitude of the populations. Indeed, there exists a highly significant correlation ($r = 0.38, P < 0.01$). The scattergram of this relationship is shown in Fig. 3.

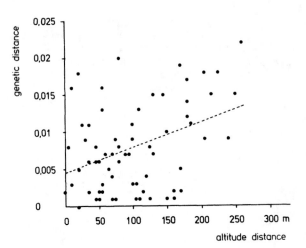

Fig. 3. Scattergram showing the relation between genetic distances and distances of altitude of the sampled populations. The line of best fit was estimated by linear regression

Since there is a general increase of altitude from southwest to northeast, the sampling stations with the longest topographic distances are also those with the highest distances in altitude. It was therefore argued that the correlation between topographic and genetic distances is only due to this correlation between topographic and altitude distances. A partial correlation analysis showed that after elimination of the effect of altitude no significant effect of distance can be detected ($r = 0.125$, $P > 0.05$). The partial correlation coefficient between differences of altitude and genetic distances after elimination of distance effects is only slightly reduced and is still highly significant ($r = 0.234$, $P < 0.01$). From this it could be concluded that isolation by distance is very strong.

The low flight activity of *A. mendica* makes drift by wind the main reason for dispersal. Wind directions, however, are neither uniform nor randomly distributed during the time span where *A. mendica* is in the adult instar which is the only instar where dispersal is possible at all (see Fig. 4). It was therefore suggested that gene flow was not uniform in all directions, but mainly in the direction of prevailing wind, i.e. from southwest to northeast.

To account for this fact, the data set was divided into groups of distances which were in a sector of $90°$. Because of the small number of data, it was not possible to use a smaller sector, since the degrees of freedom for the tests of significance became too low. This sector was rotated in $15°$ increments. The resulting overlapping data sets were used for partial regression analysis. The results are shown in Fig. 5. Correlation and regression coefficients are only significant in a sector around the prevailing direction of wind. The partial correlation coefficient reaches a maximum of $r = 0.55$ ($P < 0.01$).

The results of these analyses show the interaction of two environmental factors which act on the populations of *A. mendica*: the selecting force of different climatic conditions at different altitudes, on the one side, and the effect of a unidirectional gene flow on the other. The interaction of both leads to a strong reduction of gene flow which is expressed in the low correlation between genetic distance and topographic distance.

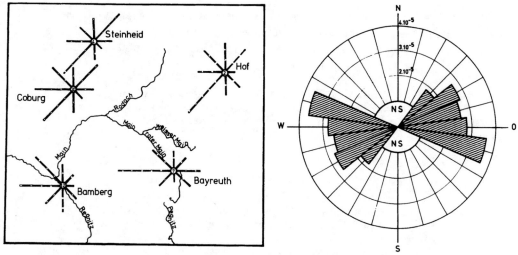

Fig. 4   **Fig. 5**

Fig. 4. Distribution of wind direction and wind force within the study area. Adapted from "Klima-Atlas von Bayern" (Deutscher Wetterdienst 1952). Averages for June from 1881 to 1925. The lengths of the *bars* correspond to percentages of wind direction. The *numbers within the circles* give percentages of calm

Fig. 5. Partial regression coefficients of genetic distances on topographic distances after elimination of the effect of altitude differences in relation to direction. The *shaded areas* show significant partial correlation ($P < 0.05$). *NS* = not significant

The situation of non-random gene flow was studied by means of a simple simulation model. It was developed in the algorithmic language PASCAL and run at a DEC VAX 11/780 of the computation centre of the University of Bayreuth. This model simulates a special case and a modification of the stepping stone model of Kimura (Kimura 1953; Kimura and Weiss 1964; Endler 1977) in a completely anisotropic environment. The setup of the model is shown in Fig. 6.

Simulations were performed with different degrees of gene flow, random genetic drift and cases with dispersal only in one or both directions of one dimension.

The results of the simulations agree with the expectations deducted from the analysis of genetic distances: passive dispersal by wind can be a source of non-random variation of genetic polymorphism in populations. The patterns which are induced in

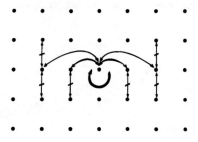

Fig. 6. Model of a metapopulation with one-dimensional gene flow. The *dots* represent local subpopulations. *Arrows* indicate direction and amount of gene flow. The situation is shown only for one subpopulation

this way are superimposed on the variations caused by genetic adaptation to special properties of the biotic and abiotic environment. Since investigations in clinal variations mainly focus on relations between climatic conditions, small-scale patterns induced by wind might be often overlooked.

Acknowledgements. Thanks are due to Mrs. A. Servant-Miosga for lab assistance and especially to G. Heusinger for providing the animals and information on the ecology of *A. mendica*. Mrs. J. Taft has checked the English. The work was partially supported by a grant of the Deutsche Forschungsgemeinschaft.

References

Ayala FJ (1975) Genetic differentiation during the speciation process. Evol Biol 8:1–78

Deutscher Wetterdienst der US-Zone 1952: Klima-Atlas von Bayern. Unter Mitarbeit von Knoch K, Bad Kissingen

Endler J (1977) Geographic variation, speciation, and clines. Monogr Popul Biol, vol 10. Princeton Univ Press

Karlin S, Nevo E (1976) Population genetics and ecology. Academic Press, London New York, p 832

Kimura M (1953) "Stepping stone" model of population. Annu Rep Nat Inst Genet Jpn 3:62–63

Kimura M, Weiss GH (1964) The stepping stone model of population structure and the decrease of genetic correlation with distance. Genetics 49:561–576

Nei M (1972) Genetic distance between populations. Am Nat 106:283–292

Nevo E (1978) Genetic variation in natural populations. Patterns and theory. Theor Popul Biol 13:121–177

Nevo E, Yang SJ (1979) Genetic diversity and climatic determinants of tree frogs in Israel. Oecologia (Berlin) 41:47–63

Nevo E, Bat-el C, Bar Z, Beiles A (1981) Genetic structure and climatic correlates of desert landsnails. Oecologia (Berlin) 48:199–208

Rezac M (1963) Zur Bionomie der *Argyresthia*-Arten auf mitteleuropaischen Obstbaumarten. Zool Listy 12:43–62

Rezac M (1964) Die Schädlichkeit und die Parasiten der auf den mitteleuropäischen Obstbaumarten lebenden *Argyresthia*-Arten und die Möglichkeit ihrer Bekämpfung. Zool Listy 13:57–72

Seitz A, Komma M (1984) Genetic polymorphism and its ecological background in Tephritid populations (Diptera: Tephritidae). In: Wöhrmann K, Loeschcke V (eds) Population biology and evolution. Springer, Berlin Heidelberg New York

Shaw CR, Prasad R (1970) Starch gel electrophoresis of enzymes – a compilation of recipes. Biochem Genet 4:297–320

Wöhrmann K (1984) Population biology of the rose aphid, *Macrosiphum rosae*. In: Wöhrmann K, Loeschcke V (eds) Population biology and evolution. Springer, Berlin Heidelberg New York

Wright SL (1943) Isolation by distance. Genetics 28:114–138

Zwölfer H, Bauer G, Heusinger G, Stechmann D (1984) Die tierökologische Bedeutung und Bewertung von Hecken. Ber Akad Natursch Landschaftspfl (Laufen/Salzach), Beiheft 3, Teil 2, p 155

The Significance of Sexual Reproduction on the Genetic Structure of Populations

J. Tomiuk and K. Wöhrmann[1]

1 Introduction

In recent years many investigations have been carried out to test the interdependence of the reproductive mode and the genetic structure of populations. It is generally considered that recombination by sexual reproduction is not necessarily advantageous in constant environments, but is of advantage under variable environmental conditions. Sexual reproduction can accelerate the evolutionary process because favourable mutations originating in different individuals can be combined (Fisher 1930; Muller 1932). In constant environments, however, the transition from sexuality to asexual reproduction is said to increase the fitness of populations (Maynard Smith 1978).

The generalization that sexuality is essential for the maintenance of genetic variation has to be revised. Parthenogenetically reproducing species can also maintain a high amount of genotypic heterogeneity (see e.g. Lynch 1985; Saura 1983). The degree of genetic variation of different species can cover a wide range, even within one genus (Graur 1985; Nei and Graur 1984). This is in accordance with the data found in aphid species. Some authors have only reported low degrees of heterozygosity (May and Holbrook 1978; Suomalainen et al. 1980; Tomiuk and Wöhrmann 1980a, 1983; Wöhrmann et al. 1986; Wool et al. 1978), whereas Loxdale et al. (1985) have found a high level of genetic variability in the grain aphid *Sitobion avenae*.

In this chapter we compare measures describing the genetic structure of populations. The mechanisms are discussed which influence the genotypic structure of rose aphid populations and of experimental yeast populations capable of reproducing either by cyclic or by permanent parthenogenesis.

2 Measures of Genetic Heterogeneity

The average genetic heterozygosity per locus and individual is usually described as

$$H_{exp} = 1 - \Sigma \frac{q_{ij}^2}{n}, \tag{1}$$

1 Lehrstuhl für Populationsgenetik, Institut für Biologie II der Universität,
   Auf der Morgenstelle 28, D-7400 Tübingen, FRG

Population Genetics and Evolution
G. de Jong (ed.)
© Springer-Verlag Berlin Heidelberg 1988

where q_{ij} is the frequency of the i-th allele at the j-th locus and j = 1,...,n (Nei 1975). This measure describes the allelic heterogeneity of a population. Some studies are more interested in the actual genotypic composition of populations than in the allelic one. In these population studies the investigators, therefore, prefer Shannon's information measure

$$H_{Sh} = - \Sigma P_i \ln (P_i) , \tag{2}$$

where P_i is the frequency of the i-th genotype (Lewontin 1972). Hedrick (1971) proposed a different approach and defined estimators of the genotypic structure on the analogy of the estimators of the genic structure given by Nei (1975). The genotypic variability is defined by

$$V = \Sigma P_i^2 , \tag{3}$$

where P_i is again the frequency of the i-th genotype. We define the genotypic heterogeneity as

$$H_v = 1 - V . \tag{4}$$

3 Material

Since 1976 Wöhrmann and co-workers have investigated the genetic structure of natural populations of the rose aphid (*Macrosiphum rosae* L.) sampled from different locations in Europe (Eggers-Schumacher et al. 1979; Tomiuk 1987; Tomiuk and Wöhrmann 1980a, 1981, 1982a,b, 1983, 1984; Tomiuk et al. 1979; Wöhrmann 1984; Wöhrmann and Tomiuk 1979; Wöhrmann et al. 1978, 1987). The life cycle of *M. rosae* is governed by environmental conditions. In mild climates the rose aphid can reproduce exclusively parthenogenetically, whereas in cold climates a sexual generation is induced in autumn, and winter eggs are layed on the host plants. In Fig. 1 some interactions between environmental and population genetics factors are given. The assessment of a single factor is obviously rendered difficult by the complexity of the system. Since cyclically parthenogenetically reproducing populations have not yet been successfully reared in the laboratory, the system cannot be investigated under controlled laboratory conditions. Alternatively, we investigated the significance of recombination by sexual reproduction in experimental populations of yeast *(Saccharomyces cerevisiae)* which can show a correponding alternate between sexual and asexual phases (Wöhr-

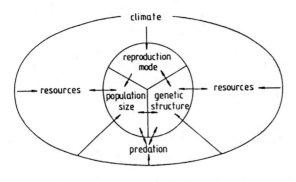

Fig. 1. The interactions between some external and internal factors that influence the population structure of aphids

mann 1982; Wolf et al. 1987). In this species the sporulation rate and the environmental conditions can be influenced by a change in the culture conditions.

In both species polymorphic enzyme systems were used as genetic markers for the
total variability of the genome. Starch-gel electrophoresis was carried out according
to Ayala et al. (1972) and Shaw and Prasad (1970). In *M. rosae* polymorphic loci
(malate dehydrogenase, phosphoglucomutase), and in *S. cerevisiae* four loci (esterase1,
esterase2, alcohol dehydrogenase and glucose-6-phosphate dehydrogenase) were investigated.

4 Results and Discussion

The expected genetic heterozygosity (H_{exp}) as defined in Eq. (1) is equal to the heterozygosity in randomly mating populations. Testing the neutral theory the expected
heterozygosity (H_{exp}) can be compared with the observed one (H_{obs}), where H_{obs} is
the relative proportion of observed heterozygotes per locus. Nei and Graur (1984)
discussed the estimates of both measures from investigations on 77 species. They were
able to refute Levins' (1968) theory that species living in heterogeneous environments
have higher genetic variability than species found in constant environments. In this
chapter interest is focussed on whether differences in variability exist within one species which are caused by the environmental conditions of its habitat. This is quite different from a comparison between species where a random sample of loci is taken. In
our case the maximum number of alleles is fixed and the polymorphism at only two
and four loci respectively, is viewed.

4.1 Aphids

In Table 1 the different heterogeneity measures (H_{exp}, H_{obs}, H_v) the X^2 values testing
deviations from the Hardy-Weinberg equilibrium of one locus (malate dehydrogenase)
and the mean temperature in January (°C) are listed from the investigated aphid
populations from Europe. The values of expected heterozygosity (H_{exp}) are close to
the observed frequencies of heterozygotes (H_{obs}). Furthermore, there is no obvious
correlation between the environmental factor (temperature) and the expected heterozygosity (H_{exp}) as well as the observed one (H_{obs}). Populations can maintain either
high or low genetic heterozygosity in warm climatic regions, and in cold climates the
estimator H_{exp} is in the same range as in mild climates. Moreover, deviations of the
population structure from the Hardy-Weinberg equilibrium can be tested with the
X^2 test. In populations where there is a disturbance of random mating, we expect the
calculated X^2 values to increase. In this respect deviations from the equilibrium must
be greater in populations from warm climates, where sexual reproduction is partially
or completely suppressed, than from cold climates where sexuality is necessary. No
correlation can be observed between the calculated X^2 values and the temperature.
There is a random distribution: populations from cold and warm climates have either
significant or non-significant values (South Germany 0.0 to 34.7, Madeira 0.7, North
Spain 41.2).

Table 1. Mean temperature in January ($^{\circ}$C), observed and expected average heterozygosity (H_{obs}, H_{exp}), X^2 values testing the deviations from the Hardy-Weinberg equilibrium at one locus (malate dehydrogenase) (d.f. = 1) and genotypic heterogeneity (H_V) in different aphid populations from Europe

Region	$^{\circ}$C	H_{obs}	H_{exp}	X^2	H_V
Canary Islands	17.8	0.00	0.00	–	0.00
Madeira	15.0	0.22	0.12	0.7	0.63
North Spain	9.4	0.40	0.32	41.2	0.32
Turkey	8.3	0.24	0.22	2.9	0.54
Great Britain	3.5–6.0	0.28–0.37	0.28–0.36	0.2– 38.5	0.67–0.71
Switzerland	1.9	0.44–0.56	0.28–0.31	11.1–160.8	0.08–0.44
Middle Europe A[a]	−1.4–0.2	0.32–0.49	0.27–0.46	0.0– 34.7	0.82–0.85
Middle Europe B[a]	−1.9–1.5	0.31–0.48	0.44–0.52	0.0– 11.2	0.78–0.85
North Norway	−9.1	0.30–0.37	0.33–0.37	0.0– 1.2[b]	0.78–0.80

[a] A: North Germany, Denmark, South Norway. B: South Germany, Austria, North Italy.
[b] Low class size of one MDH homozygote.

However, in the case of the estimator H_{exp} we can show that this measure can be influenced by the mode of reproduction resulting in a low correlation between the genic structure and environmental factors. If one considers for instance an asexual population consisting exclusively of heterozygotes: this population does not have any genotypic variability, but the expected heterozygosity reaches a maximum.

In contrast, the distribution of the heterogeneity values H_v can be easily explained in dependence on the climatic zones. In cold climates the populations restore a relatively high amount of genotypic heterogeneity. This is caused by the annual recombination of the gene pool during the sexual phase whilst in warm climatic regions the populations can be randomly fixed in a genotypic state. The latter results from the annual occurring dynamic changes in density of aphid populations. The asexual (summer) populations experience annually a series of bottlenecks due to host changes, parasitization and the physiological state of the hosts (Rhomberg et al. 1985; Steiner et al. 1985; Tomiuk 1986; Tomiuk and Wöhrmann 1981). Therefore, such populations come rapidly to fixation in a random genotypic state.

4.2 Yeast

The aphid model, according to which sexual reproduction preserves higher genotypic variability than permanent parthenogenesis, has also been corroborated in experimental yeast populations. Six populations were cultured under different environmental conditions (temperature, aeration and medium). The sporulation rate and the degree of genotypic heterogeneity were estimated every week over a period of 36 weeks. Genetic drift due to small effective population sizes could be excluded because at least 10^5 cells exist in 1 ml of a yeast suspension. There was no directional alteration of allele frequencies but in total the sporulation rate and the genotypic heterogeneity were positively correlated (Wolf et al. 1987).

5 Conclusions

From these results we conclude that the influence of a sexual phase on the genetic structure of a population depends mainly on the order of magnitude of external and/or internal factors which affect the subsequent asexually reproducing populations. There are forces operating during the asexual phase which can drastically and randomly change the genotypic structure, and sexual reproduction can only preserve the degree of heterogeneity of the population. However, we are of the opinion that Levins' theory can be modified in such a way that the environment mainly influences the mode of reproduction, for example cyclic or permanent parthenogenesis which again influences for the short-term the genotypic structure and for the long-term the genic one. There is no general strategy: in aphids sexuality is suppressed in constant (favourable) environments and we observed a low degree of heterogeneity. Whereas in other species favourable conditions could shorten the generation time of the local populations and, thus, increase their level of genotypic heterogeneity.

Acknowledgements. We wish to thank Drs. V. Loeschcke and W. Pinsker for their critical and helpful comments and Miss Janet Walsh for the correction of this manuscript. This work was supported by a grant from the Deutsche Forschungsgemeinschaft.

References

Ayala FJ, Powell JR, Tracey ML, Moura CA (1972) Enzyme variability in the *Drosophila willistoni* group. III. Genetic variation in natural populations of *Drosophila willistoni*. Genetics 70:113–119
Eggers-Schumacher HA, Tomiuk J, Wöhrmann K (1979) The estimation of growth and size of aphid populations. Z Angew Entw 88:261–268
Fisher RA (1930) The genetical theory of natural selection. Oxford Univ Press
Graur D (1985) Gene diversity in Hymenoptera. Evolution 39:190–199
Hedrick PW (1971) A new approach to measuring genetic similarity. Evolution 25:276–280
Levins R (1968) Evolution in changing environments. Princeton Univ Press
Lewontin RC (1972) The apportionment of human diversity. Evol Biol 6:381–398
Loxdale HD, Rhodes JA, Fox JS (1985) Electrophoretic study of enzymes from cereal aphid populations. 4 Detection of hidden genetic variation within populations of the grain aphid *Sitobion avenae* (F) (Hemiptera:Aphididae). Theor Appl Genet 70:407–412
Lynch M (1985) Spontaneous mutations for life-history characters in an obligate parthenogen. Evolution 39:804–818
May B, Holbrook FR (1978) Absence of genetic variability in the green peach aphid, *Myzus persicae* (Hemiptera:Aphididae). Ann Entomol Soc Am 71:809–812
Maynard Smith J (1978) The evolution of sex. Cambridge Univ Press, New York
Muller HJ (1932) Some genetic aspects of sex. Am Nat 66:118–138
Nei M (1975) Molecular population genetics and evolution. Elsevier/North-Holland Biomedical Press, Amsterdam
Nei M, Graur D (1984) Extent of protein polymorphism and the neutral mutation theory. Evol Biol 17:74–118
Rhomberg LR, Joseph S, Singh RS (1985) Seasonal variation and clonal selection in cyclically parthenogenetic rose aphids *(Macrosiphum rosae)*. Can J Genet Cytol 27:224–232
Saura A (1983) Population structure, breeding systems and molecular taxonomy. In: Oxford GS, Rollinson D (eds) Protein polymorphism and taxonomic significance. Academic Press, London New York

Shaw CR, Prasad R (1970) Starch gel electrophoresis of enzymes – a compilation of recipes. Biochem Genetics 4:297–320

Steiner WM, Voegtlin DJ, Irwin ME (1985) Genetic differentiation and its bearing on migration in North American populations of the corn leaf aphid *Rhopalosiphum maidis* (Fitsch). Ann Entomol Soc Am 78:518–525

Suomalainen E, Saura A, Lokki J, Teeri T (1980) Genetic polymorphism and evolution in parthenogenetic animals IX. Absence of variation within parthenogenetic aphid clones. Theor Appl Genet 57:129–132

Tomiuk J (1987) The neutral theory and enzyme polymorphism in populations of aphid species. In: Holman J (ed) Int Symp Population structure, genetics and taxonomy of aphids, 9.–14. Sept., Smolenice, CSSR

Tomiuk J, Wöhrmann K (1980a) Enzyme variability in populations of aphids. Theor Appl Genet 57:125–127

Tomiuk J, Wöhrmann K (1980b) Growth and population structure of natural populations of *Macrosiphum rosae* (L) (Hemiptera:Aphididae). Z Angew Entw 90:464–473

Tomiuk J, Wöhrmann K (1981) Changes of the genotypic frequencies at the MDH-locus of *Macrosiphum rosae* (L) (Hemiptera:Aphididae). Biol Zentralbl 100:631–640

Tomiuk J, Wöhrmann K (1982a) Comments on the genetic stability of aphid clones. Experientia 38:320–321

Tomiuk J, Wöhrmann K (1982b) Effect of temperature and humidity on natural populations of *Aphis pomi* De Geer and·of *Macrosiphum rosae* (L) (Hemiptera:Aphididae). Z Pflanzenkrankh Pflanzenschutz 89:157–169

Tomiuk J, Wöhrmann K (1983) Enzyme polymorphism and taxonomy of aphid species. Z Zool Syst Evol 21:266–274

Tomiuk J, Wöhrmann K (1984) Genotypic variability in natural populations of *Macrosiphum rosae* (L) in Europe. Biol Zentralbl 103:113–122

Tomiuk J, Wöhrmann K, Eggers-Schumacher HA (1979) Enzyme patterns as characteristics for the identification of aphids. Z Angew Entw 88:440–446

Wöhrmann K (1982) Population genetics of yeast. In: Jayakar SD, Zenta L (eds) Evolution and the genetics of populations. Suppl Atti Ass Genet Ital Vol XXIX

Wöhrmann K (1984) Population biology of the aphid *Macrosiphum rosae*. In: Wöhrmann K, Loeschcke V (eds) Population biology and evolution. Springer, Berlin Heidelberg New York

Wöhrmann K, Tomiuk J (1979) Investigations on natural aphid populations. Proc 1 Mediteranean Congr Genetics, Cairo, p 821

Wöhrmann K, Eggers-Schumacher HA, Tomiuk J (1978) Allozyme variations in natural populations of aphids. Proc Int Congr Genetics Moscow, Part I, p 492

Wöhrmann K, Stamp J, Tomiuk J (1987) Resistance in populations of the aphid *Macrosiphum rosae*. In: Holman J (ed) Int Symp Population structure, genetics and taxonomy of aphids, 9.–14. Sept, Smolenice, CSSR

Wöhrmann K, Tomiuk J, Weber G (1986) The search for hidden variability in the aphid *Macrosiphum rosae*. Theor Appl Genet 73:77–81

Wolf HG, Wöhrmann K, Tomiuk J (1987) Experimental evidence for the adaptive value of sexual reproduction. Genetica 72:151–159

Wool D, Bunting S, van Emden HF (1978) Electrophoretic study of genetic variation in British *Myzus persicae* (Sulzer) (Hemiptera:Aphididae). Biochem Genetics 16:987–1006

Patch-Time Allocation by Insect Parasitoids: Superparasitism and Aggregation

J. J. M. VAN ALPHEN[1]

1 Introduction

Following the seminal works of MacArthur and Pianka (1966) and Emlen (1966), many ecologists have explored the foraging behaviour of animals, assuming that natural selection will favour individuals which exploit their resources most efficiently. Female insect parasitoids seek hosts to lay their eggs in or on. The direct link between successful searching and the production of offspring suggests parasitoid searching behaviour to be strongly influenced by natural selection. Therefore, it is an ideal subject for testing optimization hypotheses.

Optimization models have been applied to parasitoid behaviour to predict sex allocation, clutch size, host (size) selection and patch-time allocation (Waage 1986; van Alphen and Vet 1986). The latter subject has received surprisingly little attention, though it is of crucial importance in understanding population dynamics of host-parasitoid systems.

Early optimal foraging models assume that optimally foraging parasitoids maximize the encounter rate with unparasitized hosts (Charnov 1976; Cook and Hubbard 1977; Hubbard and Cook 1978; Comins and Hassell 1979; Waage 1979). Though this assumption certainly does not hold for all parasitoid species, it is a likely one for pro-ovigenic species with a relatively high fecundity. Such parasitoids are unlikely to become egg limited.

This chapter reviews theory on patch-time allocation by parasitoids and discusses how parasitoids should allocate patch time with respect to simultaneously searching competitors. It shows that natural selection may shape parasitoid behaviour differently from that predicted by optimal foraging models so far published. It suggests why aggregation and mutual interference of parasitoids may be adaptive.

1 Department of Population Biology, Division of Animal Ecology, Zoological Laboratory of the University of Leiden, Kaiserstraat 63, NL-2311 GP Leiden, The Netherlands

Population Genetics and Evolution
G. de Jong (ed.)
© Springer-Verlag Berlin Heidelberg 1988

2 Patch-Time Allocation by Individually Foraging Parasitoids

The allocation of searching time by an optimally foraging parasitoid searching alone
for patchily distributed hosts is predicted in a model by Charnov (1976), known as
the marginal value theorem. This model predicts that each patch within a habitat
should be exploited until the encounter rate with unparasitized hosts within a patch
has decreased to a threshold value similar for all patches in that habitat. This threshold value depends on the average number of hosts per patch, and on the time spent
travelling between patches. The model is designed to predict the behaviour of an individual parasitoid which searches a habitat alone. However, in nature conspecific
competitors are searching for hosts simultaneously.

The predictions of the marginal value theorem are: (1) patch times should increase
with increasing host densities, (2) patch times should increase with travel time and
(3) final encounter rates in all patches should be equal. Hubbard and Cook (1978),
Waage (1979), van Lenteren and Bakker (1978) and van Alphen and Galis (1983)
found increasing patch times with increasing host density in the patch. Surprisingly,
point (2) has not been tested for parasitoids. The third prediction of the model, that
each patch in a habitat should be exploited until the same threshold encounter rate
has been reached, was tested by Hubbard and Cook (1978). They find a reasonable
qualitative agreement between the predictions of their model and the experimental
data.

Hence, the evidence for time optimization in parasitoids foraging for patchily distributed hosts is scanty.

Hubbard and Cook (1978) tested the marginal value theorem in experiments where
a number of parasitoids simultaneously exploited the hosts within a habitat. This approach neglects the effect of interactions between parasitoids. The models of Cook
and Hubbard (1977) and Comins and Hassell (1979) both maximize the overall rate
of encounters between parasitoids and unparasitized hosts, and assume that all parasitoids are equal and behave similarly. However, natural selection works on individuals,
and may select individuals that realize the highest encounter rates, but it does not
necessarily maximize the overall rate of encounter of the parasitoid population. Therefore, we must consider how the presence of conspecifics searching simultaneously in
the same habitat may affect searching behaviour.

3 Simultaneously Foraging Conspecific Parasitoids

3.1 The Ideal Free Distribution

Fretwell and Lucas (1970) predicted the stable distribution of foragers over a habitat.
Their model predicts that foragers will distribute themselves in such a way that encounter rates for all foragers are equal. They call this an ideal free distribution. Any
individual in an ideally free-distributed population of foragers which moves to another
patch decreases its encounter rate. Therefore, the ideal free distribution is an ESS.

Ideal free distributions can only be expected when travel times between patches
are negligibly short. Therefore, tests of this model have been restricted to situations

in which no effects of exploitation occur, e.g. the distribution of sticklebacks over places with different constant feeding rates (Milinski 1979), or situations where animals stay in the same place during a whole season, e.g. the distribution of territorial birds over a habitat (Fretwell 1972) or the distribution of gall-forming aphids over the leaves of a poplar tree (Whitham 1978, 1979, 1980). Insect parasitoids, however, may spend a considerable part of their life time moving between patches (Driessen et al., in prep.). This may be a reason why parasitoids in nature are not ideally free-distributed.

Comins and Hassell (1979) used an ideal free distribution of parasitoids in a model to study the effect of optimal foraging on the stability of host-parasitoid interactions. The model requires that all parasitoids start searching simultaneously in the best patch and when encounter rate with unparasitized hosts has dropped to a level similar to the initial encounter rate in the second best patch, half of the wasp population moves to this second best patch, and so on. Though this model is clearly based on unrealistic behaviour, it assumes that optimally foraging parasitoids should maintain an ideal free distribution. This is, however, only true when travel times are negligible.

3.2 Aggregation

Hassell and May (1973, 1974) explored the effect of an uneven distribution of parasitoids over the habitat, and discovered that aggregation of parasitoids in high density patches promotes stability of host-parasitoid interactions. They believe that aggregation occurs commonly in nature. Hassell and May (1974) quote some examples from field and laboratory studies showing predator and parasitoid aggregation. Hassell (1968, 1980), Waage (1983) and Driessen et al. (in prep.) have also shown that parasitoids aggregate rather than distribute themselves in a way that matches host distribution. Hence, the distribution of parasitoids in nature is unlikely to result in equal encounter rates in all patches. Does this imply that parasitoids behave in a non-adaptive way?

3.3 Mutual Interference and Aggregation

Hassell and Varley (1969) introduced this concept into parasitoid ecology by showing that in a number of studies the searching efficiency of parasitoids decreased with increasing parasitoid density. Starting from these empirical data they proceeded to show that such an effect can stabilize the otherwise unstable Nicholson-Bailey (1935) model. Because of this property, mutual interference has elicited much theoretical interest (Royama 1971; Rogers and Hassell 1974; Beddington 1975; Free et al. 1977; Sutherland 1983), but good evidence that it occurs in nature is not available (van Alphen and Vet 1986). Hassell and May (1974) assumed that aggregation of parasitoids causes interference. Interference is per definition a decrease in encounter rate with increasing parasitoid density. This decrease is believed to be caused by a loss of searching time due to time spent in interactions between parasitoids. When parasitoids lose time in interactions with conspecifics, it is even less apparent to see how individual parasitoids could benefit from staying in a patch where parasitoids have aggregated.

Two components of parasitoid behaviour influence aggregation: one is the probability that parasitoids will find a patch, which determines the number of parasitoids arriving in a patch, and the other is the amount of time that individual parasitoids decide to stay. Though larger patches may produce larger odour plumes and hence attract more parasitoids, this would not result in aggregation when parasitoids obey the marginal value theorem, because they would stay only until the threshold encounter rate with unparasitized hosts in the patch had been reached and the threshold should be reached sooner with increasing numbers of parasitoid visits. Hence, short patch times would counteract the effect of more parasitoid arrivals.

Therefore, to explain why parasitoids aggregate is to explain why they stay longer in the high density patches than we expect on the basis of optimal foraging models.

3.4 Superparasitism

The key to understand why parasitoids aggregate and why interference may occur lies in the phenomenon that a host parasitized by one parasitoid is vulnerable to further attack by a conspecific.

When a single solitary parasitoid exploits a patch and oviposits in a host already parasitized during a previous encounter she is wasting an egg and wasting the time needed to lay the egg, because only one parasitoid can develop per host, and the competition between the parasitoid larvae hatching from the eggs is competition between full sibs. One therefore would expect that parasitoids searching a patch alone would refrain from oviposition in hosts already parasitized. There is ample evidence that this indeed happens (van Alphen and Nell 1982). Waage (1986) extends this argument for gregarious parasitoids, where an increase in clutch size may decrease fitness of the other sibs.

When a number of parasitoids simultaneously search the same patch, a searching female may encounter hosts parasitized by a competing female. Laying an additional egg in such a host is not necessarily a waste of an egg and time, because there is a probability that her offspring will win the competition for the host.

If parasitoids do not interfere, do not superparasitize and encounter their hosts randomly, then the product of parasitoid density (P) and searching time (Ts) should be constant for different values of P, or, the sum of the Ts values should be constant. Van Alphen and Nell (in prep.) tested whether superparasitism in *Leptopilina heterotoma*, a larval parasitoid of *Drosophila*, increased with parasitoid density when the parasitoids were free to leave the patch. *L. heterotoma* strongly avoids superparasitism when searching alone. It encounters its hosts at random (van Batenburg et al. 1983). In the experiments the sum of the searching times increased with parasitoid density, and encounter rates (with unparasitized as well as parasitized hosts) decreased. Superparasitism occurred more at higher parasitoid densities (Table 1). Thus, these experiments demonstrate mutual interference. They also show that superparasitism increases in the presence of other females, which suggests that interference occurs as a consequence of the parasitoid's decision to stay longer and superparasitize in the presence of competing conspecifics.

Table 1. Mean encounter rates of individual wasps and total time spent searching by one, two or four females of *Leptopilina heterotoma* on patches with 32 *D. melanogaster* larvae and the resulting degree of superparasitism

	Parasitoid density		
	1	2	4
Encounter rate (s^{-1})	0.030	0.015	0.014
Mean No. of eggs per host	1.0	1.2	1.8
Mean time spent searching (s)	2523.5	4214.0	6417.2

3.5 The War of Attrition

Let us now consider what would happen when two parasitoids exploit a patch and one of them does not superparasitize and leaves the patch when unparasitized hosts are no longer found, and the other stays longer and lays additional eggs in parasitized hosts. By staying and laying additional eggs, the number of offspring obtained from this patch by the latter female will increase at the expense of the first female. Whether this will affect the number of offspring produced by the first female over her lifetime, depends on the probability of finding other unexploited patches.

Estimates of travel times in the field (Driessen et al., in prep.) indicate that these may take a significant proportion of adult lifetime. The high percentages of parasitism generally found in the field indicate that for larval parasitoids of *Drosophila* the probability of finding unexploited patches must be low. Thus, it is unlikely that eggs are a limiting resource for these parasitoids. Under these circumstances, avoiding superparasitism and leaving when the encounter rate with unparasitized hosts decreases below a certain threshold rate is not a stable strategy. One would expect parasitoids to stay longer and superparasitize. But how long should they stay? This clearly depends on how long competitors are prepared to stay. The problem of patch-time allocation is therefore one of game theory. Patch-time allocation and superparasitism are an example of the war of attrition (Maynard Smith 1982).

4 Conclusions

From the literature on parasitoid foraging behaviour, two distinct views emerge on how these insects distribute themselves over patchy environments. One view, held by those interested in factors that stabilize host-parasitoid systems stresses that parasitoids may aggregate in high density patches, and that in these patches mutual interference occurs between parasitoids. The other view emerges from simple deductive models on how parasitoids should behave when maximization of the encounter rate with unparasitized hosts is favoured by natural selection. These models predict that individual parasitoids should leave a patch when encounter rate with unparasitized hosts has dropped to a certain threshold level, and that groups of parasitoids should be distributed in such a way that encounter rates are equal in all patches in the habitat, and that mutual interference should be avoided as it lowers encounter rates.

Existing evidence for individually searching parasitoids is in agreement with the predictions of optimal foraging theory. However, there is evidence that groups of parasitoids behave differently, and that the empirical models built to study population dynamic properties of parasitoid-host systems give a better description of parasitoid distribution.

The apparent contradiction between the two views arises from the fact that transit time between patches can be long relative to total available searching time. When travel times are long and parasitoids are not omniscient about the distribution of hosts and competitors, ideal free distributions cannot be expected.

Parasitoids differ from predators in that they leave their hosts vulnerable to further attack by competitors. They may compete for hosts by staying longer in a patch and laying additional eggs in parasitized hosts when other females are present. This behaviour causes aggregation and mutual interference. Aggregation and mutual interference can thus be viewed as consequences of natural selection on individuals which try to maximize their offspring number by playing the war of attrition against competitors.

Charnov and Skinner (1984, 1985), Parker and Courtney (1984), Iwasa et al. (1984) and Waage (1986) have formulated the problem of superparasitism as an optimal foraging problem. Because optimal patch times are dependent on the patch times that other parasitoids are willing to invest, patch-time allocation and superparasitism should be analyzed with an ESS rather than with an optimal foraging approach.

Acknowledgements. I thank Jeff Waage, Charles Godfray and Peter Hammerstein for the inspiring discussion on the function of superparasitism. Hannah Nadel, Frietson Galis, Marianne van Dijken and Kees Bakker are kindly acknowledged for their comments on a previous draft of the manuscript.

References

Alphen JJM van, Galis F (1983) Patch time allocation and parasitizing efficiency of *Asobara tabida* Nees, a larval parasitoid of *Drosophila* species. J Anim Ecol 52:937–952

Alphen JJM van, Nell HW (1982) Superparasitism and host discrimination by *Asobara tabida* Nees (Braconidae, Alysiinae), a larval parasitoid of Drosophilidae. Neth J Zool 32:232–260

Alphen JJM van, Nell HW (in prep) Mutual interference and superparasitism by *Leptopilina heterotoma* (Thomson), a larval parasitoid of *Drosophila*

Alphen JJM van, Vet LEM (1986) An evolutionary approach to host finding and selection. In: Waage JK, Greathead DJ (eds) Insect parasitoids. Academic Press, London New York

Batenburg FHD van, Lenteren JC van, Alphen JJM van, Bakker K (1983) Searching for and parasitization of *Drosophila melanogaster* (Dipt:Drosophilidae) by *Leptopilina heterotoma* (Hym.: Eucoilidae): a Monte Carlo simulation model and the real situation. Neth J Zool 33:306–336

Beddington JR (1975) Mutual interference between parasites or predators and its effect on searching efficiency. J Anim Ecol 44:331–340

Charnov EL (1976) Optimal foraging, the marginal value theorem. Theor Pop Biol 9:129–136

Charnov EL, Skinner SW (1984) Evolution of host selection and clutch size in parasitoid wasps. Fla Entomol 67:5–21

Charnov EL, Skinner SW (1985) Complementary approaches to the understanding of parasitoid oviposition decisions. Environ Entomol 14:383–391

Comins HN, Hassell MP (1979) The dynamics of optimally foraging predators and parasitoids. J Anim Ecol 48:335–351

Cook RM, Hubbard SF (1977) Adaptive searching strategies in insect parasites. J Anim Ecol 46: 115–125

Driessen GJJ, Hemerik A, Teijink A (in prep) The time budget of *Leptopilina clavipes*: estimating travel times in nature

Emlen JM (1966) The role of time and energy in food preference. Am Nat 100:611–617

Free CA, Beddington JR, Lawton JH (1977) On the inadequacy of simple models of mutual interference for parasitism and predation. J Anim Ecol 46:543–554

Fretwell SD (1972) Populations in a seasonal environment. Princeton Univ Press

Fretwell SD, Lucas HL, Jr (1970) On territorial behavior and other factors influencing habitat distribution in birds. I Theoretical development. Act Biotheor 19:16–36

Hassell MP (1968) The behavioural response of a tachinid fly [*Cyzenis albicans* (Fall)] to its host, the winter moth [*Operophtera brumata* (L.)]. J Anim Ecol 37:627–639

Hassell MP (1980) Foraging strategies, population models and biological control: a case study. J Anim Ecol 49:603–628

Hassell MP, May RM (1973) Stability in insect host-parasite models. J Anim Ecol 42:693–726

Hassell MP, May RM (1974) Aggregation in predators and insect parasites and its effect on stability. J Anim Ecol 43:567–594

Hassel MP, Varley GC (1969) New inductive population model for insect parasites and its bearing on biological control. Nature (London) 223:1133–1136

Hubbard SF, Cook RM (1978) Optimal foraging by parasitoid wasps. J Anim Ecol 47:593–604

Iwasa Y, Suzuki Y, Matsuda H (1984) Theory of oviposition strategy of parasitoids. I. Effect of mortality and limited egg number. Theor Popul Biol 26:205–227

Lenteren JC van, Bakker K (1978) Behavioural aspects of the functional response of a parasite (*Pseudocoila bochei* Weld) to its host *(Drosophila melanogaster)*. Neth J Zool 28:213–233

MacArthur RH, Pianka ER (1966) On optimal use of a patchy environment. Am Nat 100:603–609

Maynard Smith J (1982) Evolution and the theory of games. Cambridge Univ Press

Milinski M (1979) An evolutionary stable feeding strategy in sticklebacks. J Tierpsychol 51:36–40

Nicholson AJ, Bailey VA (1935) The balance of animal populations. Part I. Proc Zool Soc London 102:551–598

Parker GA, Courtney SP (1984) Models of clutch size in insect oviposition. Theor Pop Biol 26: 21–48

Rogers DJ, Hassell MP (1974) General models for insect parasite and predator searching behaviour: interference. J Anim Ecol 43:239–253

Royama T (1971) Evolutionary significance of a predator's response to local differences in prey density: a theoretical study. In: Poer PJ den, Gradwell GR (eds) Dynamics of populations. Centre Agric Publ Document, Wageningen

Sutherland WH (1983) Aggregation and the "ideal free" distribution. J Anim Ecol 52:821–828

Waage JK (1979) Foraging for patchily-distributed hosts by the parasitoid, *Nemeritis canescens*. J Anim Ecol 48:353–371

Waage JK (1983) Aggregation in field parasitoid populations: foraging time allocation by a population of *Diadegma* (Hymenoptera, Ichneumonidae). Ecol Entomol 8:447–453

Waage JK (1986) Family planning in parasitoids: adaptive patterns of progeny and sex allocation. In: Waage JK, Greathead DJ (eds) Insect parasitoids. Academic Press, London New York

Whitham TG (1978) Habitat selection by *Pemphigus* aphids in response to resource limitation and competition. Ecology 59:1164–1176

Whitham TG (1979) Territorial behaviour of *Pemphigus* gall aphids. Nature (London) 279:324–325

Whitham TG (1980) The theory of habitat selection examined and extended using *Pemphigus* aphids. Am Nat 115:449–466

Developmental Constraints

The Significance of Developmental Constraints for Phenotypic Evolution by Natural Selection

G. P. WAGNER[1]

1 Introduction

During the last 10 years neo-Darwinian theory has been challenged by a number of new concepts, such as punctuation and developmental constraints. At a closer examination, however, most of these concepts may be considered compatible with neo-Darwinian thinking or as even being part of it. Especially the concept of developmental constraints was said to be already integrated into the adaptationist program (Charlesworth et al. 1982; Mayr 1983; Maynard-Smith 1983).

But what makes the discussion on the role of developmental constraints still interesting is that it stimulates attempts to take developmental constraints into account seriously. It is true that the concept of developmental constraints is implicitly contained in neo-Darwinian theory. Nevertheless, it is also true that this concept has almost never had an influence on the main stream of research that was done by neo-Darwinists. Only recently the attempt was made to elucidate the proximate causes of developmental constraints and their influence on the pattern of morphological evolution (Alberch and Gale 1985; Maynard-Smith et al. 1985).

The main deficit in recent attempts to understand developmental constraints seems to be that they are often undertaken without any reference to population genetics theory. This may become a problem, because expectations based on intuitive reasoning are often quite different compared to the results obtained from exact models (see for instance the contribution of G. de Jong, this Vol.).

The topic of the present contribution is the role of developmental constraints in adaptive phenotypic evolution. On the basis of a mathematical model of phenotypic evolution, it will be argued that developmental constraints are necessary to facilitate the adaptation of complex and functionally integrated organisms.

1 Institut für Zoologie, Universität Wien, Althanstraße 14, A-1090 Vienna, Austria

Population Genetics and Evolution
G. de Jong (ed.)
© Springer-Verlag Berlin Heidelberg 1988

2 Comparative Biology of Developmentally Constrained Characters

According to the definition given by Maynard-Smith and his co-workers, a developmental constraint is a "bias on the production of variant phenotypes or a limitation on phenotypic variability caused by the structure, character and composition or dynamics of the developmental system" (Maynard Smith et al. 1985). The primary purpose of this concept is to explain the absence of character change as for instance in the case of morphological stasis (Gould 1980), or during the course of morphological evolution (Alberch 1982; Riedl 1978; Roth and Wake 1985). Further, it was shown that constrained characters form boundary conditions for the adaptation of other characters (Stearns 1984). In addition, non-adaptive changes and characters are explained by internal constraints ("exaptations", Gould and Vrba 1982).

Usually, constraints are thought to play a negative role in relation to the mechanisms of adaptation. However, if one looks at examples of presumably constrained characters one recognizes a peculiar pattern, namely a pronounced conservativity of body plans among the most advanced and adaptively most successful groups of animals.

The best candidates for constrained characters are those which show low interspecific variation in species-rich clades. They are easily recognized because they are used by taxonomists to define supraspecific taxa, such as families and classes (Riedl 1978).

Two characteristics seem to prevail among highly fixed anatomical characters:

1. If there is a character fixed in one group of species, then there is often a related group in which this character is variable.
2. Most often the group with the constrained character condition is the more advanced and adaptively more successful group.

An example is segment number in insects. All pterygote insects are comprised of 18 segments, four comprise the head, three make up the thorax and 11 constitute the abdomen. There is little doubt that modern day insects originated from centipedelike ancestors with variable segment number and no clear tagmatation in head, thorax and abdomen. There is also little doubt, at least for holometabolic insects, that segment number is really fixed because of developmental and not functional reasons.

The developmental genetics of segmentation is well known from the work of Nüsslein-Volhard and her co-workers (Nüsslein-Volhard and Wieschaus 1980). It is known that about 25 genes contribute to segmentation in *Drosophila*. This is less than 1% of the genome. A large number of alleles have been described, most of them are lethal, but none had a phenotype with increased segment number, even though the ancestors of the insects almost certainly had variable segment number. One may conclude that constraints on character variation can be acquired during evolution (for a discussion of this topic, see Wagner 1986). Another example which shows that constrained characters are often found in the more derived groups is the evolution of the tetrapod limb from paired fins (see Riedl 1978; Hinchliffe and Johnson 1980).

Neither the insects nor the tetrapods can be considered as evolutionary dead ends, although they have a much more constrained body plan than the ancestral groups. For instance, the insects are clearly the most advanced group among the arthropods. They have evolved the most complex functional adaptations in terms of locomotion and behaviour.

Of course, the evolutionary and adaptive success of insects and tetrapods does not result from fixed characters as such, but from newly aquired characters like the auto-podial skeleton in tetrapods and the wings in pterygote insects. Functional specializations occur in the appendages and in the internal anatomy on the basis of a constrained framework. There seems to be a correlation between a constrained body plan and the evolution of highly complex functional adaptations in other characters.

3 How Constraints Influence the Rate of Phenotypic Evolution

Progressive evolution appears to be preceded by or associated with a re-allocation of variance: some characters become constrained, while other characters gain adaptive freedom. My concern in this contribution is to show what the biological reasons may be of this association between constrained body plans and progressive evolution. No comments will be made about the evolutionary mechanisms that may have caused the re-allocation of variance, i.e. the origin of acquired developmental constraints (see for instance Cheverud 1984).

The subsequent presentation is based on results which were and will be discussed in greater detail elsewhere (Bürger 1985, 1986; Wagner 1984, 1988). Bürger and I considered a simple model of multivariate phenotypic evolution according to Lande (1982). All characters have been assumed to be normally distributed, quantitative characters, $z = (z_1,...,z_n)$. The main elements of the model are the phenotypic, genetic and environmental covariance matrices.

$$P = G + E$$

and the fitness function $m(z)$, which gives Malthusian fitness of individuals with phenotype z. The fitness function will sometimes be called "fitness landscape". The basic equation is

$$d\bar{z}/dt = G \, V_{\bar{z}} \, (\bar{m}) , \tag{1}$$

where $\bar{z}$ is the vector of mean values of the characters, $\bar{m}$ is the mean fitness of the population and $V_{\bar{z}}$ is the gradient operator with respect to $\bar{z}$

$$V_{\bar{z}} = \left(\frac{\partial}{\partial \bar{z}_1},...,\frac{\partial}{\partial \bar{z}_n} \right) .$$

In order to visualize the role constraints may play in adaptive evolution we considered a special case of Eq. (1), namely a model of a developmentally *unconstrained* phenotype. Following the definition of developmental constraints of Maynard-Smith et al. (1985), an unconstrained phenotype is characterized by the absence of any bias in phenotypic and genetic variation.

Definition: The variation of a phenotype is called *unconstrained* if the phenotypic variances and the heritabilities of all characters are the same, $\sigma_i^2 = \sigma_j^2$, $h_i^2 = h_j^2$ (for all i, j), and if there are neither phenotypic nor genetic covariances among the traits.

In the case of an unconstrained phenotype the model Eq. (1) can be simplified to

$$d\bar{z}_1/dt = \sigma_1^2 \, h_1^2 \, \frac{\partial \bar{m}}{\partial \bar{z}_1} , \tag{2}$$

where $\sigma_1^2 h_1^2$ is the additive genetic variance of the trait z_1 and $\partial\bar{m}/\partial\bar{z}_1$ is the intensity of directional selection, similar to the selection differential of quantitative genetics. The interesting feature of Eq. (2) is that in general the selection intensity is not only a function of $m(z)$ and $\bar{z}$, but also of the phenotypic variances of all characters.

$$\partial\bar{m}/\partial\bar{z}_1 = f(\bar{z}, m(z), \sigma_1^2, ..., \sigma_n^2).$$

This fact is essential for the following results.

3.1 There Are Two Types of Fitness Landscapes

Properties of Eq. (2) have been analyzed, given a number of different analytical forms of the fitness function $m(z)$ (see Table 1). Only cases were considered where one character, e.g. z_1, is under directional selection and all other characters z_j are under stabilizing selection, so-called corridor models (Wagner 1984, 1988; Bürger 1985, 1986). Fitness landscapes of this type usually look like a ridge, where the direction of the ridge gives the dimension along which directional selection acts. Stabilizing selection acts perpendicular to the flanks of the ridge (see Figs. 1a or 2a).

Two basic types of fitness functions have been found, one which will be call "well behaved", and a second which will be called "malignant". The malignant types represent the majority of the examples analyzed so far.

Figure 1a shows an example of a well-behaved fitness function. As expected an increase in the amount of heritable phenotypic variation of z_1 leads to a corresponding increase in the rate of evolution $\dot{\bar{z}}_1$ (Fig. 1b), while the variances of $z_2, ..., z_n$ have no influence on the rate of evolution (Fig. 1c). In well-behaved fitness landscapes the evolution of z_1 can be considered as independent from other characters.

In this case there is no theoretical upper limit to the rate of evolution according to Eq. (2), because there is no upper limit to the amount of heritable phenotypic variation of z_1 and the variances of $z_2, ..., z_n$ do not influence the rate of evolution.

Table 1. Fitness functions analyzed so far; in well-behaved fitness landscapes there is no upper limit to the rate of evolution of a developmentally unconstrained phenotype. In malignant fitness landscapes maximal rates of evolution exist for the evolution of unconstrained phenotypes (Symbols: $/z/$ is the norm of z, $u = (z_2, ..., z_n)$ and (x, y) is the inner product of the vectors x and y)

No.	Function	Conditions	Type	Reference
1	$m_0 - f(/z/)$	$-f < 0$, strictly decreasing, concave	Well-behaved	Wagner (1988)
2	$sz_1 - \sum_i b_i z_i^2$	$b_i > 0$	Well-behaved	Wagner (1988)
3	sz_1 if $/z_{i>1}/ < b/2$ $-m_0$ otherwise	$b > 0, s > 0$	Malignant if $n > 3$	Wagner (1984)
4	$sz_1 (\exp(-(Au,u)) - b)$	A pos. def. $b > 0$	Malignant if $n > 2$	Bürger (1985)
5	$sz_1 (\exp(-(Au,u))) - b$	as in 4	Malignant if $n > 3$	Bürger (1985)
6	$sz_1 (a - \sum_i a_i z_i^2)$	$a, a_i > 0$	Malignant if $n > 2$	Bürger (1986)

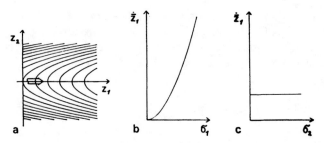

Fig. 1a–c. Example of a "well-behaved" fitness function: $m(z) = m_0 + az_1 - bz_2^2$. **a** The contour lines of the fitness landscape; the *arrow* indicates the direction of increasing fitness. **b** The rate of evolution $\dot{z}_1$ increases if the amount of phenotypic variation of z_1, σ_1, increases and the heritability remains constant. **c** The variation of z_2, σ_2, has no influence on the rate of phenotypic evolution

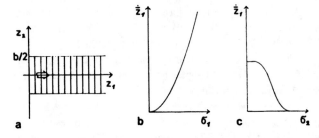

Fig. 2a–c. Example of a "malignant" fitness function: $m(z) = sz_1$ if $|z_2| < b/2$, and $= -m_0$ if $|z_2| > b/2$. **a** The contour lines of the fitness landscape. The landscape looks like a ramp in an otherwise plain land. **b** As in Fig. 1b, an increase in the amount of phenotypic variation of z_1, σ_1, increases the rate of phenotypic evolution $\dot{z}_1$, if the heritability is assumed to be constant. **c** Increasing amounts of phenotypic variation of z_2, σ_2, lead to decreasing rates of phenotypic evolution $\dot{z}_1$

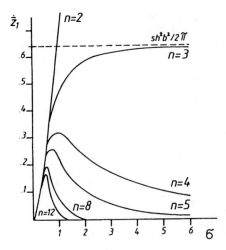

Fig. 3. The rate of phenotypic evolution $\dot{z}_1$ of an unconstrained phenotype is shown as a function of the amount of phenotypic variation per character, sigma. The fitness landscape is the same as in Fig. 2a. Note that upper limits exist to the rate of evolution as soon as three characters are involved, and that there are even maximal rates of evolution if at least four characters are taken into account. With increasing number of characters the maximal rate of evolution becomes small

In malignant fitness landscapes (e.g. Fig. 2a), the situation is a little more complicated. Similarily to the situation in well-behaved fitness landscapes an increase in the amount of heritable phenotypic variation of z_1 increases the rate of evolution (Fig. 2b). In addition, however, the variances of $z_2, ..., z_n$ have an inhibitory influence on the rate of evolution (Fig. 2c). Increasing amounts of phenotypic variance of those characters which are under stabilizing selection, lead to lower evolutionary rates.

This fact is of consequence to the evolutionary versatility of unconstrained phenotypes. Since the phenotypic variances of z_1 and of $z_2, ..., z_n$ have opposite effects on the rate of evolution and the variance of z_1 is assumed to be equal to the variances of each of the other characters, the negative effects on $\dot{\bar{z}}_1$ of $\sigma_{i>1}$ will out weight the positive effect of $\sigma_1^2 h_1^2$. If the number of characters is greater than a certain limit, then there are upper limits to the possible rate of evolution (Fig. 3). In all examples analyzed so far the complexity limit is never greater than three (Table 1).

3.2 The Types of Fitness Functions Correspond to Two Types of Functional Interactions

In this section it will be argued that the malignant fitness functions are a more adequate model for the evolution of functionally integrated organisms than the well-behaved fitness functions.

In all well-behaved fitness landscapes the selection intensity on z_1 is independent of the values of the other characters. The intensity of directional selection on the flanks of the ridge is as great as at the edge of the ridge. Therefore, variation in the characters $z_2, ..., z_n$ does not influence the rate of evolution of z_1. Mathematically this means that in well-behaved fitness functions each character contributes additively to the total fitness of the system.

In the case of well-behaved fitness functions the contribution of one character to the fitness of an individual is independent of the values of the other characters. The functional performance of each character is thus independent of the functional integrity of the other characters.

It is not easy to find characters which can serve as examples of functional independence. An example may be the contributions of sweat glands and subepidermal adipose tissue to thermoregulation in mammals. It is clear that both traits cooperate to provide an optimal thermoregulatory system for each ecological setting. However, the functional significance of each character does not depend on the other character. Variation in the number of sweat glands will lead to corresponding variation in the cooling capacity whether there is any adipose tissue or not. Equally, the fact that adipose tissue contributes to the isolation of the body does not depend on the number of sweat glands at all. Well-behaved fitness functions are models of this kind of functional independence.

In malignant fitness landscapes the selection intensity on z_1 depends on the values of $z_2, ..., z_n$. For instance in the landscape shown in Fig. 2a there is selection on z_1 only among those individuals which have all the values of $z_2, ..., z_n$ within the "corridor". If at least one character is outside the limits of the corridor, variation of z_1 ceases to influence fitness. The functional significance of z_1 depends on certain values of the other characters.

An example is the relationship between body size and respiratory capacity. There may be an advantage to increase body size, because of a pressure to improve performance in competition for mates. But variation in body size will only influence the fitness of those individuals which have an adequate respiratory capacity for their body size. Without adequate respiratory capacity one will never be in the condition to enter competition for mates. Another example is the shape of the lense in vertebrate eye and receptor density in the retina. Variation in lense shape will have functional consequences if the receptor density in the retina is high enough to mirror the differences in image quality.

It seems to be much easier to find examples of functional dependencies than examples of independent contributions to total fitness. One may consider this fact as a consequence of the "autopoietic" nature of organismic systems (see Wake et al. 1983). Therefore, I propose that malignant fitness functions are more adequate models for the evolution of functionally integrated organisms than well-behaved fitness functions.

3.3 Developmental Constraints Facilitate the Evolution of Functionally Integrated Phenotypes

As explained in Sect. 3.1, the possible rate of evolution of a certain character z_1 is limited if (1) the characters are functionally dependent, i.e. if the fitness function of the system is of malignant type; and if (2) phenotypic variation is unconstrained, i.e. if there is no bias in the production of variant phenotypes.

This fact is illustrated in Fig. 3. In this case, upper limits to the rate of evolution occur as soon as three characters are functionally integrated into a developmentally unconstrained phenotype. In addition, it can be seen that the maximal rates of phenotypic evolution decrease as the number of unconstrained characters increases. This seems to be a problem to the evolutionary versatility of complex organisms. However, there is a way out of this problem, namely an allocation of variance on those characters which are adaptively important, and a restriction of the variation of other characters, i.e. developmental constraints. Hence, developmental constraints are necessary to retain adaptive versatility in the face of functional interdependencies and high phenotypic complexity.

I do not yet know what the reasons for an adequate allocation of variance may be (for suggestions see Cheverud 1984; Riedl 1978). Nevertheless, the results indicate that the association between developmentally constrained body plans and the evolution of complex functional adaptations may not be accidental.

4 Conclusions

1. On the basis of a quantitative genetic model, one can show that there are upper limits to the rate of phenotypic evolution, if phenotypic variation is unconstrained and the characters are functionally interdependent.
2. The maximal rates of evolution become small if the number of unconstrained characters becomes large.

3. To overcome these limitations of evolutionary versatility a biased phenotypic variation is necessary, i.e. developmental constraints facilitate the evolution of functionally integrated phenotypes, if they lead to an appropriate allocation of phenotypic variance.

References

Alberch P (1982) Developmental constraints in evolutionary processes. In: Bonner JT (ed) Development and evolution. Springer, Berlin Heidelberg New York, pp 313–332
Alberch P (1985) A developmental analysis of an evolutionary trend: digital reduction in amphibians. Evolution 39:8–23
Alberch P, Gale EA (1985) A developmental analysis of an evolutionary trend: digital reduction
in amphibians. Evolution 39:8–23
Bürger R (1985) Dynamical models in quantitative genetics. In: Aubin JP, Saari D, Sigmund K
(eds) Dynamics of macrosystems. Springer, Berlin Heidelberg New York Tokyo, pp 75–89
Bürger R (1986) Constraints for the evolution of functionally coupled characters: a nonlinear
analysis of a phenotypic model. Evolution 40:182–193
Bürger R (1986) Evolutionary dynamics of functionally constrained phenotypic characters. IAM
J Math Appl Med Biol 3:265–287
Charlesworth B, Lande R, Slatkin M (1982) A neo-Darwinian commentary on macroevolution.
Evolution 36:474–498
Cheverud JM (1984) Quantitative genetics and developmental constraints on evolution by selec-
tion. J Theor Biol 101:155–171
Gould SJ (1980) Is a new and general theory of evolution emerging? Paleobiology 6:119–130
Gould SJ, Vrba ES (1982) Exaptation – a missing term in the science of form. Paleobiology 8:
4–15
Hinchliffe JR, Johnson DR (1980) The development of the vertebrate limb. Clarendon, Oxford
Lande R (1982) A quantitative genetic theory of life history evolution. Ecology 63:607–615
Maynard-Smith J (1983) The genetics of stasis and punctuation. Annu Rev Genet 17:11–25
Maynard-Smith J, Burian R, Kauffman S, Alberch P, Campbell J, Goodwin B, Lande R, Raup D,
Wolpert L (1985) Developmental constraints and evolution. Q Rev Biol 60:265–287
Mayr E (1983) How to carry out the adaptationist program? Am Nat 121:324–334
Nüsslein-Volhard C, Wieschaus E (1980) Segmentation in Drosophila: mutations affecting seg-
ment number and polarity. Nature (London) 287:795–801
Riedl R (1978) Order in living organisms, Wiley & Son, Chichester New York Brisbane Toronto
Roth G, Wake DB (1985) Trends in the functional morphology and sensory motor control of
feeding behavior in salamanders: an example of the role of internal dynamics in evolution.
Acta Biotheor 34:175–192
Stearns SC (1984) The tension between adaptation and constraints in the evolution of reproduc-
tive patterns. Adv Invert Reprod 3:387–398
Wagner GP (1984) Coevolution of functionally constrained characters: prerequisites for adaptive
versatility. Bio Systems 17:51–55
Wagner GP (1986) The systems approach: an interface between developmental and population
genetic aspects of evolution. In: Jablonski D, Raup DM (eds) Patterns and processes in the
history of life. Springer, Berlin Heidelberg New York, pp 149–165
Wagner GP (1988) The influence of variation and of developmental constraints on the rate of
multivariate phenotypic evolution. J Evol Biol 1:45–66
Wake DB, Roth G, Wake MH (1983) On the problem of stasis in organismal evolution. J Theor
Biol 101:211–224

Selection on Morphological Patterns

W. Scharloo[1]

1 Introduction

Selection acts on phenotypes. This causes genetic effects when phenotypic variation is correlated with genetic variation. The link between the phenotypic and the genotypic level is formed by gene physiology and developmental genetics. Understanding the impact of selection on genetic variation is dependent on the elucidation of this connection (Scharloo et al. 1977; Scharloo 1984).

The causal path involves genetic differences − phenotypic differences − variation in fitness − selection, and then back, − phenotypic differences − genetic differences − genetic change. The logical structure of this chain of events suggests two complementary approaches to the understanding of selection processes: (1) bottom up, investigation of single gene differences in terms of differences in gene action resulting in differences in fitness. Examples are our research on amylase variants in *Drosophila* (de Jong and Scharloo 1976; Hoorn and Scharloo 1979) and on alcohol dehydrogenase (Heinstra et al. 1983, 1987; Eisses et al. 1985). This approach can be named ecology of genes. (2) Top down, starting from phenotypic variation in characters of obvious ecological significance, making a biometrical genetic analysis of the phenotypic variance in terms of environmental and genetic components and their interaction. This must be followed up by physiological and developmental analysis of the variation. Examples are our work on the great tit (van Noordwijk et al. 1981), a study on quantitative genetics of clutch size, body size, begin of breeding, etc., and on the anal papillae of *Drosophila melanogaster* larvae and their salt resistance (te Velde 1985; te Velde et al. 1988; te Velde and Scharloo 1988). This top down approach I have called the genetics of ecology (Scharloo 1984).

It is startling to see how little work is done to connect the different levels of variation. This is in particular true for quantitative genetics. Most work is strictly limited to the phenotypic level. Analysis in terms of underlying physiological and/or developmental processes are rare (see e.g. Robertson 1959; Spicket 1963). Concepts of development genetics were involved in work on canalization of morphological charac-

1 Department of Population and Evolutionary Biology, University of Utrecht, Padualaan 8, NL-3584 CH Utrecht, The Netherlands

Population Genetics and Evolution
G. de Jong (ed.)
© Springer-Verlag Berlin Heidelberg 1988

ters (Dunn and Fraser 1959; Rendel 1959; Maynard Smith and Sondhi 1960; Scharloo 1962) and in work explaining the effect of disruptive and stabilizing selection (Rendel and Sheldon 1960; Rendel et al. 1966; Scharloo et al. 1967, 1972; Scharloo 1970).

2 Canalization of Scutellar Bristles

Rendel (1959) investigated the constancy of the pattern of scutellar bristles in *Drosophila*. This number is almost a constant four, only rarely do flies have three or four bristles. Its constancy makes it a diagnostic character for the genus *Drosophila*. Rendel found that variability occurred when the mutant *scute* decreased bristle number to two bristles in females and one in males. Rendel started selection for higher bristle number in *scute* flies. A model was derived by Rendel from the pattern of change of mean bristle number and the frequency distributions, in which the following were postulated: (1) a bristle-forming substance which is supposed to vary in the base population according to a normal distribution and is strongly affected by the *scute* mutant. (2) Thresholds separating the bristle classes: to obtain a bristle number of four the amount of morphogenetic substance must transgress the threshold between the bristle classes three and four and be lower than the threshold between the classes four bristles and five bristles. (3) A function transforming the bristle-forming substance in bristle number (Fig. 1). This mapping function is the consequence of the organization of bristle development which is canalized to produce four bristles. The horizontal part of this mapping function represents the canalization zone; it suppresses the phenotypic effects of variation in the bristle-forming substance and the small difference in the bristle-forming substance between females and males. Therefore, wild-type flies and both females and males have four bristles. When the distribution of bristle-forming substance is removed from the canalization zone by a mutant, by artificial selection or by environmental factors, there is variation in bristle number, a sex difference appears and selection has an immediate result.

In this model genetic differences, environmental differences and the stochastic variability due to imprecision of development (developmental error or developmental noise) all act via the same gene-environmental factor/phenotype mapping function (GEPM, Scharloo 1987).

It is stated that the canalization of scutellar bristle number is the consequence of the permanent pressure of stabilizing natural selection for the normal number of four bristles. Artificial selection can be used to explore the possibility that such a selection can cause canalization. Generation of canalization has indeed been observed when artificial stabilizing selection was applied to morphological characters which were affected by mutants with large effects. Morphological characters produced by such mutants have a relatively large variability when compared with wild-type morphological characters. Rendel (1967) suggested that stabilizing selection could only be successful in generating canalization when the character was affected by what he called an uncontrolled major gene. An alternative hypothesis is that which is controlled is not the action of a certain gene but more importantly the state of the developmental system leading to the formation of the bristle pattern. The presence of a major mutant would not be crucial, but rather whether the expression of the character re-

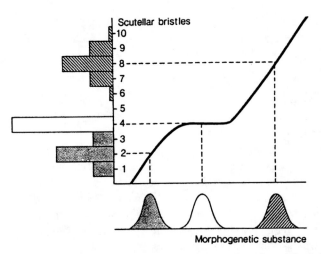

Fig. 1. Model for the determination of the number of scutellar bristles (after Rendel 1959). The number is determined by (1) a morphogenetic substance which is supposed to vary in the population according to a normal distribution. (2) Thresholds separating the bristle classes: to obtain a bristle number of four, the amount of morphogenetic substance must transgress the threshold between the bristle classes three and four and be lower than the threshold between the four- and five-bristles classes. (3) A mapping function translating the morphogenetic substance in number of bristles. The mapping function is sigmoid with a horizontal part around four bristles; in this region changes in the amount of morphogenetic substance have scarcely any effect on bristle number. Thus, the number of scutellar bristles is canalized at four bristles, i.e. there is no variability; males and females have the same bristle number and environmental factors have no effect. This mapping function, a gene-environmental factor/phenotype mapping function (GEPM) is a reflection of the organization of the developmental system underlying bristle formation. Beyond the canalized zone bristle number is approximately proportional to the amount of morphogenetic substance and therefore sensitive to genetic and environmental differences

mains within the bounds of the expression range in which the developmental system is buffered. If the alternative hypothesis is correct, canalization should be obtained from the genetic variability present in each wild population without involvement of a mutant gene with a large effect on the character concerned, the uncontrolled major gene of Rendel. Canalization based on genetic variability of an unknown number of loci with a small effect connects quantitative genetics with the developmental system.

3 Experiments on Scutellar Bristles

The problem is whether canalization can be generated from unspecified polygenic variability of a character, without the involvement of a major gene. We started therefore artificial selection for a higher number of scutellar bristles in two unrelated wild populations. We had first to look at a large number of flies to find enough with five bristles to start the selection. After some generations the number of five bristle flies started to increase and the first flies with six bristles appeared. The mean started to increase rapidly, and so did the variability. When both lines had reached a mean bristle number of eight we crossed the lines and used the third generation of this cross as a

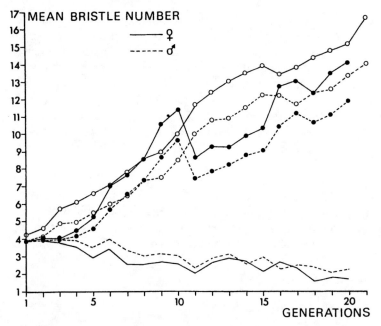

Fig. 2. Selection for higher scutellar bristle number from two wild-type cage populations which were founded independently for several years before the start of these experiments. When between generations eight and ten a mean of approximately eight bristles was obtained in both lines, they were crossed to establish the base population for experiments analyzing the effect of stabilizing and disruptive selection outside the normal zone where the number of scutellar bristles is almost invariable

base population for our selection experiments. Crossing of these lines and their further response thereafter (Fig. 2) guaranteed that there was plenty of genetic variability present in our base population. We started three lines. For the control line C we selected parents for the next generation at random. For the line under stabilizing selection S we selected from each generation flies with eight bristles. Further, we started a line under disruptive selection with mating of opposite extremes (D⁻ selection), i.e. the females with the highest number were mated with the males with the lowest number, and the females with the lowest number with the males with the highest number. After mating the males were discarded and the high and low females transferred to the same culture bottle.

3.1 Selection Results

Mean Bristle Number. Mean values in the control stock first remain at the level of the base population but decrease in later stages of the experiment. In the disruptive line the mean decreased. In the stabilizing line the mean of the females remains very constant, 0.2 bristles higher than the eight selected for; the males increased slowly from a mean of 7 to almost 8 at the end of the experiment.

Variance. The variance in the control line is rather constant until the mean starts to decline. Then there is a marked rise. In the disruptive line the phenotypic variance increases strongly in the first five generations in females, in males the increase is slower but lasts longer. In the stabilizing line there is a long decline lasting over 25 generations of selection until the variance is only one-third of its original value. Frequency distributions (Fig. 3) show a concentrated distribution for the S line with more than 80% with eight bristles. In the D⁻ line the distribution is broad. In the C line the females have an approximately normal distribution; in males which have a lower mean value flies accumulate in the four-bristle class as a consequence of the canalization of this class.

Variance Components. To obtain more insight into the causes of the variance changes under the influence of selection, components of the phenotypic variance were determined by progeny tests and by comparison of the variance in the selection and control lines with the variance of inbred lies and crosses of inbred lines. The progeny

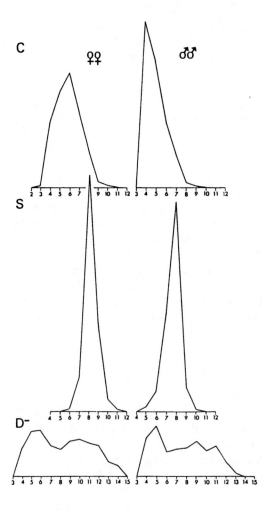

Fig. 3. Scutellar bristles. Frequency distributions in females and males of the selection lines after 15 generations. When the mean approaches four bristles as in the males of the control line, the effect of the normal canalization zone around four bristles is noticeable. In the S line there are unimodal distributions with the majority of the flies having eight bristles

tests were performed at several stages of the selection experiment. In later stages of the selection and in the inbred lines and their crosses the scutellars were counted separately on the right and the left half of the scutellum to determine within-fly variances as the variance of the differences between the two halves of the scutellum. The heritability was determined as the regression of progeny means on the midparent values. The additive genetic variance was then calculated as the product of the heritability h^2 and the phenotypic variance V_p. Deducting the additive genetic variance from the phenotypic variance gives a residual variance which contains the rest of the genetic variance and the non-genetic variance. The non-genetic variance was derived as the mean phenotypic variance of a number of genetic homogeneous stocks, i.e. inbred lines and their F_1 crosses. By counting bristle number separately on the right and the left half of the scutellum, the non-genetic variance can be split into the real environmental variance and the within-fly variance (developmental error or developmental noise). The genetic variance not included in the additive genetic variance, i.e. the variance components caused by dominance and genetic interaction are calculated as the difference between the residual variance derived from the progeny test and the non-genetic variance determined as the phenotypic variance of the genetic homogeneous stocks (Fig. 4).

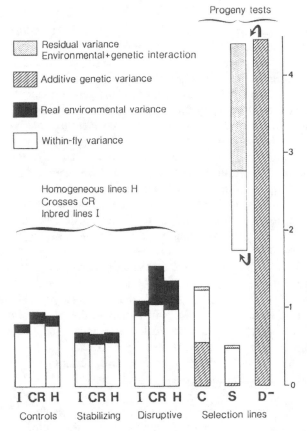

Fig. 4. Scutellar bristles. The composition of the phenotypic variance inbred lines *I* and crosses *CR*, combined in one estimate for genetic homogeneous lines *H*. The inbred lines were obtained by full sib mating from generation 15 of the control line and the stabilizing and disruptive selection lines. Selection lines as revealed by *progeny tests*. Further explanation in text

Inbred Lines and Crosses. Inbred lines were obtained by brother-sister mating from the control line and the two selection lines after 15 generations of selection. Scutellar bristle number behaves just like the other bristle characters as abdominals or sternopleurals: In crosses the mean bristle number is not higher nor is the phenotypic variance lower than in the inbred lines. It is clear that in this bristle character real environmental variance is small compared with the within fly variance measuring the precision of bristle number development on thorax halves. The difference in phenotypic variance between the genetic homogeneous lines from the control line and the stabilizing line is caused by a difference in the within-fly variance; the environmental variance has similar values. In the disruptive line both variance components increased.

Progeny Tests. The progeny tests confirm the figures from the comparison of genetic homogeneous stocks: the within-fly variance is manifold larger than the residual variance. The heritability in the base population is quite large, more than 40%. In the stabilizing line the additive genetic variance declines continuously in the course of the selection. Simultaneously, there is the decline of the within-fly variance, while the environmental variance (included in the residual variance) does not change.

3.2 Canalization

The question is now whether the concentration of flies in the eight-bristle class in the stabilizing selection line and the decrease of the non-genetic variance by a decrease of the within-fly variance is a consequence of real canalization according to Rendel. Rendel (1959, 1967) used as a crucial criterion whether the width of the canalized class, here the eight-bristle class, has been increased relative to the adjacent classes. The width of a class is determined on the basis of the supposedly normal distribution of the bristle-forming substance (*make* in Rendel's terminology). When this underlying variable has a normal distribution, the distance of the thresholds separating bristle classes to the mean of the distribution can be determined from the frequencies of the bristle classes in the population. The width of a class becomes then expressed in standard deviations of the underlying normal distribution. In our control line the width of the eight-bristle class and the adjacent seven- and nine-bristle classes have similar values (Fig. 5). In the disruptive line the width of all three bristle classes decreases in the first generations of the selection. However, this is not an indication of decreased canalization. Since the disruptive selection can be expected to increase the variance of the underlying normal distribution and because the class width is expressed in standard deviations of the underlying distribution, the width of all classes will start to decrease. But this does not occur in the stabilizing line: there is a steady increase of the width of the eight-bristle class and it obtains a value two times the value in the control population and the adjacent seven- and nine-bristle classes in the stabilizing selection line. So, Rendel's criterion is met. But is this sufficient for the conclusion that we really have obtained, by artificial stabilizing selection, a canalization of the eight-bristle class comparable to the canalization of the four-bristle class in wild populations?

   Interchange of the three large chromosome pairs between the stabilizing line and the control line shows that all large chromosomes are involved in the increase of the

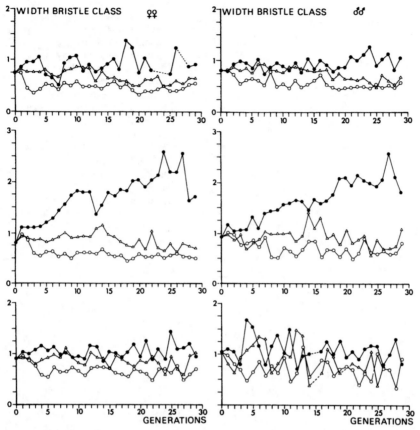

Fig. 5. Scutellar bristles. The width of the classes 7 bristles (*top*), 8 bristles (*middle*) and 9 bristles (*bottom*) in the stabilizing line (*black dots*), the control line (*triangles*) and the disruptive line (*open circles*). The class width is expressed in the standard deviation of the hypothetical normal distribution of morphogenetic substance. In the stabilizing line there is a continuous increase of the width of the 8-bristle class, while the width of the adjacent classes remains on the same level. The width of all three bristle classes has decreased slightly in the disruptive line which seems to be a consequence of increase of the variance of the morphogenetic substance which could be expected under this type of selection

width of the eight-bristle class and the decrease in sex difference (Fig. 6). The selection response is clearly polygenic.

3.3 Testing Canalization

Rendel's model implies that various factors, e.g. genetic factors, environmental factors and the random factors involved in the within-fly variance, all act via the genotype-environmental factor/phenotype mapping function (GEPM, see Scharloo 1987). Rendel and his collaborators showed that genetic differences and environmental differences, i.e. temperature, indeed act on the same scale. In our experiments we saw

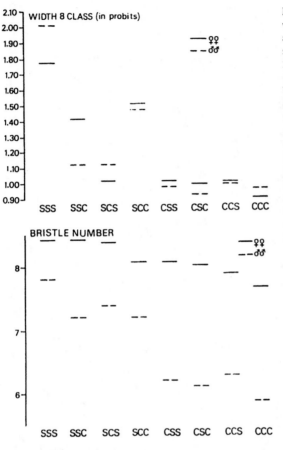

Fig. 6. Scutellar bristles. The width of the eight-bristle class and mean bristle number in the stabilizing selection line after 25 generations of selection, the control line and stocks obtained by interchange of the three large chromosomes between the stabilizing line and the control line. The chromosome combinations are indicated under the figures where S indicates chromosomes of the stabilizing line and C chromosomes of the control line. All three large chromosomes affect the width of the 8-bristle class and the difference between the means of females and males

in the stabilizing line that the decrease in the phenotypic variance is a consequence of a decrease in the within-fly variance, while the real environmental variance was not affected. This suggests that these two factors do not act via the same GEPM. The presence of canalization is indicated because in the stabilizing line during the selection, the difference between males and females decreases until it is only about one-third of its original value.

The presence of canalization was also tested with environmental factors. First, we used temperature. There is no sign that the stabilizing line is less sensitive to temperature change than the disruptive and control line. Remarkable is that the number of scutellars increases as the temperature increases in the stabilizing line and decreases in the disruptive and control lines (Fig. 7). Further, we added aminopterin to the food medium. Aminopterin inhibits the enzyme dihydrofolate dehydrogenase which is involved in the synthesis of thymidine. It causes an increase of the number of scutellar and dorsocentral bristles (Bos et al. 1969). In contrast to temperature, the reaction to aminopterin is in the same direction in all three lines. While the stabilizing and control lines show a very similar response, the disruptive line is far more sensitive. Strange enough this is most obvious for the dorsocentral bristles, a character on

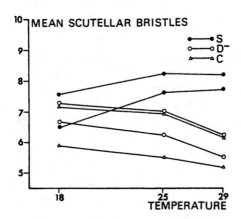

Fig. 7. Scutellar bristles. The effect of temperature on bristle number. The stabilizing line
shows a reaction opposite to the reaction in
the control line and the disruptive line, and
there is not much difference in the extent of
the reaction. For each line the reactions of
males and females are given separately. The
line with the higher bristle numbers always
represents the females

which no selection was performed. The canalization is present but is limited to genetic
and environmental differences present when the selection was performed.

3.4 Regulation Scutellar Bristle Number per Half-Thorax?

We have shown that the non-genetic variance of scutellar bristle number consists predominantly of within-fly variance. This is the variance component which is, in addition to the additive genetic variance, strongly affected by the stabilizing selection.
This component is involved in the change of the width of the eight-bristle class: there
is a negative correlation of more than 90% between width of the eight-bristle class
and the within-fly variance in the control and stabilizing selection lines and in stocks
obtained by interchange of the three large chromosome pairs of the two selection
lines.

Such a correlation between the canalization and the variance between thorax halves
could mean that this canalization of the eight-bristle class is a consequence of canalization of the development of four bristles on the two thorax halves separately. Perhaps, this could be expected because the two thorax halves originate from two separate
imaginal discs. Then a negative correlation between the presence of bristles on the
different sites on a half-thorax could be expected. This would mean that the sum of
the variances of the bristle numbers on separate sites would be higher than the variance
of the sum of the bristle numbers on half-thoraxes. But this is not the case; canalization does not proceed per thorax halve. Canalization here is an increase in precision
of the development of bristles on the separate sites. It is not a change of a gene-
environmental factor/phenotype mapping function (GEPM) for the total number of
scutellar bristles as was found for the length of the fourth vein in *ci-dominant* and
Hairless (Scharloo 1987), and in experiments canalizing the number of scutellar bristles
around two in a stock carrying the mutant *scute* (Rendel et al. 1966).

4 Bristle Patterns

That canalization in this experiment is an increase in precision of the development of
bristle number on the separate sites could be related to the establishment of a fixed
pattern in which the eight bristles are present. Therefore, we cannot avoid the involvement of the pattern aspect in this kind of selection experiment, and we cannot expect
to obtain an understanding of the mechanisms of selection response by confining the
analysis to the number of scutellar bristles only. We have seen with this character that
canalization can be obtained around an arbitrary number by artificial stabilizing selection. This means that constraints for a further selection response can be generated by
selection and that such constraints, at least in this character, are not inherently a consequence of fundamental, inevitable properties of the structure of the developmental
system underlying this character. Similar phenomena were found in our experiments
on wing venation in *Drosophila* (see Scharloo 1987).

But it is possible that this conclusion cannot be generalized. These phenomena are
perhaps generated by the very nature of our selection experiments, i.e. because of the
definition of our character as a number only, without involving the fact that these
bristles appear in fixed patterns, and by the nature of our selection programs. Perhaps
we would meet a more fundamental type of constraint if we designed selection programs impinging on the pattern aspects of such characters. Therefore, we decided to
start selection on the position of dorsocentral bristles in *Drosophila melanogaster*.

5 Selection on Pattern Characteristics

In our experiments we selected on the position of the dorsocentral bristles of *Drosophila*. The dorsocentral bristles are present on the mesonotum of *Drosophila* in a
rectangular pattern of four bristles (Fig. 8). Such a pattern can be defined by the
distances between the elements of the pattern, the dorsocentral bristles, and the distance to the borders of the body part in which the pattern is formed, i.e. the borders

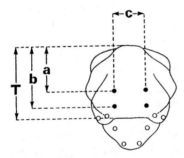

Fig. 8. The thorax (mesonotum) of *Drosophila*. The dorsocentral bristles are indicated in *black*.
Between scutum and scutellum there is a clear separation. The four bristles on the scutellum are
indicated as *open dots*. *T* represents thorax length, i.e. length of scutum and prescutum, not separated by a visible morphological boundary; *a* and *b* denote the distance of the anterior dorsocentral
bristle and the posterior dorsocentral bristle respectively, to the anterior border of the prescutum;
c denotes the distance between the two anterior dorsocentral bristles

of the scutum. Further, the absolute dimensions of these distances are not essential in a pattern; the relative positions of the elements to each other and to the borders of the body part involved are characteristic for the pattern. Therefore, our characterization of the pattern must be independent of the size of the relevant body part. For this reason we defined the position of the elements in the dorsocentral bristle pattern in the following way. The length of the thorax (scutum + prescutum) was measured (T) and the distance (a) of the anterior dorsocentrals to the anterior border of the mesonotum. The ratio a/T was used as a measurement of the position of the anterior dorsocentral bristles. Further, the distance of the posterior dorsocentral bristle to the anterior border of the mesonotum was measured. The distance between the anterior and posterior bristles was then expressed as the ratio (b-a)/T. Is this measurement really independent of body size? Thorax size is different in the two sexes, females are approximately 10% larger than males. We generated a similar difference in size between cultures by varying the population density. The comparison of the flies with different body size showed that the pattern is slightly dependent on body size. When the length of the thorax changed approximately 10% the relative position of the anterior bristle changed only about 4% and the relative distance between the anterior and posterior dorsocentrals only 2%. In smaller flies the dorsocentral bristles seem to have a position closer to the anterior and posterior margins of the scutum.

5.1 Genetic Variation

Selection response is dependent on the presence of genetic variation and its organization, i.e. the genetic architecture of the character selected for. Therefore, we first compared the phenotypic variation of genetic homogeneous lines, i.e. the pooled phenotypic variance of a number of long inbred lines and their crosses, with the phenotypic variance of a wild-type population which was used later as the base population for the selection experiment (Table 1). The pooled phenotypic variance of the three inbred lines is very similar to the pooled phenotypic variance of the six crosses. Therefore, we can pool the phenotypic variances of the inbred lines and their crosses to one pooled phenotypic variance for genetic homogeneous lines which represents here the environmental variance V_e. When we deduct this environmental variance from the phenotypic variance of our base population, we obtain an estimate of 0.10 for the total genetic variance V_g and a broad sense heritability h_{sl}^2 of 0.36. A progeny test in which the regression of progeny means on midparent values was calculated gives an estimate for the heritability of 0.40 and for the additive genetic variance V_{ag} of 0.11. This estimate for the additive genetic variance is very similar to the estimate for the total genetic variance of 0.10.

5.2 Artificial Selection

On the basis of the genetic architecture, we would perhaps expect a simple symmetric selection response in which both the shift forwards of the anterior dorsocentral bristles in the line selected for a low value of a/T, and the shift backwards

Table 1. The position of the anterior dorsocentral bristle[a]

Wild population			Homogeneous stocks		
N	V_p		N	V_e	
126	0.28	3 Inbred lines	168	0.17	0.18
		6 Crosses	336	0.19	

$V_p + V_g$	= 0.28	Progeny test:	
V_e	= 0.18	Regression of progeny means	
V_g	= 0.10	on midparent $b = h^2_{s.s}$	
$h^2_{s.1}$	= 0.36	$h^2_{s.s} = 0.40$	

[a] The phenotypic variance V_p in the base population of the selection experiment and in inbred lines and crosses which were used for an estimate of the environmental variance V_e. The total genetic variance was obtained by subtraction of the pooled phenotypic variance of the inbred lines and crosses from the phenotypic variance of the base population. The heritability (*sensu lato*) $h^2_{s.1}$ was then calculated as the ratio of the total genetic variance and the phenotypic variance and the phenotypic variance of the base population. The heritability (*sensu stricto*) $h^2_{s.s}$, i.e. the ratio of the additive genetic variance to the phenotypic variance, was obtained from progeny tests as the regression of the progeny means on the midparent values.

of the anterior bristles in the line selected for a high a/T value would occur promptly. However, one could also base some expectations on the basis of what is known and theorized in the developmental biology of *Drosophila*, in particular with respect to the development of bristle patterns and the role of compartment formation in the development of the thorax. For bristle patterns there are the models of Meinhardt (1982). He suggests that the pattern is generated because of the production of an autocatalytic bristle-forming substance on certain points in the bristle-forming tissue coupled with the production of a bristle-inhibiting substance which spreads in that tissue, thus preventing the formation of other bristles in the neighbourhood. If such a mechanism would occur, one could perhaps expect that a posterior shift of the anterior bristles would be constrained by mutual inhibitory effects of posterior and anterior bristles. Further, in the last 10 years it has become clear that an essential aspect of the development of segmentation of *Drosophila* is compartment formation. Compartment borders are often not reflected in visible, morphological landmarks. Developmental events within a compartment are separated from those in other compartments. If there would be such a compartment border between scutum and prescutum, just in front of the anterior bristles, there could be a constraint to move the anterior bristles in a forward direction.

Figure 9 shows the results of 15-generation selection. There is a strong and continuing response for a large a/T value, i.e. a shift of the anterior dorsocentral bristles in a posterior direction. There is no response in the line in which selection was applied for a shift of the bristles in an anterior direction. The a/T value in the low line is almost exactly the value in the base population. We have seen earlier that change of thorax (= scutum + prescutum) length has a small effect on the a/T value. Therefore, it is important that the T values of the two lines show only a small difference. Moreover, it is in the opposite direction to that which must be expected if this size dif-

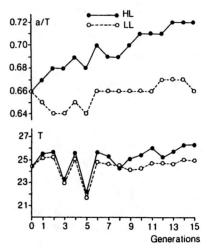

Fig. 9. Dorsocentral bristles. The response to selection on the position of the anterior dorsocentral bristle.
a The distance of the anterior dorsocentral bristle to the front of the thorax and *T* the thorax length; *a/T* the position of the anterior bristle relative to the anterior boundary of the thorax. *HL* is the line selected for a more posterior position of the anterior dorsocentral bristle; *LL* is the line selected for a more anterior position

ference is involved in the generation of the difference in a/T value. So, this selection response in a pattern parameter cannot be explained as a consequence of a change in body size.

5.3 Correlated Responses

We have already seen that our selection response is not a consequence of a correlated response of a change in body size. But there is an intriguing correlated response. When the anterior bristles shift backwards under the influence of artificial selection on their position, the posterior bristles shift in the same direction. Is this a consequence of mutual inhibition on which we were speculating earlier? Or is selection on the position of the anterior dorsocentral bristles, in fact, selection on the pattern of the four dorsocentral bristles as a whole? The last hypothesis is perhaps, to a certain extent, supported by another correlated response. We were looking too at the position of the scutellar bristles. The border between scutum and scutellum is definitely a compartment border, the differentiation of the pattern of the scutellar bristles occurs in another compartment. It is intriguing that the distal shift of the pattern of the dorsocentral bristles on the scutum is matched by a similar shift of the bristle pattern on the scutellum. This seems to implicate that what is involved here is really a shift of the whole pattern of dorsocentral bristles caused by the selection on the position of the anterior dorsocentral bristles, which then also changes the formation of a similar pattern, i.e. of the scutellar bristles in the next compartment.

5.4 Constraints on Shifting the Anterior Bristle Forwards

Striking is the complete lack of response in the line selected for a low a/T value, i.e. the line designed to displace the anterior dorsocentral bristle in a forward direction. Asymmetric responses have been described in many experiments. Various causes have been suggested and/or shown to generate such an asymmetric response, e.g. an asymmetric distribution of gene frequencies or asymmetric action of natural selection. There are no indications for these mechanisms here. Another possibility are constraints on the selection response by developmental mechanisms. Such constraints were found in Rendel's experiments when he selected for a higher bristle number in a stock with a *scute* mutant. Starting from a mean bristle number between one and two bristles response was halting when the mean value approached four bristles which is highly canalized (Rendel 1959). Similar constraints were found when the number of structures (bristles and ocelli) on the top of the head of *Drosophila subobscura* was selected towards the normal situation in lines with the mutant *ocelliless* (Maynard Smith and Sondhi 1960), and when the length of the fourth vein of ci^D was selected for and approached completion (Scharloo 1962 and unpublished). Features of the developmental system of the character selected for were certainly involved here. But in these characters the constraints were not absolute. Similar constraints could be established at other arbitrary values of the character by artificial stabilizing selection or changed by artificial selection or by mutants. The question is whether more fundamental features of development are involved here, e.g. the establishment of compartment boundaries which inhibit the selection response. Are there indications for the presence of a compartment boundary just in front of the anterior dorsocentral bristles, e.g. between scutum and prescutum? The evidence is not clear. Murphy and Tokunaga (1970) suggested subdivision of the notum in several regions which are characterized by common ancestry. Garcia Bellido et al. (1976), however, warned that the demarcation lines between these regions could represent spurious compartment borders generated by region-specific mitotic orientation, although they did not exclude the possibility that real compartment borders are present.

If there is a compartment border between scutum and prescutum, then the distance between the anterior dorsocentral bristle and this border is not smaller than the distance between the posterior dorsocentral bristle and the compartment border between scutum and scutellum, both in the adult fly and on the fate map of the imaginal discs (see Bryant 1978). Nevertheless, the posterior dorsocentral bristle is shifted backwards as a correlated response to the backward shift of the anterior bristle generated by the selection. However, it could be that compartment borders are not established at the same time, and that within compartments there is a polarity. This could perhaps affect the possibilities for shifting bristle positions.

6 Discussion and Conclusions

The renewed interest in the relation between genetic and phenotypic change expressed in morphology is stimulated by the debate between advocates of the description of the evolutionary process as punctuated equilibria (Eldredge and Gould 1972; Gould and Eldredge 1977; Gould 1980; Eldredge 1984) and population geneticists (Jones 1981; Charlesworth et al. 1982). The stability during periods of equilibrium was attributed to developmental constraints or developmental stability, which would be caused by the intrinsic structure of developmental systems (Alberch 1980, 1982; Kauffman 1983). It was stated that quantitative genetics had produced evidence supporting this theory and that thereby the explanation of the equilibrium periods would go beyond the Neo-Darwinian theory of evolution. However, it was shown (Scharloo 1987) that although quantitative genetic experiments were performed in which developmental constraints affected the expression of genetic variability and effects of environmental change, this did not mean that the constraints involved can be attributed to the intrinsic structure of developmental systems. The developmental stability or canalization (Waddington 1957) was thought to have arisen as a consequence of stabilizing natural selection. This was supported by experiments in which the expression of genetic variability, genetic change by selection, environmental differences and stochastic factors (developmental error) could be changed by artificial stabilizing or disruptive selection (see Scharloo 1987). In the first part of this chapter we described the effect of stabilizing selection on the number of scutellar bristles in a population of lines selected earlier for a bristle number higher than normal, i.e. eight bristles. We showed that canalization around the values selected for was obtained, i.e. the width of the eight-bristle class expressed in standard deviations of a hypothetical underlying distribution of a morphogenetic substance had increased. Rendel (1959) explained canalization of the number of scutellar bristles around the wild-type number of four in terms of a model (Fig. 1), in which the total number of scutellar bristles was determined by a hypothetical, underlying, normally distributed morphogenetic substance or "make" which was translated by a sigmoid gene-environmental factor/phenotype mapping function in the number of scutellar bristles. A basic assumption is that the number of scutellar bristles is determined in its totality. The canalization of the four-bristle class is revealed by the width of this class, i.e. the change in the amount of morphogenetic substance necessary to change from three to five bristles expressed in the standard deviation of the hypothetical normal distribution of the morphogenetic substance. This class width can be derived from the frequency distribution of the bristles and the theory of normal distribution. A. Robertson (1965) designed a model which showed that such phenomena could also be expected according to a model in which bristle number is not determined as a total, but when bristle number is determined at each site separately with a certain probability. When the precision of determination increases, i.e. the probability of the realization of one bristle per site, then the canalization of the four-bristle class expressed as class width also increases. This seems to occur in our stabilizing selection line. However, Rendel (1965) found support of his model in a study of bristle patterns which showed that the appearance of bristles on the separate sites is not independent. New complications were introduced by Fraser (1967), who suggested that different genetic systems are involved in the selection response to a higher bristle number and the response to a lower number. Evidence for separate effects on anterior and posterior scutellar bristles was found by several authors (Scowcroft et al. 1968; Whittle 1969; Latter and Scowcroft 1970; Scowcroft 1973). They found that an increase in scutellar bristle number was generally initiated by an increase in the number of anterior bristles and a decrease due to loss of posterior bristles and that canalization can be changed for separate bristle sites. These complications, which do not fit Rendel's simple model, are not surprising in view of the complicated developmental pathways underlying the realization of a bristle pattern. Similar complications are found in relation to our model for determining the length of the fourth vein in ci^{DG} stocks (Scharloo 1970; Scharloo et al. 1972;

Scharloo 1987). In this model, too, hypothetical normal distributions of a vein-form-ing substance are translated via a genotype-environmental factor/phenotype mapping function (GEPM) in phenotypic values, i.e. length of the fourth vein. In that case there was plenty of evidence that genetic differences, temperature and stochastic variability expressed in the within-fly variance were acting on this same scale, thus via the same GEPM. However, in two lines under artificial stabilizing selection the environmental variance decreased strongly without change of the sensitivity to temperature or the effect of the developmental error expressed in the within-fly variance (Scharloo et al. 1967). The environmental factors of which the effect was reduced by the stabilizing selection, clearly act on other steps in development than the temperature and the fac-tors involved in the error variance. While the action of different factors, which change phenotypes via the same GEPM, clearly implicates a possible common path in the realization of their effects, it does not mean that other factors will necessarily act via the same GEPM, i.e. the same developmental path. In the complicated network of developmental pathways, there will always be numerous opportunities to change morphological phenotypes. Therefore, it is not surprising that there are various pos-sibilities for different mechanisms maintaining together the canalization of the wild-type number of four bristles. Perhaps it must be expected that when canalization is generated in a short-term experiment with artificial stabilizing selection, only one of these mechanisms is realized. This was observed in a long-term experiment in which disruptive selection was performed on the length of the fourth vein of a ci^{DG} stock. In the first stage of the experiment the phenotypic variance increased with an in-crease of the genetic variance only; in the second stage there was a change of a GEPM which led to a simultaneous increase of the phenotypic expression of genetic dif-ferences, environmental differences and the differences involved in the within-fly variance (de Waal Malefijt and Scharloo, in preparation).

While Rendel in his selection experiments in his *scute* stocks found predominantly evidence for one GEPM as mediator between several different factors which changed the phenotypic character, i.e. scutellar bristle number as a whole, in our experiment the effects were on bristle sites separately. But our second experiment, in which selec-tion was performed on the position of dorsocentrals, showed that there is mutual in-fluence between bristle sites in the dorsocentral pattern. The pattern was moved as a whole. Moreover, the mutual influence was not limited to the dorsocentral pattern. The pattern of scutellar bristles also moved backwards as a correlated response to the change of position of the anterior dorsocentral bristle caused by selection on this pat-tern character. The correlated shift of the scutellar bristle pattern as a whole shows again the interdependence of bristle formation at the different sites on the suctellum.

These different possibilities of changing complex morphological characters either as a whole or in their separate elements suggest that it will not be easy to predict pat-terns of selection response on the basis of developmental relations between elements of such a character. This was also indicated by the correlated response of the length of the fifth wing vein in three lines with the mutant *Hairless* selected for a short fourth vein. In one line the fifth vein became shorter parallel to the shortening of the fourth vein; in a second line the fifth vein did not change at all when the fourth vein became shorter, and in a third experiment the fifth vein changed in the first part of the experi-ment parallel with the fourth vein but increased in length in the second part until it reached its original length again (Scharloo 1987). No fixed, unalterable developmental correlations which could determine the selection response existed.

Experiments in which selection was performed on the pattern characteristics of morphological patterns are rare compared with experiments with artificial selection on quantitative characters.

Reeve (1961) succeeded in shifting the small sternopleural bristles below the imaginary line between two large bristles to the dorsal side of this line.

Maynard Smith and Sondhi (1960) selected to change the number and pattern of bristles and ocelli on the head of *Drosophila subobscura* in lines with the mutant *ocelliless*. In addition to the canalization of the wild-type number of the ocellar structures, they showed that artificial selection can promote a pattern in which only the two posterior ocelli are present. However, it was not possible to obtain a handed asymmetric pattern in which either the left or the right ocellus was present.

An interesting phenomenon in relation to the role of developmental constraints in evolution is the appearance of new structures in experiments with artificial selection on morphological characters in a fixed position. Sondhi (1962) described in a line selected for a higher number of ocellar structures how a new bristle appears on the top of the head, always in the same position. He showed that such a bristle is always present in a related genus and suggested that the two species have the same prepattern for bristles on the head but that this bristle is lost in the further evolution of one of the species because of a decrease of the ocellar organ-forming substance. When the level of the substance increases as a consequence of selection for an increased number of ocellar organs, the position of the extra bristle is already fixed by the presence of the preparation descended from an earlier stage in the evolution of the species.

A similar appearance of a new bristle was found in transplantation experiments with imaginal discs of halteres. When discs differentiated during the metamorphosis of the host larva a new bristle appeared in a position in which a bristle is always present in another genus. A similar explanation as for the appearance of a new bristle in the case of the selection on the ocellar organs seems appropriate (Loosli 1959).

What we see here are constraints present for historical reasons and without obvious function. This contrasts with the constraints due to continuing stabilizing selection which fixes a canalized region around the value favoured by the selection. It is also in contrast to the constraints due to fundamental features of the developmental system which originated in the evolutionary history of the species but are still necessary for obtaining a functional adult organism, e.g. body segmentation and compartment formation.

Our first example of the stabilizing and disruptive selection on the scutellar bristles falls in the second category. The experiments with selection on the position of the anterior dorsocentral bristles belong to the third category, fundamental features, because in that experiment we are involved with fundamental aspects of pattern formation, compartmentalization and interaction of processes of pattern formation in different compartments.

Replication of these experiments must elucidate whether the phenomena discovered in this selection experiment are reflections of inevitable rules of the developmental system involved. We have already pointed out that when dealing with correlated responses of selection on wing veins in *Hairless* stocks, three replicates produced three different results. But we could explain this as the consequence of the complicated network of developmental relations which offers the possibility of manifold solutions for changing the same character, either specifically the character selected for alone or in relation with other characters. Moreover, the character on which the selection was performed showed a similar reaction in all three lines, in particular similar effects of the developmental system were revealed. The difference in the responses concerned the correlated responses only. The more fundamental and deep-seated the properties on which selection impinges, the more difficult it will be to bypass constraints in the developmental pathway. Be this as it may, this kind of selection experiment on pattern characteristics gives fascinating results of a type which are urgently needed for understanding the role of development in evolution.

Acknowledgement. Thanks are due to Ad Velders for performing the selection on the position of the dorsocentral bristles, and to him and Dr. Gerdien de Jong for stimulating discussion.

References

Alberch P (1980) Ontogenesis and morphological diversification. Am Zool 20:653–667
Alberch P (1982) Developmental constraints in evolutionary processes. In: Bonner JT (ed) Evolution and development. Springer, Berlin Heidelberg New York, pp 313–332
Bos M, Scharloo W, Bijlsma R, de Boer IM, den Hollander J (1969) Induction of morphological aberrations by enzyme inhibition in *Drosophila melanogaster*. Experientia 24:811–812
Bryant P (1978) Pattern formation in imaginal discs. In: Ashburner M, Wright TRF (eds) The genetics and biology of *Drosophila*, vol 2c. Academic Press, London New York, pp 230–336
Charlesworth B, Lande R, Slatkin M (1982) A neo-Darwinian commentary on macroevolution. Evolution 36:474–498
de Jong G, Scharloo W (1976) Environmental determination of the significance or neutrality of amylase variants in *Drosophila*. Genetics 84:77–94
Dunn RB, Fraser AS (1958) Selection for an invariant character, "vibrissae number" in the house mouse. Nature (London) 181:1018–1019
Eisses KTh, Schoonen WG, Aben W, Scharloo W, Thörig GEW (1985) Dual function of the alcohol dehydrogenase of *Drosophila melanogaster*: ethanol and acetaldehyde oxidation by two allozyme ADH-71K and ADH-F. Mol Gen Genet 199:76–81
Eldredge N (1985) Time frames. Simon & Schuster, New York
Eldredge N, Gould SJ (1972) Punctuated equilibria, an alternative to phyletic gradualism. In: Schopf TJM (ed) Models in paleobiology. Freeman, Cooper, San Francisco, pp 82–115
Fraser A (1967) Variation of scutellar bristles in *Drosophila* XV. Systems of modifiers. Genetics 57:919–934
Garcia Bellido A, Ripoll P, Morata G (1976) Developmental compartmentalization in the dorsal mesothoracic disc of *Drosophila*. Dev Biol 48:132–147
Gould SJ (1980) Is a new and general theory of evolution emerging? Paleobiology 11:27–41
Gould SJ, Eldredge N (1977) Punctuated equilibria: the tempo and mode of evolution reconsidered. Paleobiology 3:115–151
Heinstra PWH, Eisses KTh, Schoonen WGEJ, Aben W, de Winter AJ, Horst DJ van der, Marrewijk WJA van, Beenakkers AMTh, Scharloo W, Thörig GEW (1983) A dual function of alcohol dehydrogenase in *Drosophila*. Genetica 60:129–136
Heinstra PWH, Scharloo W, Thörig GEW (1987) Physiological significance of the alcohol-dehydrogenase polymorphism in larvae of *Drosophila*. Genetics (in press)
Hoorn AJW, Scharloo W (1979) Selection on enzyme variants in *Drosophila*. Aquilo Ser Zool 20: 41–48
Jones JS (1981) An uncensored page of fossil history. Nature 293:427–428

Kauffman SA (1983) Developmental constraints: internal factors in evolution. In: Goodwin BC, Holder N, Wylie CC (eds) Development and evolution, 6th Symp British society for developmental biology. Cambridge Univ Press, pp 195–225

Latter BDH, Scowcroft WR (1970) Regulation of anterior and posterior scutellar bristle number in *Drosophila*. Genetics 66:685–694

Loosli R (1959) Vergleichung von Entwicklungspotenzen in normalen, transplantierten und mutierten Halteren-Imaginaal-Scheiben von *Drosophila melanogaster*. Dev Biol 1:24–64

Maynard Smith J, Sondhi KC (1960) The genetics of a pattern. Genetics 45:1039–1050

Meinhardt H (1982) Models of biological pattern formation. Academic Press, London New York

Murphy C, Tokunaga C (1970) Cell lineage in the dorsal mesothoracic disk of *Drosophila*. J Exp Zool 175:197–220

Noordwijk AJ van, Balen JH van, Scharloo W (1980) Heritability of ecologically important traits in the great tit. Ardea 68:193–203

Reeve ECR (1961) Modifying the sternopleural hair pattern in *Drosophila* by selection. Genet Res 2:158–160

Rendel JM (1959) Canalization of the scute phenotype of *Drosophila*. Evolution 13:425–439

Rendel JM (1965) Bristle pattern in scute stocks of *Drosophila melanogaster*. Am Nat 99:25–32

Rendel JM (1967) Canalization and gene control. Plenum, London New York

Rendel JM, Sheldon BL (1960) Selection for canalization of the scute phenotype in *Drosophila melanogaster*. Austr J Biol Sci 13:36–45

Rendel JM, Sheldon BL, Finlay DL (1966) Selection for canalization of the scute phenotype. II. Am Nat 100:13–31

Robertson A (1965) Variation in scutellar bristle number – an alternative hypothesis. Am Nat 99:19–24

Robertson FW (1959) Studies in quantitative inheritance. XII. Cell size and number in relation to genetic and environmental variation in body size. Genetics 44:869–896

Scharloo W (1962) The influence of selection and temperature on a mutant character (ci^D) in *Drosophila melanogaster*. Arch Neerl Zool 14:431–512

Scharloo W (1970) Stabilizing and disruptive selection on a mutant character in *Drosophila melanogaster*. III. Polymorphism caused by a developmental switch mechanism. Genetics 65:693–705

Scharloo W (1984) Genetics of adaptive reactions. In: Wöhrmann K, Loeschcke V (eds) Population biology and evolution. Springer, Berlin Heidelberg New York, pp 5–15

Scharloo W (1987) Constraints in selection response. In: Loeschcke V (ed) Genetic constraints on adaptive evolution. Springer, Berlin Heidelberg New York Tokyo (in press)

Scharloo W, Hoogmoed MS, ter Kuile A (1967) Stabilizing and disruptive selection on a mutant character in *Drosophila*, I. The phenotypic variance and its components. Genetics 56:709–726

Scharloo W, Zweep A, Schuitema KA, Wijnstra JG (1972) Stabilizing and disruptive selection in a mutant character in *Drosophila*. IV. Selection on sensitivity to temperature. Genetics 71:551–566

Scharloo W, Dijken FR van, Hoorn AJW, de Jong G, Thörig GEW (1977) Functional aspects of genetic variation. In: Christiansen FB, Fenchel TM (eds) Measuring selection in natural populations. Springer, Berlin Heidelberg New York, pp 131–148

Scowcroft WR (1973) Scutellar bristle components and canalization in *Drosophila melanogaster*. Heredity 30:289–301

Scowcroft WR, Green MM, Latter BD (1968) Dosage at the scute locus, and canalization of anterior and posterior scutellar bristles in *Drosophila melanogaster*. Genetics 60:373–388

Sondhi KC (1962) The evolution of a pattern. Evolution 16:186–191

Spickett SG (1963) Genetic and developmental studies of a quantitative character. Nature (London) 199:870–873

Velde J te (1985) The significance of the anal papillae in salt adaptation of *Drosophila melanogaster*. Thesis, Univ Utrecht

Velde JH te, Scharloo W (1987) Natural and artificial selection on a deviant character of the anal papillae in *Drosophila melanogaster* and their significance in salt adaptation. J Evol Biol 1(2):155–164

Velde JH te, Molthoff CFM, Scharloo W (1987) The function of anal papillae in salt adaptation of *Drosophila melanogaster*. J Evol Biol 1(2):139–153

Waddington CH (1957) The strategy of the genes. Allen & Unwin, London

Whittle JRS (1969) Genetic analysis of the control of number and pattern of scutellar bristles in *Drosophila melanogaster*. Genetics 63:167–181

Extrapolations

The Evolutionary Potential of the Unstable Genome[1]

A. FONTDEVILA[2]

1 The Mobile, Dispersed, Middle Repetitive DNA Sequences May Be Responsible for Sterility and Genetic Instability in Interspecific Hybrids

The eukaryotic genome contains a large fraction of repetitive DNA, which is considered of primary importance for the expression of genetic information. Most of the moderately repetitive DNA is dispersed in the genome. In *Drosophila melanogaster* it has been estimated that disperse, middle repetitive DNA represents about 12% of its whole genomic DNA. At least, a minimum of 50 different families belonging mainly to three classes (copia, P and foldback) have been characterized so far (Rubin 1983). The degree of repetition varies among families but it may be estimated between 30 and 50 and the maximum copy number seems to be under control (Syvanen 1984).

Perhaps the most notorious attribute of these dispersed elements is their capacity of transposition. Their transposition is effected by "transposase" activities encoded in some cases by the elements themselves and, at least in copialike elements, this mobility is thought to be achieved by a system of replication and transposition similar to that of retroviruses, via reverse transcription. The evidence of their retroviral origin stems from DNA structural studies (for a review see Finnegan 1983), from some observations of endogenous proviruses in vertebrates (Todaro et al. 1980) and from the isolation of circular copies of copia DNA in cell cultures (Flavell 1984) and in adults of *D. melanogaster* (McDonald et al. 1986) as well. In summary, not only the structure of mobile elements but also their dynamics are strikingly similar to those of retroviruses. Whether they originated from actual retroviruses or represent host DNA sequences that have escaped host control (Miller and Miller 1982) is open to discussion.

From an evolutionary point of view, their capacity of producing genetic changes is worth discussing. Recent data indicate that most mutations from natural populations are due to insertions of mobile elements (Rubin et al. 1982). These mutational events range from the inactivation of a functional gene (Modolell et al. 1983; Levis et al.

1 This work is dedicated to Professor Antonio Prevosti on the occasion of his retirement and as an homage to his scientific contribution to Population Genetics
2 Departamento de Genética y Microbiología, Universidad Autónoma de Barcelona, Bellaterra (Barcelona), Spain

Population Genetics and Evolution
G. de Jong (ed.)
© Springer-Verlag Berlin Heidelberg 1988

1984) to important changes in its regulation (Campuzano et al. 1986; Parkhurst and Corces 1986). Genetic changes induced by mobile elements are not only limited to point mutations but they can also include long chromosomal rearrangements such as duplications, deficiencies, inversions and translocations. Early evidence of chromosomal rearrangement association with transposable elements comes from studies by McClintock (1951) in maize. Later, data on yeast suggest that the Ty elements may recombine intra- or interchromosomally and give rise to inversions and translocations (Roeder and Fink 1980). More recently, the evidence has increased significantly through studies in *Drosophila* rearrangements using mainly P elements (Engels and Preston 1984).

The biological consequences of P transposition have been studied in much detail because they may illustrate what could be considered a powerful mechanism of speciation. More than 10 years ago it was observed that certain crosses between wild males and laboratory females displayed a series of abnormal traits in F_1 progenies, including sterility, male recombination and a high frequency of chromosomal rearrangements and mutations (Kidwell et al. 1977). Later on, it has been proved that this set of abnormalities, designated as "hybrid dysgenesis", is produced only when a male containing P elements is crossed to a female free from this element (called M) (Engels 1983). The molecular mechanism for the P transposition repression in P cytoplasm and its derepression in M cytoplasm is not known. Yet, several molecular studies have shown that a P element encodes its own transposase function (Spradling and Rubin 1982) and that its germ tissue specificity is regulated at the level of mRNA splicing (Laski et al. 1986). There is another interacting system (I-R) that produces hybrid dysgenesis (Picard et al. 1978). There are several differences between the I-R and P-M systems, namely that in the former only female progeny is affected by the syndrome, whereas in the latter both sexes are affected. Yet, their relationship to mobile elements seems to be well substantiated in both cases (for a review see Bregliano and Kidwell 1983).

The biological similarities between intraspecific hybrid dysgenesis and interspecific hybrid sterility have been pointed out several times, but studies with interspecific hybrids do not provide, in general, enough information to be compared with all the events observed in hybrid dysgenesis, besides sterility. Nonetheless, recent studies in my laboratory with hybrids between *D. buzzatii* and *D. serido* are shedding much light on the genetic syndrome that accompanies hybridization.

1.1 Studies on Hybrid Sterility

These studies are based on the analysis of backcross progenies of interspecific hybrids. In *Drosophila*, several kinds of hybrid sterility have been reported. Namely, the F_1 progeny may consist of (1) both sterile males and sterile females (Muller and Pontecorvo 1940, 1941; Pontecorvo 1943); (2) sterile males and fertile females; (3) both fertile males and fertile females, but where sterile males appear among the F_2 and/or backcrosses due to some special combinations of chromosomes of both species, frequently involving Y-autosomal interactions (Alexander et al. 1952; Henning 1985; Schäfer 1978; Stone 1947; Zouros 1981), and (4) progeny exemplified by *D. paulisto-*

rum, in which the sterility of backcross hybrid males between races, or incipient species, is due to some microorganisms (Somerson et al. 1984). The second class (2) has traditionally provided the most informative material to unveil the genetic architecture of reproductive isolation. By backcrossing fertile hybrid females to each parental species and using genetic markers in all chromosomes, it has been possible to obtain males with all chromosomal combinations. The fertility analysis of these males has shown that in the species pairs *D. pseudoobscura-D. persimilis* (Dobzhansky 1936) and *D. simulans-D. mauritiana* (Coyne 1984) fertility factors are distributed among all chromosomes. These and other similar studies (Pontecorvo 1943; Wu and Beckenbach 1983) suggest that the number of genes affecting hybrid sterility must be very large, yet their results only indicated the existence of a minimum of one genetic factor (or two, at most) per chromosome. This is so because fine genetic dissection of fertility factors requires a large number of markers all over the genomes of closely related species and those studies are based on one or a few markers per chromosome, which makes it impossible to distinguish between one and several genetic factors per chromosome arm. Thus, the nature, distribution, actual number and mode of interaction of those "segregating units" still remains a matter of high speculation.

Recently, the observation that certain pairs of sibling species produce hybrids, in which asynapsis is present almost along the whole length of all chromosomes (Naveira et al. 1986), has provided a unique opportunity to use this cytological character as a genetic marker. *D. buzzatii* and *D. serido*, two closely related species of the *D. repleta* group, produce hybrids when the direction of crossing is *D. buzzatii* males x *D. serido* females. Hybrid males are sterile, but hybrid females are fertile. Using a planned scheme of successive backcrosses between hybrid females and parental males it is possible to select recombinant strains of each parental species, such that each one is introgressed with a different chromosomal segment in heterozygous condition (segmental hybrids). Identification and selection of segmental hybrids is feasible due to the use of interspecific asynapsis as a marker technique.

The experimental procedure of introgression has led to a large set of segmental hybrids with introgressed chromosomal lengths, ranging from a few polytene bands to the whole chromosome. These segments cover the whole genome in an overlapping way (Fig. 1), allowing for the exact localization of any point genetic factor with a precision of three to four polytene bands (Naveira et al. 1986). The analysis of male sterility performed by Naveira and Fontdevila (1986) using this material leads to the following conclusions: (1) The genetic factors producing hybrid sterility are not localized in specific chromosomal zones, but they are dispersed along the genome. (2) In autosomes, it is impossible to map any of these genetic factors because each one produces no detectable effect individually. (3) In autosomes, the size of the introgressed segment is the critical factor that elicits male sterility, irrespective of its location in the genome. Thus, segments under a critical size do not produce male sterility when introgressed. On the other hand, segments above this critical size always produce sterile males (Table 1). (4) Any X-chromosome introgressed segment produces either sterility or inviability, irrespective of its size. (5) Sterility effects are chromosome-dependent. Thus, for example, a segment of chromosome 4 has a larger effect that a chromosome 3 segment of equal size when introgressed. (6) Introgression of a *D. serido* segment into a *D. buzzatii* background is more effective than the reciprocal introgres-

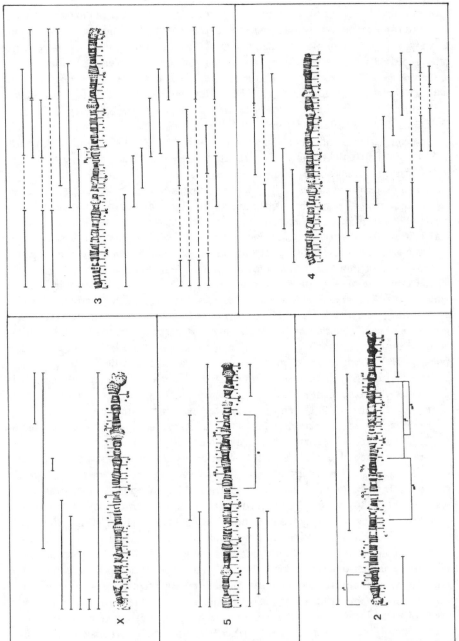

Fig. 1

Table 1. Critical sizes (%) of introgressed segments relative to chromosome 2, showing the differential effect of the chromosomes in the determination of hybrid male sterility

	Chromosome 3	Chromosome 4	Chromosome 5
Maximum for fertility	34.5	23.2	27.9
Minimum for sterility	34.1	29.8	34.1

Naveira and Fontdevila (1986)

sion for a segment of equal size. (7) Two autosomal segments can cooperate in an additive way regardless of whether they are introgressed in the same (*cis*-action) or in different autosomes (*trans*-action) to produce sterility once their cumulative size exceeds a critical value.

These results represent a further understanding of the genetic basis of hybrid sterility. They prove that the number of sterility factors is enormous and can be classified into two types. First, there are many X-linked specific factors distributed all over the X chromosome, any one of which produces sterility by itself. Second, there are many non-specific factors spread all over the autosomes, which produce dominant sterility only when accumulated in critical amounts. The unambiguous dispersive nature of sterility factors requires as a counterpart a dispersive molecular architecture of genetic functions acting in concert during the spermatogenesis. In autosomes this concerted action is only impaired when a minimum of these functions are substituted, but any substitution in the X chromosome is prone to produce severe alterations either in spermatogenesis or in the general ontogeny of the males.

Detailed physiological studies of these hybrids have also shown that hybrid spermiogenesis abnormalities are dependent not only on the introgressed chromosome but also on the length of the introgressed segment (Naveira et al. 1984). Unfortunately, the majority of similar studies performed in *Drosophila* are concerned with F_1 hybrids and at most with F_2 hybrids with combinations of whole chromosomes of both species. The former show a full range of abnormal spermatogenesis, from nearly functional gamete formation (Mainland 1942) to grossly abnormal patterns of development (Kerkis 1933; Dobzhansky 1934), and the latter indicate that not all combinations affect spermatogenesis in a similar way or with a comparable intensity (Stone 1947; Schäffer 1978). Although these results are less informative than those obtained by Naveira et al. (1984), they also seem to fit a general pattern which has been compared and found compatible with hybrid dysgenesis (Engels and Preston 1979; Kidwell and Novy 1979). Moreover, the unambiguous dispersive nature of sterility

---

Fig. 1. Graphic distribution of some *D. serido* chromosomal segments introgressed into a *D. buzzatii* genomic background. *Bars* (⊢——⊣) represent length and position of introgressed segment relative to each *D. buzzatii* chromosome. Segments producing male hybrid sterility are drawn above each *D. buzzatii* chromosome, while those not producing sterility are drawn below. Each row with a single segment corresponds to an individually introgressed segment. Two segments united by *dots in a row* means that they are introgressed simultaneously. *D. buzzatii* chromosomes are drawn and inverted parts due to cytological evolution are shown upside down. In chromosomes 2 and 5 the extent of inverted segments due to species-specific inversions for *D. serido* (2j^9, 2m^9, 2l^9) and for *D. buzzatii* (5 g) is shown (Naveira and Fontdevila 1986)

factors requires as a counterpart a dispersive molecular architecture of genetic function and middle repetitive DNA is a plausible candidate.

1.2 Studies on Hybrid Genetic Instability

Naveira and Fontdevila (1985) have also found high frequencies of new chromosome rearrangements induced by introgressive hybridization. Figure 2 shows the variety and complexity of these rearrangements that occasionally may be the result of multiple breaks. These data on interspecific hybrids can be summarized as follows. First, the rate of rearrangement production has been computed as 2% per gamete per generation in hybrid males. If we take the hybrid female rate as the expected frequency in non-dysgenic events, the figure drops to 0.08% or 8×10^{-4} per gamete per generation, two orders of magnitude lower (Table 2). Second, the distribution of breakage points is non-random. This heterogeneity inside and between chromosomes is in accordance with the observation that mobile elements insert non-randomly and are represented more in heterochromatic zones (Young and Schwartz 1981). Engels and Preston (1984) also found a non-random distribution for breakage points induced by P elements.

This new information on interspecific hybrid genetic instability is so similar to hybrid dysgenesis in intraspecific crosses that both may be considered as the result of the same genetic cause: namely the transposition of some kind of mobile element. Transposition of mobile elements has been shown to occur in interspecific hybrids of *Drosophila* (Evgenev et al. 1982).

2 Genetic Instability as a Promoter of Speciation

Most population circumstances described as critical to speciation may be linked to episodes of genetic instability (Lewin 1983). As an example, Wright (1969) has for a long time considered a realistic model in which populations are subdivided into small demes subjected to environmental fluctuations, where new genetic novelties can be established by a combined action between genetic drift and natural selection (shifting-

Table 2. Frequency of new rearrangements in progenies of segmental hybrids (5 b/s) and of *D. buzzatii* (5 b/b)

Sex	Genotype	Number of parental individuals	Number of progeny larvae (l)	Number of new rearrangements (n)	Frequency of new rearrangements per gamete and generation (n/l)
♂	5 b/s	59	1668	35	2.1×10^{-2}
♂	5 b/b	62	1382	0	0
♀	5 b/s	50	1243	1	8.0×10^{-4}
♀	5 b/b	40	800	0	0

Data from Naveira and Fontdevila (1985)

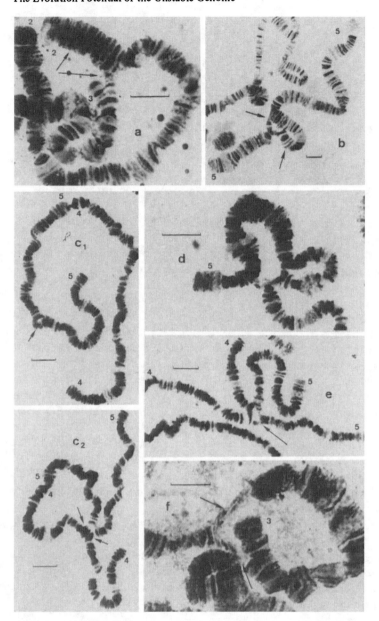

Fig. 2. Some new chromosomal rearrangements observed in the male progeny of *D. buzzatii* introgressed with a telomeric segment of *D. serido* (5A-Bld). a Translocation between chromosomes 2 and 3; b inversion in chromosome 5; c_1, c_2 duplication in chromosomes 4 and 5; d duplication in chromosome 5; e translocation between chromosomes 4 and 5; f duplication in chromosome 3. *Numbers* identify different chromosomes and *bars* represent 10 μm (Naveira and Fontdevila 1985)

balance theory). Once a novelty has been fixed in a deme, it has to be expanded to the whole population range through interdemic selection. This is a complicated process unless we assume an extremely unrealistic high gene flow associated with the new mutant, as Lande (1985) has pointed out. However, temporal stochastic changes in population structure may break through the barrier to gene flow and allow diffusion of the mutant (Lande 1985). This population instability can be achieved reasonably in two ways: (1) by large synchronic fluctuations in population density that produces alternative periods of subdivision and continuous distribution (habitat fragmentation model) and (2) by permanent subdivision with fast local extinctions and subsequent colonizations (extinction-recolonization model) (Wright 1940). In any of these occasions demes are facing new, unexpected environmental and genetic conditions, often stressful to their original adaptations.

Since environmental stress is capable of mobilizing and inducing transcription of repetitive nomadic elements (for a review see McDonald et al. 1986), stressful conditions created in the process of habitat fragmentation, due to the environmental shift, may give rise to many marginal zones in which gene activation and transposition are induced. Rates of mutation can increase significantly in these demes and insufflate new genetic variability of evolutionary value. Some classic observations on changes in mutation rate may be explained by this phenomenon of stress-induced genetic instability (for a review see Woodruff et al. 1983). In this process, the differentiation of middle repetitive sequences is of utmost interest. Genetic turnover for this class of DNA is very high and several mechanisms such as unequal crossovers and gene conversion result in rapid genic amplification (Flavell 1982). Besides, several surveys in *Drosophila* species have often shown that closely related species do not share more similarly dispersed repetitive sequences than species phylogenetically more distant (Dowsett and Young 1982; Martin et al. 1983; Lansman et al. 1985). Of particular interest is the observation that the P element of *D. melanogaster* is not present in any of the *melanogaster* subgroup species but it has been found in many of the *willistoni* and *saltans* group species (Daniels et al. 1984; Lansman et al. 1985; Daniels and Strausbaugh 1986). These studies have raised the possibility of horizontal transmission by interspecific infection (Syvanen 1986).

Regardless of the origin of these dispersed sequences, they evolve at a high rate and may produce a high level of differentiation among closely related species whose genetic similarity is otherwise very large. Since low differentiated incipient species show very often a significant level of reproductive isolation it can be hypothesized that genetic hybrid incompatibilities due to divergence in mobile-dispersed elements may be the initial cause of reproductive isolation.

3 Genetic Instability as a Helper to Maintain and Enhance Reproductive Isolation

Environmental stress may not be the unique event triggering genetic instability. Several authors (Sturtevant 1939; Dayal 1979; Sage and Selander 1979; Walters 1952) have witnessed enhanced mutation rates and chromosomal breakages when inbreeding and/or hybridization occurs. Inbreeding is a common event in initial episodes of speciation in peripatric (Mayr 1982), founder-flush (Carson 1982) and genetic transilience

(Templeton 1981) models. Hybridization is produced when incipient species or divergent populations isolated for some time come into contact due to migration or habitat expansions produced by climatic changes. In the case of intraspecific hybridization, the phenomenon of hybrid dysgenesis has been well documented (Kidwell et al. 1977) and its relevance for speciation as a promotor of genetic isolation has been discussed several times (Kidwell 1983; Woodruff et al. 1983).

Studies on genetic instabilities produced in interspecific hybrids are much less documented. Some early studies suggest that foreign heterochromatin may be associated with phenotypic and chromosomal abnormalities in interspecific hybrids (Burns and Gerstel 1969, 1972, 1973; Gerstel and Burns 1967, 1972). Naveira and Fontdevila (1985) have shown that segmental hybrids for a small telomeric section of chromosome 5 of D. serido introgressed in the D. buzzatii genome are able to produce in the male germ line a high frequency of chromosome breakages and new arrangements. These results might be the manifestation of a phenomenon similar to hybrid dysgenesis.

After a period of habitat fragmentation, climatic and/or biogeographic changes may produce a following period of habitat expansion. Populations are no longer in isolates and expand their range. Incipient species may meet in zones of secondary contact before they have completed their genetic isolation and hybrids are only partially sterile. This is a crucial moment for the species fate. Usually, hybrid males are sterile and hybrid females are fertile in incipient species. Recombination in the semisterile sex hybrids and successive backcrosses to the parental species will lead to the genomic absorption, usually from the less to the more abundant incipient species.

It has been suggested that stabilization of a new rare recombinant or a new mutant or karyotypic hybrid form may lead to speciation (see Templeton 1981 for a discussion), but this depends on the simultaneous occurrence of hybrid fitness inferiority and enhanced reproductive isolation from parental species. It is much more plausible to envisage new mutations and chromosomal rearrangements to be selected against in the heterozygous state. These new genetic novelties may be produced by the genetic instability created in the hybrids and contribute to maintain the species integrity by stabilizing a hybrid zone. Some chromosomal inversions may be selected and fixed in one species for several reasons. First, they may prevent recombination in chromosomal segments longer than the critical size for sterility, thus maintaining in their carriers reproductive isolation at a maximum. Second, some of these segments covered by inversions may also include coadapted gene complexes. Any inversion sharing both properties would be selected because homozygotes will be perfectly adapted to a previous habitat and their genetic isolation totally maintained through the lowest levels of hybrid fertility.

These inversions have a low probability of fixation in large populations. However, contact zones during habitat expansion will create the perfect conditions of instability and low effective breeding size to increase randomly their frequency to values where selection may efficiently operate to homozygous fixation.

Recent evidence with the pair of sister species D. buzzatii and D. serido has shown that it is possible to fix as homozygous some newly induced inversions (Naveira 1985). Moreover, most of the species-specific inversions could have been fixed due to their dual role in reproductive isolation and coadaptation. D. serido is fixed for inversion $2j^9$ and D. buzzatii for 5 g, both inversions larger than the critical size. Moreover, although the adaptive value of these fixed inversions is unknown, Ruiz and Fontdevila (1985) have shown that intraspecific polymorphism for the second chromosome is adaptive to trophic resources in *D. buzzatii*.

In summary, species-specific paracentric inversions in *Drosophila* are not regarded as primary inducers of speciation since sterility factors are not specifically located in them, but rather they may be seen as essential tools for maintaining and enhancing incipient reproductive isolation.

4 Concluding Remarks

Evolutionists are paying much attention to the recent results on the possible adaptive significance of mobile genetic elements. The initial idea of considering these genetic elements as non-adaptive selfish DNA (Dover and Doolittle 1980; Orgel et al. 1980) is giving way to the already documented view that enhanced response to selection is coupled with the presence and mobility of these elements by means of their great ability to produce genetic variability (Mackay 1985; Pasyukova et al. 1986). This new variability is produced not only at the genic level, but also at the chromosomal level by generating gross rearrangements (Engels and Preston 1984). Thus, this genomic instability may be one of the major factors in evolution if coupled with other necessary factors of population instability discussed above. In this respect, it is of paramount importance to define what the external and internal environments are that may elicit this kind of mutator potential. Among them environmental stress (Strand and McDonald 1985) and hybridization (Kidwell 1983; Naveira and Fontdevila 1986) are the best documented.

Acknowledgements. This work has been supported in part by research grants 0910/81 and 2825/83 awarded to the author by the Comisión Asesora de Investigación Cientifica y Técnica, Spain and also by a travel grant awarded to the author by the Ministerio de Educatión y Ciencia, Spain. The first version of this work was written while the author was on sabbatical at the University of Georgia (USA) in Athens (USA) sponsored by a personal grant from the US-Spain Joint Committee for Scientific and Technological Cooperation.

References

Alexander ML, Lea RB, Stone WS (1952) Interspecific gene variability in the *virilis* group. Texas Univ Publ 5204:106–113
Bregliano JC, Kidwell MG (1983) Hybrid dysgenesis determinants. In: Shapiro (ed) Mobile genetic elements. Academic Press, London New York, pp 363–410
Burns JA, Gerstel DU (1969) Consequences of spontaneous breakage of heterochromatic segments in *Nicotiana* hybrids. Genetics 63:427–439
Burns JA, Gerstel DU (1972) Inhibition of chromosome breakage and of megachromosomes by intact genomes of *Nicotiana*. Genetics 69:211–220
Burns JA, Gerstel DU (1973) Formation of megachromosomes from heterochromatic blocks of *Nicotiana tomentosiformis*. Genetics 75:497–502
Campuzano S, Balcells L, Villares R, Carramolino L, Garcia-Alonso L, Modolell J (1986) Excess function *Hairy-wing* mutations caused by *gypsy* and *copia* insertions within structural genes of the achaete-scute locus of *Drosophila*. Cell 44:303–312

Carson HL (1982) Speciation as a major reorganization of polygenic balances. In: Barigozzi C (ed) Mechanisms of speciation. Liss, New York, pp 411–433 (Proc Int Meet Mechanisms of speciation)

Coyne JA (1984) Genetic basis of male sterility in hybrids between two closely related species of Drosophila. Proc Natl Acad Sci USA 81:4444–4447

Daniels SB, Strausbaugh LD (1986) The distribution of P element sequences in Drosophila: the willistoni and saltans species groups. J Mol Evol 23:138–148

Daniels SB, Strausbaugh LD, Ehrman L, Armstrong R (1984) Sequences homologous to P elements occur in Drosophila paulistorum. Proc Natl Acad Sci USA 81:6794–6797

Dayal N (1979) Cytogenetic studies in the inbred lines of radish (Raphanus sativus L. var. radicola pers.) and their hybrids. III. Meiotic abnormalities. Cytologia 44:1–5

Dobzahnsky Th (1934) Studies on hybrid sterility. I. Spermatogenesis in pure and hybrid Drosophila pseudoobscura. Z Zellforsch Mikrosk Anat 21:169–223

Dobzahnsky Th (1936) Studies on hybrid sterility. II. Localization of sterility factors in Drosophila pseudoobscura hybrids. Genetics 21:113–135

Dover GA, Doolittle RF (1980) Modes of genome evolution. Nature (London) 288:646

Dowsett AP, Young MW (1982) Differing levels of dispersed repetitive DNA among closely related species of Drosophila. Proc Natl Acad Sci USA 79:4570–4574

Engels WR (1983) The P family of transposable elements in Drosophila. Annu Rev Genet 17: 315–344

Engels WR, Preston CR (1979) Hybrid dysgenesis in Drosophila melanogaster: the biology of male and female sterility. Genetics 92:161–174

Engels WR, Preston CR (1984) Formation of chromosome rearrangements by P factors in Drosophila. Genetics 107:657–678

Evgenev MB, Yenikolopov GN, Peunova NI, Ilyin V (1982) Transposition of mobile genetic elements in interspecific hybrids of Drosophila. Chromosoma (Berlin) 85:375–386

Finnegan DJ (1983) Transposable elements in eukaryotes. Int Rev Cytol 93:281–326

Flavell RB (1982) Sequence amplification, deletion and rearrangement: major sources of variation during species divergence. In: Dover GA, Flavell RB (eds) Genome evolution. Academic Press, London New York, pp 301–323

Flavell RB (1984) Role of reverse transcription in the generation of extrachromosomal copia mobile genetic elements. Nature (London) 310:514–516

Gerstel DU, Burns JA (1967) Phenotypic and chromosomal abnormalities associated with the introduction of heterochromatin from Nicotiana otophora into N. tabacum. Genetics 56: 483–502

Gerstel DU, Burns JA (1972) On the absence of cytoplasmic determination of the formation of megachromasomes in Nicotiana. J Heredity 63:256–258

Hennig W (1985) Y chromosome function and spermatogenesis in Drosophila hydei. In: Caspari EW, Scandalios JG (eds) Advances in genetics, vol 23. Academic Press, London New York, pp 179–234

Kerkis J (1933) Development of gonads in hybrids between Drosophila melanogaster and D. simulans. J Exp Zool 66:477–509

Kidwell MG (1983) Evolution and hybrid dysgenesis determinants in Drosophila melanogaster. Proc Natl Acad Sci USA 80:1655–1659

Kidwell MG, Novy JB (1979) Hybrid dysgenesis in Drosophila melanogaster: sterility resulting from gonadal dysgenesis in the P-M system. Genetics 92:1127–1140

Kidwell MG, Kidwell JF, Sved JA (1977) Hybrid dysgenesis in Drosophila melanogaster: a syndrome of aberrant traits including mutation, sterility and male recombination. Genetics 86:813–833

Lande R (1985) The fixation of chromosomal rearrangements in a subdivided population with local extinction and colonization. Heredity 54:323–332

Lansman RA, Stacey SN, Grigliatti TA, Brock HW (1985) Sequences homologous to the P mobile element of Drosophila melanogaster are widely distributed in the subgenus Sophophora. Nature (London) 318:561–563

Laski FA, Rio DC, Rubin GM (1986) Tissue specificity of Drosophila P element transposition is regulated at the level of mRNA splicing. Cell 44:7–19

Levis R, O'Hare K, Rubin G (1984) Effects of transposible element insertions on RNA encoded by the *white* gene of *Drosophila*. Cell 38:471−481

Lewin R (1983) Origin of species in stressed environments. Science 222:1112

Mackay TFC (1985) Transposable element-induced response to artificial selection in *Drosophila melanogaster*. Genetics 111:351−374

Mainland GB (1942) Genetic relationships in the *Drosophila funebris* group. Univ Texas Publ 4228:74−112

Martin G, Wiernasz D, Schedl P (1983) Evolution of *Drosophila* repetitive dispersed DNA. J Mol Evol 19:203−213

Mayr E (1982) Processes of speciation in animals. In: Barigozzi C (ed) Mechanisms of speciation. Liss, New York, pp 1−19

McClintock B (1951) Chromosome organization and genetic expression. Cold Spring Harbor Sym Quant Biol 16:13−47

McDonald JF, Strand DJ, Lambert ME, Weinstein IB (1986) The responsive genome: evidence and evolutionary implications. In: Rauff R, Rauff E (eds) Development as an evolutionary process. Liss, New York

Miller DW, Miller LK (1982) A virus mutant with an insertion of a copia-like transposable element. Nature (London) 299:562−564

Modolell J, Bender W, Meselson M (1983) *Drosophila melanogaster* mutations suppressible by the suppressor of hairy-wing are insertions of a 7.3 kilobase mobile element. Proc Natl Acad Sci USA 80:1678−1682

Muller HJ, Pontecorvo G (1940) Recombinants between *Drosophila* species the F1 hybrids of which are sterile. Nature (London) 146:199−200

Muller HJ, Pontecorvo G (1941) Recessive genes causing interspecific sterility and other disharmonies between *D. melanogaster* and *D. simulans*. Genetics 27:157

Naveira H (1985) Base genética de las barreras de aislamiento reproductivo en el cluster *buzzatii* de *Drosophila*. PhD Thesis, Univ Autónoma Barcelona, Spain

Naveira H, Fontdevila A (1985) The evolutionary history of *Drosophila buzzatii*. IX. High frequencies of new chromosome rearrangements induced by introgressive hybridization. Chromosoma (Berlin) 91:87−94

Naveira H, Fontdevila A (1986) The evolutionary history of *Drosophila buzzatii*. XII. The genetic basis of sterility in hybrids between *D. buzzatii* and its sibling *D. serido* from Argentina. Genetics 114:841−857

Naveira H, Hauschteck-Jungen E, Fontdevila A (1984) Spermiogenesis of inversion heterozygotes in backcross hybrids between *Drosophila buzzatii* and *D. serido*. Genetica 65:205−214

Naveira H, Plá C, Fontdevila A (1986) The evolutionary history of *D. buzzatii*. XI. A new method for cytogenetic localization based on asynapsis of polytene chromosomes in interspecific hybrids of *Drosophila*. Genetica 71:199−212

Orgel LE, Crick FHC, Sapienza C (1980) Selfish DNA. Nature (London) 288:645−646

Parkhurst SM, Corces V (1986) Interactions among the gypsy transposable element and the yellow and the supressor of hairy-wing loci in *Drosophila melanogaster*. Mol Cell Biol 6:47−53

Pasyukova EG, Belayeva ESP, Kogan GL, Zaidanov LZ, Gvozdev VA (1986) Concerted transpositions of mobile genetic elements coupled with fitness changes in *Drosophila melanogaster*. Mol Biol Evol 3(4):299−312

Picard G, Bregliano JC, Bucheton A, Lavige JM, Pelisson A, Kidwell MG (1978) Non-Mendelian female sterility and hybrid dysgenesis in *Drosophila melanogaster*. Genet Res 32:275−287

Pontecorvo G (1943) Hybrid sterility in artificially produced recombinants between *Drosophila melanogaster* and *Drosophila simulans*. Proc R Soc Edinburgh Ser B 61:385−397

Roeder GS, Fink GR (1983) Transposable elements in yeast. In: Shapiro JA (ed) Mobile genetic elements. Academic Press, London New York, pp 299−328

Rubin GM (1983) Dispersed repetitive DNAs in *Drosophila*. In: Shapiro JA (ed) Mobile genetic elements. Academic Press, London New York, pp 329−361

Rubin GM, Kidwell MG, Bingham PM (1982) The molecular basis of P-M hybrid dysgenesis: the nature of induced mutations. Cell 29:987−994

Ruiz A, Fontdevilla A (1985) The evolutionary history of *Drosophila buzzatii*. VI. Adaptive chromosomal changes in experimental populations with natural substrates. Genetica 66:63−71

Sage RD, Selander RK (1979) Hybridization between species of the *Rana pipiens* complex in central Texas. Evolution 33:1069–1088

Schäffer U (1978) Sterility in *Drosophila hydei* X *D. neohydei* hybrids. Genetica 49:205–214

Somerson NL, Ehrman L, Kocka JP, Gottlieb FJ (1984) Streptococcal L-forms isolated from *Drosophila paulistorum* semispecies cause sterility in male progeny. Proc Natl Acad Sci USA 81:282–285

Spradling AC, Rubin GM (1982) Transposition of cloned P elements into *Drosophila* germ line chromosomes. Science 218:341–347

Stone WS (1947) Gene replacement in the *virilis* group. Univ Texas Publ 4720:161–166

Strand DJ, McDonald JF (1985) Copia is transcriptionally responsive to environmental stress. Nucl Acids Res 13:4401–4410

Sturtevant AH (1939) High mutation frequency induced by hybridization. Proc Natl Acad Sci USA 25:308–310

Syvanen M (1984) The evolutionary implications of mobile genetic elements. Annu Rev Genet 18:271–293

Syvanen M (1986) Cross-species gene transfer: a major factor in evolution? Trends Genet:63–66

Templeton AR (1981) Mechanisms of speciation. A population genetic approach. Annu Rev Ecol Syst 12:23–48

Todaro GJ, Callahan R, Rapp VR, de Larco JE (1980) Genetic transmission of retroviral genes and cellular oncogenes. Proc R Soc London Ser B 210:367–385

Walters MS (1952) Spontaneous chromosome breakage and atypical chromosome movement in meiosis of the hybrid *Bromus marginatus* X *B. pseudolaevipes*. Genetics 37:8–25

Woodruff RC, Slatko BE, Thompson JN Jr (1983) Factors affecting mutation rate in natural populations. In: Ashburner M, Carson HL, Thompson JN, Jr (eds) The genetics and biology of *Drosophila*, vol 3c. Academic Press, London New York, pp 37–124

Wright S (1940) Breeding structure of populations in relation to speciation. Am Nat 74:232–248

Wright S (1969) Evolution and the genetics of populations, vol 2. The theory of gene frequencies. Univ Chicago Press

Wu CI, Beckenbach AT (1983) Evidence for extensive genetic differentiation between the sex-ratio and the standard arrangement of *Drosophila pseudoobscura* and *Drosophila persimilis*, and identification of hybrid sterility factors. Genetics 105:71–86

Young MW, Schwartz HE (1981) Nomadic gene families in *Drosophila*. Cold Spring Harbor Symp Quant Biol 45:629–640

Zouros E (1981) The chromosomal basis of sexual isolation in two sibling species of *Drosophila*: *D. arizonensis* and *D. mojavensis*. Genetics 97:703–718

Consequences of a Model of Counter-Gradient Selection

G. DE JONG[1]

1 Introduction

It might be an old adage that the genotype tracks the environment. Given persistent change in the environment, populations have to evolve constantly to stay adapted (Fisher 1930). Evolution under environmental change must be one of the more common processes in history.

But another question is whether genotypic change under environmental change must always mean phenotypic change. An often unspoken assumption seems to be that environmental change must always imply a change of the optimum phenotype; and as adaptation unfailingly produces the optimum phenotype, environmental change must mean phenotypic change. Given this reasoning, morphological stasis under environmental change becomes a major problem (Wake et al. 1983). Yet the assumption is not necessary. The factors determining what the optimum phenotype in any environment is might either not be related to the observed environmental change, or, the same environmental variables might have another relation to the optimum phenotype than to phenotypic changes within one genotype. If the optimum phenotype is constant over changes in the environment, we face two problems: what determines the phenotypic optimum, and how well genotypes are selected to approach the phenotypic optimum.

Following a classical approach, an example from natural history is adduced where the observed phenotypic change in the field over a range of environments is less than the phenotypic change expected on the basis of phenotypic plasticity. On such natural history observations evolutionary explanations should be based. The name of this type of ecological morphological stasis is counter-gradient selection. The natural history observations provide a framework of reflection on genetic change in a changing environment, with any degree of morphological change. Selection need not follow environmental change: it might counter it, perhaps deleting its effect.

---

1 Vakgroep Populatie- en Evolutiebiologie University of Utrecht, Padualaan 8,
   NL-3584 CH Utrecht, The Netherlands

Population Genetics and Evolution
G. de Jong (ed.)
© Springer-Verlag Berlin Heidelberg 1988

2 Counter-Gradient Selection in the Green Frog

In 1979, Berven et al. published a combined field and laboratory study of the green frog, *Rana clamitans*. Interest centered on observed differences in developmental time, weight at metamorphosis and stage-specific growth rate, comparing natural montane and lowland populations.

The four mountain populations lived in permanent ponds at an elevation of about 1000 m. The breeding season in the mountain localities was restricted to a 2-month period, July and August. The minimum larval period in mountain ponds is between 365 and 425 days, the maximum 670 days; development cannot be completed within one summer, and all tadpoles have to go through one winter at least, and two winters for eggs laid late in the breeding season.

The two lowland populations came from a pond at the foot of the mountains, and from a pond at sea level. Apart from elevation, the ponds are in the same general area as the mountain ponds. The breeding season extended from mid-May to mid-September.

The minimum larval period for green frogs in the lowlands is 90 days, and the maximum 334 days: development is completed either in the first summer, for eggs laid earlier in the breeding season, or after one winter, for eggs laid later in the breeding season. Differences in duration of the breeding season and in developmental time of the tadpoles were ascribed to temperature differences due to elevational differences (see Berven et al. 1979 for details).

In the field, in any month and at any developmental state, body size is larger for mountain tadpoles. At similar stages in development, mountain tadpoles are larger, and concordantly mountain tadpoles are larger at metamorphosis. The developmental time of mountain tadpoles is longer. In the field, the environment in the mountains allows only slow development, which results in a large size at metamorphosis.

In order to determine whether the observed differences in growth and development were induced ontogenetically by the ambient environment or controlled genetically, Berven et al. complemented the field observations by laboratory experiments. Raising tadpoles from several ponds in the laboratory at a range of constant temperatures showed that at any temperature the mountain tadpoles grew faster, and were smaller at metamorphosis. In Fig. 1 a diagram is presented summarizing those laboratory experiments and indicating the field situation. For both mountain and lowland popula-

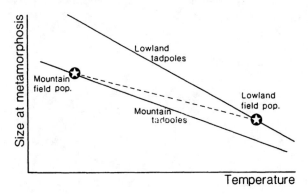

Fig. 1. Diagrammatic representation of the results of Berven et al. (1979). On the reaction norms in the laboratory the position of the field populations is indicated. The *dotted line* gives a possible trajectory for the optimum phenotype under changing temperature in nature

tions of *Rana clamitans*, the body size at metamorphosis (as well as developmental time and stage-specific growth rate) is influenced by the temperature during tadpole development. Mountain and lowland populations show differences that are easiest interpreted as genetic differences. The population that develops slowest in the field, the mountain population, is in fact genetically a fast developer. The population that metamorphoses at a smaller size in the field is in fact genetically a larger size at metamorphosis. Selection for fast development in an environment leading to slow growth and selection for large size at metamorphosis in an environment leading to fast growth must have had as a result that the observed phenotype over the environmental range is more constant than it would be according to any of the reaction norms. The direction of the genetic variation is opposite to that of the observed variation.

This has been termed contra- or counter-gradient selection by Levins (1968, 1969). Examples of counter-gradient selection are rare. This does not necessarily imply that counter-gradient selection is rare in nature, as the combination of field observations and laboratory experiments necessary to detect it might not be common.

3 Questions Raised

This splendid natural history story of Berven et al. (1979), on the one hand, raises a number of intriguing questions and, on the other hand, provides the essential features for a model of counter-gradient selection. The field work concerned itself with two environments out of a potential range. If pools all along the mountain side had been sampled, would the body size at metamorphosis have gradually changed with elevation, i.e. lie along the dashed line in Fig. 1? And if so, would that be because every population from every pool was genetically distinct, with each population showing a separate reaction norm for body size at metamorphosis, in between the two reaction norms found? Or can the present lowland and montane populations be regarded as homozygotes for alternate alleles, and would all intermediate populations along the elevation be polymorphic? And what would the mean body size at metamorphosis be in such a series of polymorphic populations?

Those are questions of natural history, but the work on the green frog suggests the ingredients of a model for counter-gradient selection. The model can tell us what the several alternatives would look like, whether we can expect the mean phenotype to follow the dashed line in Fig. 1 smoothly, both in a series of polymorphic and in a series of monomorphic populations.

4 A Model of Counter-Gradient Selection

A model for counter-gradient selection starts by describing the genotypes as reaction norms. The simplest choice is to assume the phenotype to be a quantitative character under the usual model:

$P = G + E$

with $\bar{E} = 0$, and cov $(G,E) = 0$. For any locus, each with two alleles, the genotypic values can be represented by linear functions of an environmental variable x. The

Table 1. One locus in n locus model (mean genotypic value over all other loci is G_0)

Genotype	Genotypic value	Genotypic frequency
$A_1 A_1$	$g_{11} = -2 a^a + cx + G_0$	p^2
$A_1 A_2$	$g_{12} = \quad d \quad + cx + G_0$	$2 pq$
$A_2 A_2$	$g_{22} = +2 a \quad + cx + G_0$	q^2

^{a} $a > 0$.

genotypic values, and genotypic frequencies before selection, can be found in Table 1. In the simplest case the reaction norms of the three genotypes vary in the same way with the environment; the slope c is independent of the genotype. We will restrict ourselves to this simple case of reaction norms as equidistant lines, with gene effects −a and +a of alleles A_1 and A_2.

The next feature of the model that is suggested by the data for *Rana clamitans* is that the selected phenotype changes less over the range of environments than the reaction norms. Selection will be described as optimizing, with a constant or slowly changing optimum. The choice of fitness function for optimizing selection is here the (truncated) quadratic:

$$w(P) = 1 - a'(b' - P)^2 \quad b' - \sqrt{\frac{1}{a'}} < P < b' + \sqrt{\frac{1}{a'}}$$

$$w(P) = 0 \quad b' - \sqrt{\frac{1}{a'}} > P, P > b' + \sqrt{\frac{1}{a'}}$$

for fitness as function of the phenotype.

In a one-locus model, the genotypic fitness becomes:

$$w_{ij} = 1 - a'(b' - g_{ij})^2 - a'V_E$$

for

$$b' - \sqrt{\left(\frac{1}{a'} - V_E\right)} < g_{ij} < b' + \sqrt{\left(\frac{1}{a'} - V_E\right)}.$$

We will assume that the genotypic values are always within the allowed range.

The optimum phenotype, and genotypic value, is at $P = b'$. This optimum value can itself be a function of the same environmental variable x that plays its role in the reaction norm. For simplicity, let the optimum phenotype, too, be a linear function of x : $b' = bx$.

One point of interest is the role of the slopes b of the optimum phenotype and c of the reaction norm in determining gene frequencies and average genotypic values in a population.

The last assumption concerns a series of independent populations along the range of the environmental variable x. There is no migration, and selection in each population proceeds independently. We are interested in the gene frequency and average genotypic value at the end of selection in each population. With optimizing selection and genotypes as reaction norms, we have, over the environmental range, selection

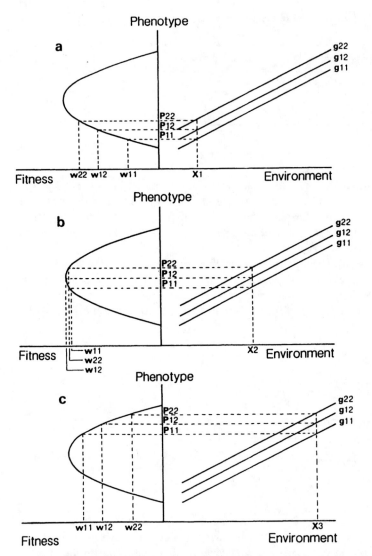

Fig. 2. Change in fitness of three genotypes with a change in the environment. All three diagrams show the same three reaction norms for the three genotypes at one locus, and the same fitness function with a constant optimum phenotype. The figures have three axes:

1. *Right-hand side,* phenotype as a function of the environment:
   abscissa, to the right: environmental variable: x,
   ordinate, up: phenotype = genotypic value: P.
2. *Left-hand side,* fitness as a function of phenotype:
   abscissa, up: phenotype = genotypic value: P,
   ordinate, to the left: fitness: w.

In any given environment, the three genotypic values can be found from the reaction norms; the relative fitnesses can be found from the genotypic values. In **a**, at low x, $w_{11} < w_{12} < w_{22}$, leading to fixation of allele A_2. In **b**, at intermediate x, $w_{11} < w_{12} > w_{22}$, leading to a stable polymorphism. In **c**, at high x, $w_{11} > w_{12} > w_{22}$, leading to fixation of allele A_1.

for the one homozygote, polymorphism, and selection for the other homozygote. In Fig. 2 the three situations are shown graphically.

The problem is therefore to find the range of x that allows polymorphism, and the equilibrium gene frequency within that range.

4.1 Additivity, One Locus

Let the genotypic values at locus A be additive (d = 0). The genotypic fitnesses are:

$$w_{ij} = 1 - a' \, (bx - g_{ij})^2 - a'V_E \, . \tag{1}$$

For the range of values of the environmental variable x

$$-\frac{a}{|c - b|} < x < \frac{a}{|c - b|} \tag{2}$$

populations are polymorphic, at an equilibrium gene frequency of

$$\hat{p} = \frac{1}{2} + \frac{c - b}{2a} \, x \, . \tag{3}$$

The mean genotypic and phenotypic value in the polymorphic populations is:

$$\bar{P} = \bar{G}_p = x \, (2b - c) \, . \tag{4}$$

In Fig. 3 a one-locus example of an environmental trajectory is shown, with populations homozygotic for the one allele at one end of the range, polymorphic populations in the middle of the range and populations homozygotic for the other allele at the other end of the range. At no point over this environmental range is the mean phenotypic value equal to the optimal value, b = 0. This might not be surprising for

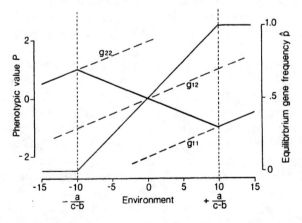

Fig. 3. Reaction norms (---) of three genotypes at one locus, mean phenotypic values $\bar{P}$ (——) (*left-hand ordinate*) and equilibrium gene frequency (——) (*right-hand ordinate*) in a series of separate populations over an environmental range. Here a = 2, c = 0.2, b = 0. Since the optimum phenotype is b' = 0 in all environments, the slope of the mean phenotypic value $\bar{P}$ in the polymorphic range is −cx, while the slope of P is the slope of the reaction norm +cx over the monomorphic ranges

the monomorphic populations, but even in the polymorphic populations the mean phenotype under selection is not at the optimum. This as the optimal phenotypic value, b=0, does not coincide with the phenotypic value, cx, from which genetic variation is measured (Table 1). The resulting pattern for the phenotypic mean is a pronounced zigzag pattern, contrary to naive expectation.

Note that the intensity of selection a' plays no role at all. This is a major difference between using quadratic and Gaussian optimizing fitness functions. The gene effect a plays a role in the boundaries of the polymorphic range and in the equilibrium value of the gene frequency. A large gene effect means a large range. But in the phenotypic mean the gene effect appears only at fixation of alleles, at $\overline{P} = -2a + cx$ or $\overline{P} = +2a + cx$, for monomorphic populations of $A_1 A_1$ and $A_2 A_2$ respectively. In the polymorphic populations the gene effect has no influence on the mean phenotype.

It might seem obvious that the difference of the slopes, $c-b$, appears in the boundaries and the equilibrium gene frequency. This is what one would expect, only the difference in slope playing a role as an indication of the difference between ideal and reality. Again, the phenotypic mean in a polymorphic population does not conform: the slopes play a different role. It is as if the realized mean phenotype tracks the optimum phenotype (bx) as far as it is allowed by the difference in slope between optimum phenotype and reaction norm:

$$bx - (c-b)x = (2b-c)x = \overline{P}.$$

4.2 Additivity, N Loci

In Fig. 1, an a priori idea about the behaviour of the phenotypic mean over a range of separate populations was given.

As shown in Fig. 3, the results from a one-locus model are quite different from that. This zigzag pattern might be an extreme case, as only one locus is considered. Is it possible to get better tracking of the phenotypic optimum by the phenotypic mean if more loci are involved?

We will consider again a large number of separate populations along a range of an environmental variable x. Now the character we are interested in is coded for by n independent loci. The choice of submodel has been for considering n loci of equal effect, where the difference between the most extreme homozygotes in any one environment is 4 an, and the gene effects at the individual loci are $-a$ for any 1 allele and $+a$ for any 2 allele. The slope of the reaction norm per locus is c, and this means that the total slope in the linear relation giving phenotype as a function of the environment is ncx for all genotypes. The quantities of importance are total slope nc and the distance between homozygotes at one locus, 4a.

The reaction norm at any one focal locus becomes for additive loci:

$$g_{ij} = (-1)^i (a) + (-1)^j (a) + cx + \overline{G}_0,$$

where $\overline{G}_0$ is the genotypic mean over all other loci. The course of selection at the focal locus now directly depends upon $\overline{G}_0$- as $\overline{G}_0$ determines the position of the genotypic values at the focal locus with respect to the optimum phenotypic value. Let the reaction norms in Fig. 2a now be the reaction norms at the focal locus at low $\overline{G}_0$. In environment x_1, selection at the focal locus is for allele 2 with low $\bar{G}_0$, but with higher $\bar{G}_0$ a polymorphism at the focal locus would be indicated, and at very high $\bar{G}_0$ selection would be for allele 1. This is of course represented in the fitnesses at the focal locus:

$$w_{ij} = 1 - a' (bx - g_{ij} - \bar{G}_0)^2 - a'V_E - a'V_{G_0} , \tag{5}$$

where V_{G_0} is the genetic variance at the other loci.

In any system of n additive loci, at most one can be polymorphic in the equilibrium situation (Wright 1935, 1969, pp. 107-109). Computer simulation bears this out. It means that for a focal locus to be polymorphic, all other loci must be fixed at either allele 1 or allele 2. Let k of the n loci be fixed at allele 1, the $k+1^{th}$ locus being the focal locus, and $n-(k+1)$ loci be fixed at allele 2. The mean genotypic value at all other loci is then:

$$\bar{G}_0 = k(-2a+cx) + (n-k-1)(2a+cx) = cx(n-1) + 2a(n-2k-1) . \tag{6}$$

The equilibrium gene frequency at a polymorphic focal locus is:

$$\hat{p} = \frac{1}{2} + \frac{\bar{G}_0 + x(c-b)}{2a} = \frac{1}{2} \{2n-4k-1\} + \frac{x(cn-b)}{2a} . \tag{7}$$

The polymorphism at locus k + 1 exists for:

$$- \frac{a}{cn-b} \{2n-4k-3\} < x < - \frac{a}{cn-b} \{2n-4k-1\} \quad \text{at } cn-b < 0 \tag{8a}$$

$$- \frac{a}{cn-b} \{2n-4k-1\} < x < - \frac{a}{cn-b} \{2n-4k-3\} \quad \text{at } cn-b > 0 \tag{8b}$$

giving a polymorphism distance, and an interpolymorphism distance, of:

$$2a/|cn-b| .$$

The equality of polymorphism distances and interpolymorphism distances follows from the equality of the effects of all n loci. The mean phenotypic value in an interpolymorphism trajectory with k loci fixed at allele 1 and $n-k$ loci fixed at allele 2, is:

$$\bar{P} = \bar{G} = ncx + 2a(n-2k) , \tag{9}$$

while the mean phenotypic value in the polymorphism trajectory with k loci fixed at allele 1 and $n-(k+1)$ loci fixed at allele 2 is:

$$\bar{P} = \bar{G} = \bar{G}_0 + \bar{G}_p$$
$$= \bar{G}_0 + \hat{p}^2 (-2a + cx) + 2 \hat{p}\hat{q}(cx) + \hat{q}^2 (+2a + cx)$$
$$= \bar{G}_0 + n(2b - c) - 2\bar{G}_0 = -2a(n-2k-1) + x(2b - cn) .$$

The slope of $\bar{P}$ in the interpolymorphism trajectory is cn; in the polymorphism trajectory the slope is $2b-cn$. This leads to a marked zigzag pattern in the phenotypic mean. The value of the phenotypic mean is shown in Fig. 4a for three loci, together with the optimal phenotype over the same environmental range and some reaction norms. In Fig. 4b equilibrium gene frequencies for populations at each environmental value are given. From Fig. 4, it becomes clear that in some way the phenotypic mean tracks the phenotypic optimum. This is for a range of optimum phenotypic values

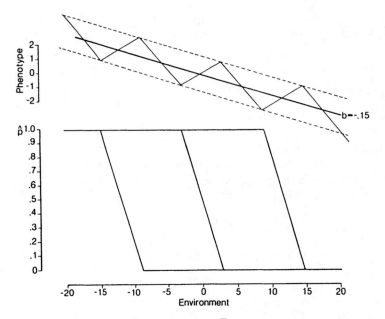

Fig. 4. Mean equilibrium phenotypic value $\bar{P}$ (——) and optimum phenotype bx (——) (*upper graph*) and equilibrium gene frequency (*lower graph*) in an additive system of three loci of equal effect. Per locus: a = 4/3, c = 0.2, b = 0.15. To the *left* all three loci are fixed at allele 1; to the *right* all three loci are fixed at allele 2. At most one locus is polymorphic in any environment. In the graph of the equilibrium mean phenotypic value $\bar{P}$ monomorphic ranges are shown by lines going northwest to southeast; polymorphic ranges are shown by lines going southwest to northeast

demonstrated in Fig. 5, where phenotypic means for five loci are given for six sets of values of b and cn. Since b and cn do not play the same role in the phenotypic mean, the zigzag pattern of the phenotypic mean is quite different when the values of b and cn are exchanged (Fig. 5d and f).

The alternating pattern of the phenotypic mean in the monomorphic and polymorphic populations has a general direction, a certain width and a certain length. Let us first consider the general direction. The corner points of the zigzag are at those values of the environmental variable where a focal locus becomes polymorphic or monomorphic. For cn – b < 0, as in Fig. 5, the corner points below the optimum phenotype are given by

$$x = \frac{-a}{cn - b} \{2n - 4k - 3\}$$

for successive values of k, the number of loci fixed at allele 1. The corner points above the optimum are given by

$$x = \frac{-a}{cn - b} \{2n - 4k - 1\} \ .$$

The distance between two corners on the same side of the optimum phenotype is made up of one polymorphic and one monomorphic trajectory, i.e. the distance be-

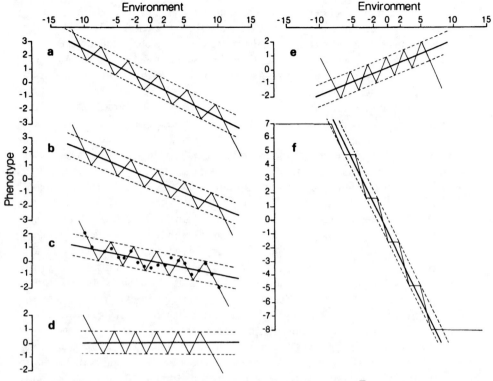

Fig. 5. Effect of the slopes b and c. Mean equilibrium phenotypic value $\bar{P}$ (——) and optimum phenotype bx (——) in an additive system of five loci of equal effect. To the *left* all five loci are fixed at allele 1; to the *right* all five loci are fixed at allele 2. At most one locus is polymorphic in any environment. In the graph the line of the equilibrium mean phenotypic value $\bar{P}$ goes north-west to southeast in monomorphic ranges; in polymorphic ranges the line of $\bar{P}$ goes southwest to northeast. The boundary lines P_{upper} and P_{lower}, both with slope b, are indicated by *dashed lines*. In all graphs a = 4/5.

a c = −0.20, b = −0.25; b c = −0.20, b = −0.20; c c = −0.20, b = −0.15; d c = −0.20, b = 0; e c = −0.20, b = +0.1; f c = 0, b = −1.0.

In **c** the separate points give the mean phenotype after 500 generations of selection from random starting frequencies at the five loci

tween corner points in the direction of the abscissa is $4a/|cn-b|$. The distance along the ordinate between any two successive corners on the same side of the optimum phenotype is

$$\bar{P}_k - \bar{P}_{k-1} = \frac{4a \cdot b}{|cn-b|} .$$

This means that the two lines connecting the corner points on each side of the optimum phenotype both have slope:

$$\left(\frac{4a \cdot b}{|cn-b|}\right) \Big/ \left(\frac{4a}{|cn-b|}\right) = b .$$

Since the distance between sequential homozygotes, i.e. between the reaction norms of k loci fixed at allele 1 and of k + 1 loci fixed at allele 1, is 4a at any x, the boundary lines through the corners of the zigzag (Figs. 4, 5) have equations $P_{upper} = bx + 2a$ and $P_{lower} = bx - 2a$. The width of the tracking zone is therefore 4a in any environment, and $4a/\sqrt{1 + b^2}$ across. As can be seen from this expression and from the examples in Fig. 5, the width across the tracking zone is highest at b = 0; a quickly changing optimum leads to a narrow tracking zone. The value of the slope of the reaction norm has no influence on the width of the tracking zone.

The influence of the number of loci on the width of the tracking zone is nil; width depends upon the gene effect of the individual loci. The length of the tracking zone depends upon the number of loci. The total length of the tracking zone, from the first crossing of actual phenotypic mean and optimum phenotype to the last crossing, is $2n * 2a/|cn-b| = 4na/|cn-b|$.

Naively, one might expect that the larger the number of loci involved, the longer the tracking zone. This, however, is not true: there are complications due to the slope b and the per locus effects. Here, opposite homozygotes at one locus differ by 4a, and each locus constributes a slope c. The influence of the number of loci on the length $4an/|cn-b|$ of the tracking zone can easiest be seen at b = 0, since at b = 0 there is none; the effect of the larger difference between extreme homozygotes is compensated for by the increase of the slope. All that happens is a faster change between loci; the polymorphic and interpolymorphic distances, $2a/|cn|$ for b = 0, just get smaller. Only if the total slope of the reaction norm is constant, c, and the per locus effect on the slope is c/n, while the gene effect a is independent of the number of loci, does the tracking zone increase in length with increasing number of loci; then length is $4an/|c|$ at b = 0.

Conversely, if the distance between extreme homozygotes is always 4a, and the per locus gene effect a/n, while the slope c is constant and determined per locus, the total length of the tracking zone $(4a/|cn-b|)$ decreases with n. This can be seen by comparing Figs. 4 and 5. (One should imagine bundles of reaction norms of the same width but with increasing slope being crossed by the same optimum phenotype line.) The length of the tracking zone, therefore, cannot be determined without knowing the effects of all n loci upon the reaction norm.

Any presupposition of a larger tracking zone with more loci seems based upon an implicit assumption of a changing optimum but constant genetic expression: the case b ≠ 0, but c = 0, while a is independent of n. The length of the tracking zone is then $4an/|-b|$, and there is a direct increase of the length of the tracking zone with the number of loci. One should be careful not to regard this as the general case.

In Fig. 5c the phenotypic mean after 500 generations of selection, starting from random gene frequencies at all five loci, is shown as well as the equilibrium phenotypic mean. Optimizing selection on five loci has not led to the equilibrium mean phenotype in 500 generations. Selection is slow, as the total selection is spread out over five loci; this can be seen in Eq. (5) for fitness at one locus, as the presence of the term including the genetic variance over the other loci effectively equalizes relative fitnesses. Non-fixation in ultimately monomorphic populations and non-equilibrium gene frequencies in polymorphic populations give a fuzzy pattern of the phenotypic mean around the phenotypic optimum, rather than the sharp zigzag pattern. In natural populations such deviations would be attributed to sampling error, or to intermittent selection. The overall impression would be of the phenotypic mean closely following the phenotypic optimum.

Whether the change of the phenotypic mean over the whole range of the environment will appear gradual or not depends on the range of observation. Over the whole range of the environmental variable we see three regions: a monomorphic region at extremely high and at extremely low values of the environmental variable, and a tracking zone in between. In the monomorphic regions, the phenotypic mean changes with the overall slope of the reaction norm, cn. In the tracking zone the overall slope is that of the optimum phenotype, b. Whether the total pattern will appear as a gradual or abrupt change will depend upon the difference between cn and b, and on the length of the tracking zone. Depending upon the number of loci, the details of gene action, and the optimum phenotype, any pattern of the phenotypic mean can be generated.

Constancy of the optimum for the phenotype leads to a changing genotype when the per locus effect of the environment on the genotype is a reaction norm. The genotype tracks the environment, without any phenotypic change.

5 Gradualism, Punctuations and Stasis

"Traditional gradualism imagined a graded response to gradually shifting environments". The quotation is from Gould and Eldredge (1986) in an amusing article with the title: "Punctuated equilibria at the third stage" — the stage, that is, when a new idea is recognized as not only not controversial, but as the one that people always subscribed to. In this article, Gould and Eldredge discuss whether punctuated equilibrium is really arrivée, somewhat trying to rescue it as a nice controversy. But this brings them to pay attention to population geneticists' remarks about punctuated equilibria — and they find that the population geneticists' proposals have not addressed their principal problem: gradualism.

There are several models by population geneticists on punctuated equilibria (Kirkpatrick 1982; Newman et al. 1985; Lande 1985). Gould and Eldredge (1986) give explicit comment on one of those: "Thus, Newman et al. (1985) did not render punctuated equilibrium as the expected outcome of those projected environmental histories that engendered the expectation of gradualism". Quite true: the population geneticists' problems were the punctuations, and they addressed themselves to explaining both stability (Charlesworth et al. 1982) and punctuation.

The models (Kirkpatrick 1982; Newman et al. 1985; Lande 1985) differ in their details, and somewhat in their objective, but all three models are more or less explicit elaborations of Wright's shifting balance theory (Wright 1982). In the models two phenotypic optima exist at some distance from each other. The phenotypic optimum P_{opt} at which the population finds itself has a lower fitness maximum in the adaptive landscape than a pre-existing phenotypic optimum P^*_{opt} across a valley. The models give ways in which the population can cross the valley. At both phenotypic optima stabilizing selection removes noise, genotypes lower down the slope. By some random element (that differs between the models) some genotypes come to cross the valley, and on the basis of those genotypes a selective runover to the higher fitness peak comes about. Genetic variation is low during stasis, and high during the selective punctuation (Fig. 6). No environmental influence seems present. The genotypes are not thought of as reaction norms, and the higher fitness peak is already present in the same environment without any animals occupying it. The adaptive landscape seems to have taken on an existence by itself, fixed from eternity.

But to return to the quotation starting this section, gradualists (who ever they may be) might have imagined a gradual response to a gradually shifting environment. In population genetics terms, that would be a different model from those models of punctuated equilibrium; one would think of a quantitative character coded for by many loci, and a constant selection differential underlying a constant response. The gradual impression would be given by the presence of many loci with a small effect. (An allele substitution at a major locus would look like a sudden change in phenotype!) Perhaps a classical model of gradualism is that of the tracking zone in Fig. 5f: genotypic values independent of the environment, a changing phenotypic optimum closely tracked. The gradual impression depends upon restricting oneself to the tracking zone. The total pattern in Fig. 5f shows a punctuation. The punctuated appearance depends upon the existence of monomorphic populations at very high and very low values of the environmental variable.

By an analogy argument that is often applied, we can use spatial patterns to explain patterns in time. The gradual change of the environmental variable is now in time, mutation is always present and environmental change is slow enough for selection to lead to the equilibrium gene frequency. The phenotypic pattern of reaction norm, zigzags in a tracking zone and reaction norm becomes a pattern of morphology in time. In Fig. 6b a gradual change of the phenotypic mean with a change of the phenotypic optimum is shown (it is a regraphing of Fig. 5f). This pattern gives a classic punctuation, more or less corresponding to the patterns found in the

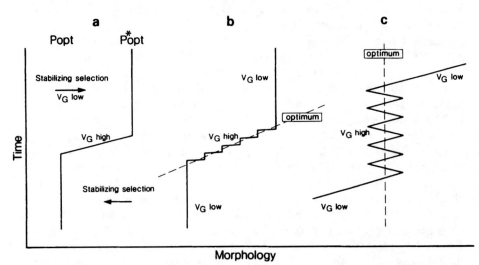

Fig. 6. Diagrams of morphology in time. **a** Published population genetics punctuation models with two fixed optima. **b** In this model, five loci are at $c = 0$, $b = -1.0$, $a = 4/5$. **c** In this model, five loci are at $c = 0.20$, $b = 0$, $a = 4/5$

models based upon a static adaptive landscape. But in Fig. 6c (regraphing of Fig. 5d) another pattern is shown. This too shows punctuation and stasis.

In contrast to the models of Fig. 6a,b stasis now corresponds to optimizing selection with gene replacement and high genetic variance, and punctuation corresponds to fixation of alleles and low genetic variance. The genetic variation is necessary to produce morphological stasis in a gradually changing environment, while a changing morphology means fixation at a certain reaction norm far from the phenotypic optimum. There is only one fitness peak in the adaptive landscape, and the changing environment forces the population downhill after the genetic variation is exhausted. By simply observing morphological change in time, it is not possible to know whether the morphological change is based upon genetic variation or on fixation of genes, on tracking a changing optimum phenotype or on the constraint of the reaction norm. The natural history of the green frog provides an example in which the optimum phenotype is more constant than the reaction norms. Such ecological examples should assist in the explanation of morphological stasis under a changing environment.

References

Berven KA, Gill DE, Smith-Gill SJ (1979) Countergradient selection in the green frog, *Rama clamitans*. Evolution 33:609–623

Charlesworth B, Lande R, Slatkin M (1982) A neo-Darwinian commentary on macro-evolution. Evolution 36:474–498

Fisher RA (1930) The genetical theory of natural selection. Oxford Univ Press

Gould SJ, Eldredge N (1986) Punctuated equilibrium at the third stage. Syst Zool 35:143–148

Kirkpatrick M (1982) Quantum evolution and punctuated equilibria in continuous genetic characters. Am Nat 119:833–848

Lande R (1985) Expected time for random genetic drift of a population between stable phenotypic states. Proc Natl Acad Sci USA 82:7641–7645

Levins R (1968) Evolution in changing environments. Princeton Univ Press (Monographs in population biology 2)

Levins R (1969) Thermal acclimation and heat resistance in *Drosophila* species. Am Nat 103:483–499

Newman CM, Cohen JE, Kipnis C (1985) Neo-Darwinian evolution implies punctuated equilibria. Nature (London) 315:400–401

Wake DB, Roth G, Wake MH (1983) On the problem of stasis in organismal evolution. J Theor Biol 101:211–224

Wright S (1935) Evolution in populations in approximate equilibrium. J Genet 20:257–266

Wright S (1969) Evolution and the genetics of populations, vol 2. The theory of gene frequencies. Univ Chicago Press

Wright S (1982) Character change, speciation and higher taxa. Evolution 36:427–443

Subject Index